Lecture Notes in Computer Science   9197

Commenced Publication in 1973
Founding and Former Series Editors:
Gerhard Goos, Juris Hartmanis, and Jan van Leeuwen

More information about this series at http://www.springer.com/series/7407

My T. Thai · Nam P. Nguyen
Huawei Shen (Eds.)

Computational Social Networks

4th International Conference, CSoNet 2015
Beijing, China, August 4–6, 2015
Proceedings

 Springer

Editors
My T. Thai
Computer and Information Science
   and Engineering
University of Florida
Gainesville, FL
USA

Huawei Shen
Chinese Academy of Sciences
Beijing
China

Nam P. Nguyen
Towsen University
Towson, MD
USA

ISSN 0302-9743   ISSN 1611-3349   (electronic)
Lecture Notes in Computer Science
ISBN 978-3-319-21785-7   ISBN 978-3-319-21786-4   (eBook)
DOI 10.1007/978-3-319-21786-4

Library of Congress Control Number: 2015944431

LNCS Sublibrary: SL1 – Theoretical Computer Science and General Issues

Springer Cham Heidelberg New York Dordrecht London

Printed on acid-free paper

Springer International Publishing AG Switzerland is part of Springer Science+Business Media
(www.springer.com)

Preface

The International Conference on Computational Social Network (CSoNet) 2015 provided a premier interdisciplinary forum bringing together researchers and practitioners from all fields of social networks – such as network computing, social network/media analysis and mining – for the presentation of original research results as well as the exchange and dissemination of innovative, practical development experiences. CSoNet 2015 addressed emerging yet important computational problems with a focus on the fundamental background, theoretical technology development, and real-world applications associated with social network analysis, modeling and data mining.

CSoNet 2015 was inherently interdisciplinary as it attempted to integrate across different disciplines such as social science, computer science, networks science, and mathematics in pursuit of a fundamental understanding of computational social networks. The conference welcomed all submissions with a focus on common principles, algorithms, and tools that govern social network structures/topologies, network functionalities, security and privacy, network behaviors, information diffusions and influence, and social recommendation systems that are applicable to all types of social networks and social media. The conference received 102 qualified submissions from which 28 papers were accepted as regular and short papers, as well as extended abstracts. Each submission was reviewed by at least three reviewers and some received meta reviews from Program Committee members.

June 2015

My T. Thai
Nam P. Nguyen
Huawei Shen

Organization

Steering Committee

My T. Thai	University of Florida (Chair), USA
Zhi-Li Zhang	University of Minnesota, USA
Weili Wu	University of Texas–Dallas, USA

Program Committee Co-chairs

Nam P. Nguyen	Towson University, USA
Huawei Shen	Chinese Academy of Science, China

Technical Program Committee

Thang N. Dinh	Virginia Commonwealth University, USA
Illes Farkas	Hungarian Academy of Sciences, Hungary
Niloy Ganguly	Indian Institute of Technology Kharagpur, India
Konstantinos Georgiou	University of Waterloo, Canada
Preetam Ghosh	Virginia Commonwealth University, USA
Steve Gregory	University of Bristol, UK
Vasileios Karyotis	National Technical University of Athens, Greece
Donghyun Kim	North Carolina Central University, USA
Sang-Wook Kim	Hanyang University, Korea
Guanfeng Liang	LinkedIn Research and Development, USA
Christopher McCarty	University of Florida, USA
Ramasuri Narayanam	IBM Research, India
Hien Nguyen	Ton Duc Thang University, Vietnam
Panos Pardalos	University of Florida, USA
Han-Woo Park	YeungNam University, Korea
Georgios Piliouras	California Institute of Technology, USA
Nishanth Sastry	King's College London, UK
Yilin Shen	Samsung Research and Development, USA
Ravi Tiwari	eBay Research, USA
Raffaele Vacca	University of Florida, USA
Mario Ventresca	Purdue University, USA
Anil Kumar Vullikanti	Virginia Tech, USA
Li Wang	Taiyuan University of Technology, China
Yu Wang	University of North Carolina at Charlotte, USA

Measuring Originality in Knowledge Networks

Ádám Szántó-Várnagy, Péter Pollner, and Illés J. Farkas

Department of Biological Physics, Eötvös University, and MTA-ELTE Statistical
and Biological Physics Group, Hungarian Academy of Sciences,
Pázmány Péter sétány 1A, 1117 Budapest, Hungary
fij@elte.hu

Abstract. Human knowledge is accumulated in several ways: through patents, scientific publications, encyclopedias, news, etc. In each case the involved "knowledge items" form a directed network that shows which item is built on which others. For example, patents (nodes) cite (link to) other patents (nodes). The usefulness of knowledge is most often measured on single knowledge items by article-level metrics (ALMs). In science the most common ALM is the citation number, n, quantifying impact. Instead of the impact here we discuss originality. We compute the probability, p, of directed links pointing from a node's in-neighbors to its out-neighbors. Low values of p mean high originality. For several large real knowledge networks we find a very low correlation between n and p. Thus, we suggest that p provides qualitatively novel information about single knowledge items of human knowledge, such as patents, scientific publications, encyclopedia and news articles, etc.

Quantitative Function and Algorithm for Community Detection in Bipartite Networks

Zhenping Li[1], Rui-Sheng Wang[2], Shihua Zhang[3,*],
Xiang-Sun Zhang[3,*]

[1] School of Information, Beijing Wuzi University, Beijing, China
[2] Department of Medicine, Brigham and Women's Hospital,
Harvard Medical School, Boston, USA
[3] National Center for Mathematics and Interdisciplinary Sciences,
Academy of Mathematics and Systems Science, CAS, Beijing, China
zsh@amss.ac.cn, zxs@amt.ac.cn

Abstract. In this paper, we propose a new quantitative function for community detection in bipartite networks, and demonstrate that this quantitative function is superior to the widely used Barber's bipartite modularity and other functions. Based on the new quantitative function, we develop an integer programming model and a heuristic and adapted label propagation algorithm (BiLPA). We demonstrate their efficiency by applying them onto artificial networks and real-world networks.

Contents

Real-Time Topic-Aware Influence Maximization Using Preprocessing

Wei Chen[1], Tian Lin[1,2] (✉), and Cheng Yang[1]

[1] Microsoft Research, Beijing, China
weic@microsoft.com
[2] Tsinghua University, Beijing, China
lint10@mails.tsinghua.edu.cn, albertyang33@gmail.com

Abstract. Influence maximization is the task of finding a set of seed nodes in a social network such that the influence spread of these seed nodes based on certain influence diffusion model is maximized. Topic-aware influence diffusion models have been recently proposed to address the issue that influence between a pair of users are often topic-dependent and information, ideas, innovations etc. being propagated in networks are typically mixtures of topics. In this paper, we focus on the topic-aware influence maximization task. In particular, we study preprocessing methods to avoid redoing influence maximization for each mixture from scratch. We explore two preprocessing algorithms with theoretical justifications. Our empirical results on data obtained in a couple of existing studies demonstrate that one of our algorithms stands out as a strong candidate providing microsecond online response time and competitive influence spread, with reasonable preprocessing effort.

Keywords: Influence maximization · Topic-aware influence modeling · Information diffusion

1 Introduction

In a social network, information, ideas, rumors, and innovations can be propagated to a large number of people because of the social influence between the connected peers in the network. *Influence maximization* is the task of finding a set of *seed nodes* in a social network such that the influence propagated from the seed nodes can reach the largest number of people in the network. More technically, a social network is modeled as a graph with nodes representing individuals and directed edges representing influence relationships. The network is associated with a stochastic diffusion model (such as independent cascade model and linear threshold model [14]) characterizing the influence propagation dynamics starting from the seed nodes. Influence maximization is to find a set of k seed nodes in the network such that the *influence spread*, defined as the expected number of nodes influenced (or activated) through influence diffusion starting from the seed nodes, is maximized [6,14].

© Springer International Publishing Switzerland 2015
M.T. Thai et al. (Eds.): CSoNet 2015, LNCS 9197, pp. 1–13, 2015.
DOI: 10.1007/978-3-319-21786-4_1

Influence maximization has a wide range of applications including viral marketing [9,14,18], information monitoring and outbreak detection [15], competitive viral marketing and rumor control [5,13], or even text summarization [22] (by modeling a word influence network). As a result, influence maximization has been extensively studied in the past decade. Research directions include improvements in the efficiency and scalability of influence maximization algorithms [8,12,21], extensions to other diffusion models and optimization problems [3,5,13], and influence model learning from real-world data [10,19,20].

Most of these works treat diffusions of all information, rumors, ideas, etc. (collectively referred as *items* in this paper) as following the same model with a single set of parameters. In reality, however, influence between a pair of friends may differ depending on the topic. For example, one may be more influential to the other on high-tech gadgets, while the other is more influential on fashion topics, or one researcher is more influential on data mining topics to her peers but less influential on algorithm and theory topics. Recently, Barbieri et al. [2] propose the topic-aware independent cascade (TIC) and linear threshold (TLT) models, in which a diffusion item is a mixture of topics and influence parameters for each item are also mixtures of parameters for individual topics. They provide learning methods to learn influence parameters in the topic-aware models from real-world data. Such topic-mixing models require new thinking in terms of the influence maximization task, which is what we address in this paper.

In this paper, we adopt the models proposed in [2] and study efficient topic-aware influence maximization schemes, i.e., finding a set of k seed nodes to trigger the information cascade whenever a diffusion item composed of multiple topics is given. It has a wide application in viral marketing for online scenarios, where the system should recommend candidate sets instantly to different queries. One can still apply topic-oblivious influence maximization algorithms in online processing of every diffusion item, but it may not be efficient when there are a large number of items with different topic mixtures or real-time responses are required. Thus, our focus is on how to utilize the preprocessing of individual topic influence so that when a diffusion item with certain topic mixture comes, the online processing of finding the seed set is fast. To do so, our first step is to collect two datasets in the past studies with available topic-aware influence analysis results on real networks and investigate their properties pertaining to our preprocessing purpose. Our data observation shows that in one network users and their relationships are largely separated by different topics while in the other network they have significant overlaps on different topics. Even with this difference, a common property we find is that in both datasets most top seeds for a topic mixture come from top seeds of the constituent topics, which matches our intuition that influential individuals for a mixed item are usually influential in at least one topic category.

Motivated by our findings from the data observation, we explore two preprocessing based algorithms (Section 3). The first algorithm, *Best Topic Selection* (BTS), minimizes online processing by simply using a seed set for one of the constituent topics. Even for such a simple algorithm, we are able to provide a theoretical approximation ratio (when a certain property holds), and thus BTS serves as a baseline for preprocessing algorithms. The second algorithm, *Marginal Influence Sort* (MIS), further uses pre-computed marginal influence of seeds on each topic to avoid slow greedy computation. We provide a theoretical justification showing that MIS can be as good as the offline greedy algorithm when nodes are fully separated by topics.

We then conduct experimental evaluations of these algorithms and comparing them with both the greedy algorithm and a state-of-the-art heuristic algorithm PMIA [21], on the two datasets used in data analysis as well as a third dataset for testing scalability (Section 4). From our results, we see that MIS algorithm stands out as the best candidate for preprocessing based real-time influence maximization: it finishes online processing within a few microseconds and its influence spread either matches or is very close to that of the greedy algorithm. Full technical details including data analysis, proofs and experimental results are available in the technical report [7].

Our work, together with a recent independent work [1], is one of the first that study topic-aware influence maximization with focus on preprocessing. Comparing to [1], our contributions include: (a) we include data analysis on two real-world datasets with learned influence parameters, which shows different topical influence properties and motivates our algorithm design; (b) we provide theoretical justifications to our algorithms; (c) the use of marginal influence of seeds in individual topics in MIS is novel, and is complementary to the approach in [1]; (d) although MIS is simple, it achieves competitive influence spread within microseconds of online processing time satisfying real-time application requirement.

2 Preliminaries

In this section, we introduce the background and problem definition on the topic-aware influence diffusion models. We focus on the independent cascade model [14] for ease of presentation, but our results also hold for other models parameterized with edge parameters such as the linear threshold model [14].

Independent cascade model. We consider a social network as a directed graph $G = (V, E)$, where each node in V represents a user, and each edge in E represents the relationship between two users. For every edge $(u, v) \in E$, denote its *influence probability* as $p(u, v) \in [0, 1]$, and we assume $p(u, v) = 0$ for all $(u, v) \notin E$ or $u = v$. The *independent cascade (IC)* model, defined in [14], captures the stochastic process of contagion in discrete time. Initially at time step $t = 0$, a set of nodes $S \subseteq V$ called *seed nodes* are activated. At any time $t \geq 1$, if node u is activated at time $t - 1$, it has one chance of activating each of its inactive outgoing neighbor v with probability $p(u, v)$. A node stays active after it is activated. This process stops when no more nodes are activated.

We define *influence spread* of seed set S under influence probability function p, denoted $\sigma(S, p)$, as the expected number of active nodes after the diffusion process ends. As shown in [14], for any fixed p, $\sigma(S, p)$ is monotone (i.e., $\sigma(S, p) \leq$

$\sigma(T,p)$ for any $S \subseteq T$) and submodular (i.e., $\sigma(S \cup \{v\}, p) - \sigma(S,p) \geq \sigma(T \cup \{v\}, p) - \sigma(T,p)$ for any $S \subseteq T$ and $v \in V$) in its seed set parameter. For two influence probability functions p and p' on graph $G = (V, E)$, we denote $p \leq p'$ if for any $(u,v) \in E$, $p(u,v) \leq p'(u,v)$. Another well-known fact is that $\sigma(S,p)$ is *monotone in* p (i.e. $\sigma(S,p) \leq \sigma(S,p')$ if $p \leq p'$ edge-wise).

Influence maximization. Given a graph $G = (V, E)$, an influence probability function p, and a budget k, *influence maximization* is the task of selecting at most k seed nodes in V such that the influence spread is maximized, i.e., finding the optimal seeds $S^* = S^*(k, p) \subseteq V$ such that $S^* = \text{argmax}_{S \subseteq V, |S| \leq k}\, \sigma(S, p)$.

Kempe et al. [14] show that the influence maximization problem is NP-hard in both the IC and LT models, and they propose the following *greedy algorithm*. Given influence probability function p, the *marginal influence (MI)* of any node $v \in V$ under any seed set S is defined as $MI(v|S, p) = \sigma(S \cup \{v\}, p) - \sigma(S, p)$. The greedy algorithm selects k seeds in the following k iterations: (a) let $S_0 = \emptyset$; (b) for each iteration $j = 1, 2, \ldots, k$, find node $v_j = \text{argmax}_{v \in V \setminus S_{j-1}}\, MI(v|S_{j-1}, p)$, and adds v_j into S_{j-1} to obtain S_j; (c) output seed set $S^g(k, p) = S_k$.

It is shown in [14] that the greedy algorithm selects a seed set $S^g(k, p)$ with approximation ratio $1 - \frac{1}{e} - \varepsilon$ for any small $\varepsilon > 0$ (i.e., $\sigma(S^g(k, p), p) \geq \left(1 - \frac{1}{e} - \varepsilon\right)\sigma(S^*, p)$), where ε accommodates the inaccuracy in Monte Carlo simulations to estimate the marginal influence.

Topic-aware independent cascade model and topic-aware influence maximization. Topic-aware independent cascade (TIC) model [2] is an extension of the IC model to incorporate topic mixtures in any diffusion item. Suppose there are d base topics, and we use set notation $[d] = \{1, 2, \cdots, d\}$ to denote topic $1, 2, \cdots, d$. We regard each diffusion item as a distribution of these topics. Thus, any item can be expressed as a vector $I = (\lambda_1, \lambda_2, \ldots, \lambda_d) \in [0,1]^d$ where $\sum_{i \in [d]} \lambda_i = 1$. We also refer such a vector I as a *topic mixture*. Given a directed social graph $G = (V, E)$, influence probability on any topic $i \in [d]$ is $p_i : V \times V \to [0,1]$, and we assume $p_i(u,v) = 0$ for all $(u,v) \notin E$ or $u = v$. In the TIC model, the influence probability function p for any diffusion item I is defined as $p(u,v) = \sum_{i \in [d]} \lambda_i p_i(u,v)$, for all $u, v \in V$ (or simply $p = \sum_{i \in [d]} \lambda_i p_i$). Then, the stochastic diffusion process and influence spread $\sigma(S, p)$ are exactly the same as defined in the IC model by using the influence probability p on edges.

Given a social graph G, base topics $[d]$, influence probability function p_i for each base topic i, a budget k and an item $I = (\lambda_1, \lambda_2, \ldots, \lambda_d)$, the *topic-aware influence maximization* is the task of finding optimal seeds $S^* = S^*(k, p) \subseteq V$ such that $S^* = \text{argmax}_{S \subseteq V, |S| \leq k}\, \sigma(S, p)$, where $p = \sum_{i \in [d]} \lambda_i p_i$.

3   Preprocessing Based Algorithms

Topic-aware influence maximization can be solved by using existing influence maximization algorithms such as the ones in [14,21]: when a query on an item $I = (\lambda_1, \lambda_2, \cdots, \lambda_d)$ comes, the algorithm first computes the mixed influence probability function $p = \sum_j \lambda_j p_j$, and then applies existing algorithms using parameter p. This, however, means that for each topic mixture influence maximization has to be carried out from scratch. It may take from half a minute to several hours to find the seed sets in large-scale networks, which could be inefficient or impractical for online scenarios.

In this paper, we are able to obtain datasets from two prior studies, one is on social movie rating network Flixster [2] and the other is on academic collaboration network Arnetminer [20], to help design our algorithms. Due to the space limit, the full data analysis can be found in [7], and we briefly summarize two key observations we made as follows: (1) Topic separation in terms of influence probabilities is network dependent: In the Arnetminer network, topics are mostly separated among different edges and nodes in the network, while in the Flixster network there are significant overlaps on topics among nodes and edges; (2) Most seeds for topic mixtures come from the seeds of constituent topics, in both Arnetminer and Flixster networks. In this section, motivated by the above observations, we introduce two preprocessing based algorithms that cover different design choices.

3.1   Best Topic Selection (BTS) Algorithm

Our first algorithm is to minimize online processing by simply selecting a seed set from one of the constituent topics that has the best influence spread in the topic mixture, and thus we call it *Best Topic Selection (BTS)* algorithm. Since the query of item $I = (\lambda_1, \lambda_2, \cdots, \lambda_d)$ may be arbitrary, our key idea is to apply a bucketing technique to establish landmarks for each topic in the preprocessing stage, and use properties of upper and lower landmarks to bound the error in the online stage, as we explain in more detail now.

Preprocess Stage. Denote constant set $\Lambda = \{\lambda_0^c, \lambda_1^c, \cdots, \lambda_m^c\}$ as a set of *landmarks*, where $0 = \lambda_0^c < \lambda_1^c < \cdots < \lambda_m^c = 1$. For each $\lambda \in \Lambda$ and each topic $i \in [d]$, we pre-compute $S^g(k, \lambda p_i)$ and $\sigma(S^g(k, \lambda p_i), \lambda p_i)$ in the preprocessing stage, and store these values for online processing. In our experiments, we use uniformly selected landmarks because they are good enough for influence maximization and can adopt parallel optimization. More sophisticated landmark selection method may be applied, such as the machine learning based method in [1].

Online Stage. We define two rounding notations that return one of the neighboring landmarks in $\Lambda = \{\lambda_0^c, \lambda_1^c, \cdots, \lambda_m^c\}$: given any $\lambda \in [0,1]$, let $\underline{\lambda} = \lambda_j^c$ such that $\lambda_j^c \leq \lambda < \lambda_{j+1}^c$, and $\overline{\lambda} = \lambda_{j+1}^c$ such that $\lambda_j^c < \lambda \leq \lambda_{j+1}^c$. Given $I = (\lambda_1, \lambda_2, \cdots, \lambda_d)$, let $D_I^+ = \{i \in [d] \mid \lambda_i > 0\}$. With the pre-computed $S^g(k, \lambda p_i)$ and $\sigma(S^g(k, \lambda p_i), \lambda p_i)$ for every $\lambda \in \Lambda$ and every topic i, the BTS algorithm is given in Algorithm 1. The algorithm basically rounds down the mixing coefficient on every topic to $(\underline{\lambda}_1, \cdots, \underline{\lambda}_d)$, and then returns the seed set $S^g(k, \underline{\lambda}_{i'} p_{i'})$ that gives the largest influence spread at the round-down landmarks.

In this paper, BTS is used as a baseline for preprocessing based algorithms. Although BTS is rather simple, we show below that it could provide theoretical guarantee with a certain condition.

Algorithm 1. Best Topic Selection (BTS) Algorithm

Require: $G = (V, E)$, k, $\{p_i \mid i \in [d]\}$, $I = (\lambda_1, \cdots, \lambda_d)$, Λ, $S^g(k, \lambda p_i)$ and $\sigma(S^g(k, \lambda p_i), \lambda p_i)$, $\forall \lambda \in \Lambda, \forall i \in [d]$.

1: $I' = (\underline{\lambda}_1, \cdots, \underline{\lambda}_d)$
2: $i' = \mathrm{argmax}_{i \in D_I^+} \sigma(S^g(k, \underline{\lambda}_i p_i), \underline{\lambda}_i p_i)$
3: **return** $S^g(k, \underline{\lambda}_{i'} p_{i'})$

We say that $\sigma(S, p)$ is *c-sub-additive in* p for some constant c if for any $S \subseteq V$ with $|S| \leq k$ and any $I = (\lambda_1, \ldots, \lambda_d)$, $\sigma(S, \sum_{i \in D_I^+} \lambda_i p_i) \leq c \sum_{i \in D_I^+} \sigma(S, \lambda_i p_i)$. The sub-additivity property above means that the influence spread of any seed set S in any topic mixture will not exceed constant times of the sum of the influence spread for each individual topic. It is easy to verify that, when each topic in the network does not interfere with each other, $\sigma(S, p)$ is 1-sub-additive. The counterexample we could find that violates the c-sub-additivity assumption is a tree structure where even layer edges are for one topic and odd layer edges are for another topic. Such structures are rather artificial, and we believe that for real networks the influence spread is c-sub-additive in p with a reasonably small c.

We define $\mu_{\max} = \max_{i \in [d], \lambda \in [0,1]} \frac{\sigma(S^g(k, \bar{\lambda} p_i), \bar{\lambda} p_i)}{\sigma(S^g(k, \underline{\lambda} p_i), \underline{\lambda} p_i)}$, which is a value controlled by preprocessing. A fine-grained landmark set Λ could make $\mu_{\max}$ close to 1. The following Theorem 1 guarantees the approximation ratio of Algorithm 1.

Theorem 1. *If the influence spread function* $\sigma(S, p)$ *is c-sub-additive in* p, *Algorithm 1 achieves* $\frac{1 - e^{-1}}{c|D_I^+|\mu_{\max}}$ *approximation ratio for item* $I = (\lambda_1, \lambda_2, \cdots, \lambda_d)$.

The approximation ratio given in the theorem is a conservative bound for the worst case (e.g., a common setting may be $c = 1.2$, $\mu_{\max} = 1.5$, $|D_I^+| = 2$). Tighter online bound in our experiment section based on [15] shows that Algorithm 1 performs much better than the worst case scenario.

3.2 Marginal Influence Sort (MIS) Algorithm

Our second algorithm derives the seed set from constituent topics, and moreover it utilizes pre-computed marginal influence from different topics to select seeds. Our idea is partially motivated by our data observation, especially for the Arnetminer dataset, which shows that in some cases the network could be well separated among different topics. Intuitively, if nodes are separable among different topics, and each node v is only pertinent to one topic i, the marginal influence of v would not change much whether it is for a mixed item or the pure topic i, as formally characterized in the following. Given threshold $\theta \geq 0$, define node set $\nu_i(\theta) = \{v \in V \mid \sum_{u:(v,u) \in E} p_i(v, u) + \sum_{u:(u,v) \in E} p_i(u, v) > \theta\}$ for every topic i, and *node overlap coefficient* for topic i and j as $R_{ij}^V(\theta) = \frac{|\nu_i(\theta) \cap \nu_j(\theta)|}{\min\{|\nu_i(\theta)|, |\nu_j(\theta)|\}}$. If θ is small and the overlap coefficient is small, it means that the two topics are

---

Algorithm 2. Marginal Influence Sort (MIS) Algorithm

Require: $G = (V, E)$, k, $\{p_i \mid i \in [d]\}$, $I = (\lambda_1, \cdots, \lambda_d)$, Λ, $S^g(k, \lambda p_i)$ and
    $MI^g(v, \lambda p_i)$, $\forall \lambda \in \Lambda$, $\forall i \in [d]$.
1: $I' = (\underline{\lambda}_1, \cdots, \underline{\lambda}_d)$
2: $V^g = \cup_{i \in [d], \underline{\lambda}_i > 0} S^g(k, \underline{\lambda}_i p_i)$
3: **for** $v \in V^g$ **do**
4:     $f(v) = \sum_{i \in [d], \underline{\lambda}_i > 0} MI^g(v, \underline{\lambda}_i p_i)$
5: **end for**
6: **return** top k nodes with the largest $f(v), \forall v \in V^g$

---

fairly separated in the network. In particular, we say that the network is *fully separable* for topics i and j if $R_{ij}^V(0) = 0$, and it is fully separable for all topics if $R_{ij}^V(0) = 0$ for any pair of i and j with $i \neq j$.

Lemma 1. *If a network is fully separable among all topics, then for any $v \in V$ and topic $i \in [d]$ such that $\sigma(v, p_i) > 1$, for any item $I = (\lambda_1, \lambda_2, \ldots, \lambda_d)$, for any seed set $S \subseteq V$, we have $MI(v|S, \lambda_i p_i) = MI(v|S, p)$, where $p = \sum_{j \in [d]} \lambda_j p_j$.*

Lemma 1 suggests that we can use the marginal influence of a node on each topic when dealing with a topic mixture. Algorithm MIS is based on this idea.

Preprocess stage. Recall the detail of greedy algorithm, given probability p and budget k, for iteration $j = 1, 2, \cdots, k$, it calculates v_j to maximize marginal influence $MI(v_j|S_{j-1}, p)$ and let $S_j = S_{j-1} \cup \{v_j\}$ every time, and output $S^g(k, p) = S_k$ as seeds. Denote $MI^g(v_j, p) = MI(v_j|S_{j-1}, p)$, if $v_j \in S^g(k, p)$, and 0 otherwise. Therefore, $MI^g(v_j, p)$ is the marginal influence of v_j according to the greedy selection order. Suppose the landmark set $\Lambda = \{\lambda_0^c, \lambda_1^c, \lambda_2^c, \cdots, \lambda_m^c\}$. For every $\lambda \in \Lambda$ and every single topic $i \in [d]$, we pre-compute $S^g(k, \lambda p_i)$, and cache $MI^g(v, \lambda p_i), \forall v \in S^g(k, \lambda p_i)$ in advance.

Online stage. Marginal Influence Sort (MIS) algorithm is described in Algorithm 2. Given an item $I = (\lambda_1, \cdots, \lambda_d)$, it first rounding down the mixture, and then use the union of seed sets as candidates. If a seed node appears multiple times in pre-computed topics, we approximate by summing the marginal influence in each topic together. Then we sort all candidates according to the computed marginal influence, and select top-k nodes as seeds.

Theorem 2. *Suppose $I = (\lambda_1, \lambda_2, \cdots, \lambda_d)$, where each $\lambda_i \in \Lambda$, and $S^g(k, \lambda_1 p_1)$, $\cdots$, $S^g(k, \lambda_d p_d)$ are disjoint. If the network is fully separable for all topics, the seed set calculated by Algorithm 2 is one of the possible sequences generated by greedy algorithm under the mixed influence probability $p = \sum_{i \in [d]} \lambda_i p_i$.*

Although MIS is a heuristic algorithm, this theorem implies that the seed set S from MIS satisfies $\sigma(S, p) \geq (1 - e^{-1} - \epsilon)\sigma(S^*, p)$ (for any $\epsilon > 0$) compared with the optimal S^* in fully separable networks. It suggests that MIS would work well for networks that are fairly separated among different topics, which are verified by our test results on the Arnetminer dataset. Moreover, even for networks that are not well separated, it is reasonable to assume that the marginal influence of nodes in the mixture can be approximated by the sum of the marginal influence in individual topics, and thus we expect MIS to work also competitively in this case, which is verified by our test results on the Flixster dataset.

4 Empirical Evaluation

We test the effectiveness of our algorithms by using multiple real-world datasets, and compare them with state-of-the-art influence maximization algorithms.

Data descriptions. The first dataset is on social movie rating network Flixster [2], an American social movie site for discovering new movies, learning about movies, and meeting others with similar tastes in movies. The Flixster network represents users as nodes, and two users u and v are connected by a directed edge (u, v) if they are friends both rating the same movie and v rates the movie shortly later after u does so. The network contains 29357 nodes, 425228 directed edges and 10 topics. We eliminate individual probabilities that are too weak ($\forall i \in [d], \lambda_i < 0.01$). We also obtain 11659 topic mixtures, from which we found that predominant ones are single topic (96.79%) or two-topic mixtures (3.04%). Mixtures with three or four topics are already rare and there are no items with five or more topics.

The second dataset is on the academic collaboration network Arnetminer [20], which is a free online service used to index and search academic social networks. The Arnetminer network represents authors as nodes and two authors have an edge if they coauthored a paper. It contains 5114 nodes, 34334 directed edges and 8 topics, and all 8 topics are related to computer science, such as data mining, machine learning, information retrieval, etc.

The above two datasets act as the baseline to verify the effectiveness of the algorithms. Furthermore, we use a larger academic collaboration network data DBLP maintained by Michael Ley (650K nodes and 2 million edges) only to test the scalability of the algorithms.

Influence probabilities. We first test our algorithms on the Flixster and Arnetminer datasets, whose influence probabilities are learned from real action trace data or node topic distribution data. The basic statistics for the learned influence probabilities show similar behavior between the two datasets, such as mean probabilities for each topic are mostly between 0.1 and 0.2, standard deviations (SD) are mostly between 0.1 and 0.3, etc. (Take the average over all topics: Arnetminer mean=0.173, SD=0.227; Flixster mean=0.131, SD=0.187.)

As DBLP does not have influence probabilities, we simulate two topics according to the joint distribution of topics 1 and 2 in the Flixster, and follow the practice of the TRIVALENCY model in [21] to rescale it into $\{0.1, 0.01, 0.001\}$ (i.e., strong, medium, and low influence).

Topic mixtures. In terms of topic mixtures, in practice and also supported by our data, an item is usually a mixture of a small number of topics thus our tests focus on testing topic mixtures from two topics. First, we test random samples to cover most common mixtures. We draw 50 topic mixtures from the uniform distribution over the polytope of any two topics. Second, since we have the data of real topic mixtures in Flixster dataset, we also test additional 50 cases following the same sampling technique described in Section 3.1 of [1], which estimates the Dirichlet distribution that maximizes the likelihood first and then generates topic mixtures by sampling from the distribution.

Algorithms for comparison. In our experiments, we test our topic-aware preprocessing based algorithms MIS and BTS comprehensively. Three classes of algorithms are selected for comparison: (a) Topic-aware algorithms: The topic-aware greedy algorithm (TA-Greedy) and a state-of-the-art fast heuristic algorithm PMIA (TA-PMIA) [21]; (b) Topic-oblivious algorithms: The topic-oblivious greedy algorithm (TO-Greedy), degree algorithm (TO-Degree) and random algorithm (Random); (c) Simple and fast heuristic algorithms that do not need preprocessing: The topic-aware PageRank (TA-PageRank) [4] and WeightedDegree (TA-WeightedDegree) [21] algorithms.

In this paper, we employ the greedy algorithm [15] with lazy evaluation and the same approximation ratio to provide hundreds of time of speedup to the original one [14]. PMIA is a fast heuristic algorithm based on trimming influence propagation to a tree structure, and it achieves thousand fold speedup comparing to optimized greedy algorithms with a small degradation on influence spread [21] (we set a small threshold $\theta = 1/1280$ to alleviate the degradation).

Topic-oblivious algorithms work under previous IC model that does not identify topics (the uniform topic mixture). TO-Greedy runs greedy algorithm for previous IC model. TO-Degree outputs the top-k nodes with the largest degree based on the original graph. Random simply chooses k nodes at random.

Finally, we study the possibility of acceleration for large graphs by comparing PMIA with greedy algorithm in preprocessing stage, and denote MIS and BTS algorithms as MIS[Greedy], BTS[Greedy] and MIS[PMIA], BTS[PMIA], respectively.

In the preprocessing stage, we use two algorithms, Greedy and PMIA, to pre-compute seed sets for MIS and BTS, except that for the DBLP dataset, which is too large to run the greedy algorithm, we only run PMIA. In our tests, we use 11 equally distant landmarks $\Lambda = \{0, 0.1, 0.2, \ldots, 1\}$ for MIS and BTS. Each landmark is independent and can be pre-computed concurrently in different processes. We choose $k = 50$ seeds in all our tests and compare the influence spread and running time, and take the average of 10000 Monte Carlo simulations to obtain the influence spread for each seed set in the greedy algorithm. In addition, we apply offline bound (the influence spread of any greedy seeds multiplied by factor $1/(1 - e^{-1})$) and online bound (Theorem 4 in [15]) to estimate influence spread of optimal solutions.

All experiments are conducted on a computer with 2.4GHz Intel(R) Xeon(R) E5530 CPU, 2 processors (16 cores), 48G memory, and Windows Server 2008 R2 (64 bits). The code is written in C++ and compiled by Visual Studio 2010.

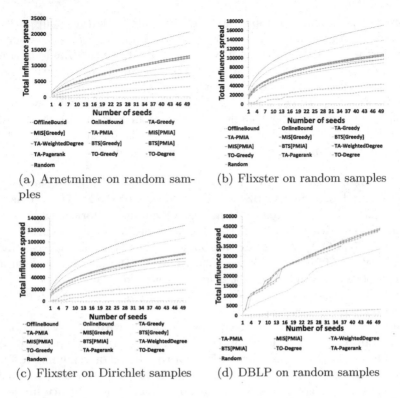

(a) Arnetminer on random samples

(b) Flixster on random samples

(c) Flixster on Dirichlet samples

(d) DBLP on random samples

Fig. 1. Influence spread of algorithms. Legends are ordered (left to right, top to bottom) according to influence spread.

Influence spread. Figure 1 shows the total influence spread results on Arnetminer with random samples (a); Flixster with random and Dirichlet samples, (b) and (c), respectively; and DBLP with random samples (d). For the Arnetminer dataset, it clearly separates all algorithms into three tiers (all percentages reported in parentheses are the gap of ratio compared with the best algorithm after taking average from one seed to 50 seeds): the top tier is TA-Greedy, TA-PMIA (0.61%), MIS[Greedy] (0.32%) and MIS[PMIA] (1.08%) whose gaps are negligible; the middle tier is TA-WeightedDegree (4.06%), BTS[Greedy] (4.68%), BTS[PMIA] (4.67%) and TA-PageRank (26.84%); and the lower tier is topic-oblivious algorithms TO-Greedy (28.57%), TO-Degree (56.75%) and Random (81.48%). Besides, MIS[Greedy] and BTS[Greedy] are 76.9% and 72.5% of the online bound, which are better than their conservative theoretical bounds ($1 - e^{-1} \approx 63.2\%$). For Flixster dataset we see that the influence spread of TA-PMIA, MIS[Greedy], MIS[PMIA], BTS[Greedy] and BTS[PMIA] are 1.78%, 3.04%, 4.58%, 3.89% and 5.29% smaller than TA-Greedy for random samples, and 1.41%, 1.94%, 3.37%, 2.31% and 3.59% smaller for Dirichlet samples, respectively, indicating that our preprocessing based algorithms can perform quite well.

Table 1. Running time statistics

(a) Preprocessing time

Arnetminer ($|A| = 8 \times 11$)		Flixster ($|A| = 10 \times 11$)		DBLP ($|A| = 2 \times 11$)	
Total	Max	Total	Max	Total	Max
Greedy 8.8 hrs	1.2 hrs	26.3 days	3.5 days	$\geq$ 100 days	$\geq$ 7 days
PMIA 37 secs	7.1 secs	2.28 hrs	12.6 mins	9.6 mins	4.2 mins

(b) Average online response time

	Arnetminer	Flixster		DBLP
		random	Dirichlet	
TA-Greedy	9.3 mins	1.5 days	20 hrs	N/A
TA-PMIA	0.52 sec	5.5 mins	3.8 mins	58 secs
MIS	2.85 s	2.37 s	3.84 s	2.09 s
BTS	1.20 s	2.35 s	1.42 s	0.49 s
TA-PageRank	0.15 sec	2.08 secs	2.30 secs	41 secs
TA-WeightedDegree	8.5 ms	29.9 ms	30.7 ms	0.32 sec

Running time. We summarize both of the preprocessing time and average online response time in Table 1. Table 1(b) shows the average online response time of different algorithms in finding 50 seeds (topic-oblivious algorithms always use the same seeds and thus are not reported). Our proposed MIS emerges as a strong candidate for fast real-time processing of topic-aware influence maximization task: it achieves microsecond response time, which does not depend on graph size or influence probability parameters, while its influence spread matches or is very close to the best greedy algorithm and outperforms other simple heuristics (Figure 1). Table 1(a) shows the preprocessing time based on greedy algorithm and PMIA algorithm on three datasets. It indicates that the greedy algorithm is suitable for small graphs but infeasible for large graphs like DBLP. PMIA is a viable choice for preprocessing, and our MIS using PMIA as the preprocessing algorithm achieves almost the same influence spread as MIS using the greedy algorithm for preprocessing (Figure 1).

5   Related Work

Domingos and Richardson [9,18] are the first to study influence maximization in an algorithmic framework. Kempe et al. [14] first formulate the discrete influence diffusion models including the independent cascade model and linear threshold model, and provide algorithmic results on influence maximization.

A large body of work follows the framework of [14]. One line of research improves on the efficiency and scalability of influence maximization algorithms [8,11,12,21]. Others extend the diffusion models and study other related optimization problems [3,5,13]. A number of studies propose machine learning methods to learn influence models and parameters [10,19,20]. A few studies look into the interplay of social influence and topic distributions [16,17,20,23]. They focus on inference of social influence from topic distributions or joint inference of influence diffusion and topic distributions. They do not provide a dynamic topic-aware influence diffusion model or study the influence maximization problem. Barbieri et al. [2] introduce the topic-aware influence diffusion models TIC and TLT as extensions to the IC and LT models. They provide maximum-likelihood based learning method to learn influence parameters in these topic-aware models. We use their proposed models and datasets with the learned parameters.

A recent independent work by Aslay et al. [1] is the closest one to our work. Their work focuses on index building in the query space while we use pre-computed marginal influence to help guiding seed selection, and thus the two approaches are complementary. Other differences have been listed in the introduction and will not be repeated here.

6 Future Work

One possible follow-up work is to combine the advantages of our approach and the approach in [1] to further improve the performance. Another direction is to study fast algorithms with stronger theoretical guarantee. An important work is to gather more real-world datasets and conduct a thorough investigation on the topic-wise influence properties of different networks, similar to our preliminary investigation on Arnetminer and Flixster datasets. This could bring more insights to the interplay between topic distributions and influence diffusion, which could guide future algorithm design.

References

1. Aslay, C., Barbieri, N., Bonchi, F., Baeza-Yates, R.: Online topic-aware influence maximization queries. In: EDBT (2014)
2. Barbieri, N., Bonchi, F., Manco, G.: Topic-aware social influence propagation models. In: ICDM (2012)
3. Bhagat, S., Goyal, A., Lakshmanan, L.V.S.: Maximizing product adoption in social networks. In: WSDM (2012)
4. Brin, S., Page, L.: The anatomy of a large-scale hypertextual Web search engine. Computer Networks and Isdn Systems 30, 107–117 (1998)
5. Budak, C., Agrawal, D., Abbadi, A.E.: Limiting the spread of misinformation in social networks. In: WWW (2011)
6. Chen, W., Lakshmanan, L.V., Castillo, C.: Information and Influence Propagation in Social Networks, vol. 5. Morgan & Claypool (2013)
7. Chen, W., Lin, T., Yang, C.: Real-time topic-aware influence maximization using preprocessing (2014). arXiv preprint arXiv:1403.0057
8. Chen, W., Wang, Y., Yang, S.: Efficient influence maximization in social networks. In: KDD (2009)
9. Domingos, P., Richardson, M.: Mining the network value of customers. In: KDD (2001)
10. Goyal, A., Bonchi, F., Lakshmanan, L.V.: Learning influence probabilities in social networks. In: WSDM (2010)
11. Goyal, A., Lu, W., Lakshmanan, L.V.: Celf++: optimizing the greedy algorithm for influence maximization in social networks. In: WWW (2011)
12. Goyal, A., Lu, W., Lakshmanan, L.V.: Simpath: an efficient algorithm for influence maximization under the linear threshold model. In: ICDM (2011)
13. He, X., Song, G., Chen, W., Jiang, Q.: Influence blocking maximization in social networks under the competitive linear threshold model. In: SDM (2012)
14. Kempe, D., Kleinberg, J., Tardos, É.: Maximizing the spread of influence through a social network. In: KDD (2003)
15. Leskovec, J., Krause, A., Guestrin, C., Faloutsos, C., VanBriesen, J., Glance, N.: Cost-effective outbreak detection in networks. In: KDD (2007)

16. Lin, C.X., Mei, Q., Han, J., Jiang, Y., Danilevsky, M.: The joint inference of topic diffusion and evolution in social communities. In: ICDM (2011)
17. Liu, L., Tang, J., Han, J., Jiang, M., Yang, S.: Mining topic-level influence in heterogeneous networks. In: CIKM (2010)
18. Richardson, M., Domingos, P.: Mining knowledge-sharing sites for viral marketing. In: KDD (2002)
19. Saito, K., Nakano, R., Kimura, M.: Prediction of information diffusion probabilities for independent cascade model. In: Lovrek, I., Howlett, R.J., Jain, L.C. (eds.) KES 2008, Part III. LNCS (LNAI), vol. 5179, pp. 67–75. Springer, Heidelberg (2008)
20. Tang, J., Sun, J., Wang, C., Yang, Z.: Social influence analysis in large-scale networks. In: KDD (2009)
21. Wang, C., Chen, W., Wang, Y.: Scalable influence maximization for independent cascade model in large-scale social networks. DMKD **25**(3), 545–576 (2012)
22. Wang, C., Yu, X., Li, Y., Zhai, C., Han, J.: Content coverage maximization on word networks for hierarchical topic summarization. In: CIKM (2013)
23. Weng, J., Lim, E.-P., Jiang, J., He, Q.: Twitterrank: finding topic-sensitive influential twitterers. In: WSDM (2010)

Pricing in Social Networks
with Negative Externalities

Zhigang Cao[1], Xujin Chen[1], Xiaodong Hu[1], and Changjun Wang[2](✉)

[1] Academy of Mathematics and Systems Science,
Chinese Academy of Sciences, Beijing 100190, China
{zhigangcao,xchen,xdhu}@amss.ac.cn
[2] Beijing Center for Scientific and Engineering Computing,
Beijing University of Technology, Beijing, China
wcj@amss.ac.cn

Abstract. We study the problems of pricing an indivisible product to consumers who are embedded in a given social network. The goal is to maximize the revenue of the seller. We assume impatient consumers who buy the product as soon as the seller posts a price not greater than their valuations of the product. The product's value for a consumer is determined by two factors: a fixed consumer-specified intrinsic value and a variable externality that is exerted from the consumer's neighbors in a linear way. We study the scenario of negative externalities, which captures many interesting situations, but is much less understood in comparison with its positive externality counterpart. We assume complete information about the network, consumers' intrinsic values, and the negative externalities. The maximum revenue is in general achieved by iterative pricing, which offers impatient consumers a sequence of prices over time. We prove that it is NP-hard to find an optimal iterative pricing, even for unweighted tree networks with uniform intrinsic values. Complementary to the hardness result, we design a 2-approximation algorithm for finding iterative pricing in general weighted networks with (possibly) nonuniform intrinsic values. We show that, as an approximation to optimal iterative pricing, single pricing works rather well for many interesting cases, such as forests, Erdős-Rényi networks and Barabási-Albert networks, although its worst-case performance can be arbitrarily bad.

1 Introduction

People interact with and influence each other to a degree that is beyond most of us can imagine. The magnitude of this connection has been upgraded to a brandnew level by the proliferation of online SNS (Social Network Services, e.g., Facebook, Twitter, and SinaWeibo). Numerous business opportunities are being incubated by this upgrading. Yet, its consequences are far from being

Supported in part by NNSF of China under Grant No. 11222109 and 11471326, 973 Project of China under Grant No. 2011CB80800, and CAS Program for Cross & Cooperative Team of Science & Technology Innovation.

© Springer International Publishing Switzerland 2015
M.T. Thai et al. (Eds.): CSoNet 2015, LNCS 9197, pp. 14–25, 2015.
DOI: 10.1007/978-3-319-21786-4_2

fully unfolded or understood, leaving many fascinating questions for scientists in a variety of disciplines to answer. One incredible fact in the SNS era is that we are now able to know the complete network of who is connected with whom. Network marketing and pricing, with this assistance, could be much more precise and flexible than traditional counterparts, and are attracting increasing attention from both industry and academia. In this paper, we study, from an algorithmic point of view, how a monopolist seller should price to the consumers connected by a known social network.

Consumption is often not a completely private thing. As opposed to standard economic settings, the utilities that a consumer obtains from consuming many kinds of goods, are not determined merely by her private needs and the functions and qualities of the goods, but also greatly affected by the consumptions of her social network neighbors. For example, the reason that we wear clothes is not only to cover ourselves from cold, but usually also to make other people think that we look great. This social side of consumption is becoming more and more prominent with the unification of E-commerce and SNS. It is now very convenient for us to share with our friends our shopping results. By clicking one more button at the time we pay for the skirt online, all our SNS friends may know immediately the information of this skirt. This effect could be much stronger and faster than face-to-face sharing. Our ladybros may think the skirt terrific and get one too; or oppositely, they may prefer later a different style to avoid outfit clash. The former case is typical *positive externality*: the incentive that a consumer wants to buy a product increases as more of her social network neighbors have it. The latter opposite scenario where the incentive decreases when more neighbors possess the product, referred to as *negative externality*, is the focus of this paper. Positive externalities are prevalent in many aspects of the society and have been extensively studied under various terminologies (herding, cascading, Matthew effect, viral marketing, to name a few). Negative externalities, in contrast, although widely exist too, are much less investigated.

Pricing with Negative Externalities. We concentrate on the negative externality among consumers of consuming a single kind of product, which is usually luxury or fashionable one. An important reason that a consumer buys this product is to showoff in front of her friends (also referred to as *invidious consumption* in literature). Naturally, a consumer buys the product if the price is not higher than her valuation of the product, which is the sum of her constant *intrinsic value* and varying *external value*. We propose and study the typical network pricing model, where the external value is the (weighted) number of people to whom the consumer can showoff (i.e., her social network neighbors who do not possess this product). We study, to obtain a maximum revenue, how a monopolist seller should price such a product with negative externality to consumers connected by a link-weighted social network, where the revenue is the total payment the seller receives, and the nonnegative integer link weights represent the influences between consumers. We assume that the seller have complete information of both the social network and the consumers' intrinsic values. Our study falls into the framework of *uniform pricing*, where at any time point the same take-it-or-leave-it price is offered (posted) to all consumers who have not bought the product. The seller adopts a strategy of *iterative pricing* – posting different prices sequentially at discrete time points, to maximize her revenue (we assume that production costs are zero). We also assume that the consumers are *myopic* (a.k.a. *impatient*) in the sense that they, when making purchase decisions, do not take into account their neighbors' future actions (which will change their external values of the product).

Contributions. Comparing with their positive counterparts, negative external-ities possess more irregularity and pose more challenges for research on product diffusion, especially from the perspective of pricing. The intuitive hardness is confirmed by the following theoretical intractability. By a reduction from the 3SAT problem we show that finding an optimal iterative pricing is NP-hard even for the extremely simple case of unweighted tree network with uniform intrinsic values (Theorem 1).

Complementary to the hardness result, we design a 2-approximation algo-rithm for iterative pricing in general weighted networks with general intrinsic values (Theorem 2). An exact $O(n^2)$-time algorithm is designed for unweighted split networks with uniform intrinsic values, where n is the number of consumers (Theorem 3).

We also study single pricing as an approximation of iterative pricing, and obtain the following negative and positive results. We prove that optimal sin-gle pricing can be arbitrarily worse (at a rate of $\ln \ln n$) than the optimal iterative pricing (Example 1); and on the other hand, the best single pricing provides nice approximations to the optimal iterative pricing for several well-known unweighted networks with uniform intrinsic values: $(\ln n)$-approximation for general networks (Theorem 7), 1.5-approximation for forest networks (Theo-rem 4), $(1 + \epsilon)$-approximation a.a.s for Erdős-Rényi networks (Theorem 5), and 2-approximation a.a.s. for Barabási-Albert networks (Theorem 6).

Organization. The remainder of the paper is organized as follows. Section 2 provides a literature review. Section 3 gives the mathematical formulation of our iterative pricing model. Section 4 is devoted to general iterative pricing. Section 5 discusses the relation between single pricing and iterative pricing. Section 6 concludes the paper with remarks on future research. The proofs and details omitted can be found in the full version [11].

2   Related Work

In the economics literature, the importance of network effects and network exter-nalities in business began to attract serious attention around three decades ago ([15, 20]). Under the most popular frameworks, network effects are assumed to be global instead of local. Namely, only complete networks are considered. Con-sumers may also act sequentially as in this paper, but are usually assumed to be completely rational in the way that they are able to forecast the decisions of later ones and make their purchase decisions accordingly. There are quite a lot of followups, most of which are beyond the scope of this paper. We refer the reader to [23] for a most recent development in this paradigm with relaxations of assumptions on consumers.

In the literature of computer science, network pricing stems mainly from the study of diffusion and cascading. One of the most important differences between this strand of research and that of economics is arguably that network structures are explicitly and seriously addressed. Over the last decade, under the framework of viral marketing, the algorithmic study of diffusing products with positive externalities is especially fruitful for influence maximization, see, e.g., [12, 21, 22]. To the best of our knowledge, Hartline et al. [18] was the first to study the diffusion problem from a network pricing perspective. They investigated marketing strategies for revenue maximization with positive externalities. Consumers are visited in a sequence (determined by the seller), and asked whether to buy or not under some price (different consumers may receive different prices, referred to as differential pricing or discriminative pricing). They showed that for myopic consumers, a reasonable approximation of the optimal marketing strategy can be achieved in a simple way of influence-and-exploit. While complete information was assumed in [18], Chen et al. [13] studied the incomplete information model with rational players and positive externalities. They provided a polynomial time algorithm that computes all the pessimistic (and optimistic) equilibria and the optimal single price. When discriminative pricing is allowed, they proved the NP-hardness of optimal equilibrium computation, and gave an FPTAS for the case that consumers are already partitioned into groups such that those within the same group must receive the same price.

Iterative pricing, with a very limited literature, was discussed by Akhlaghpour et al. [1] for positive externalities. The authors studied two iterative pricing models in which consumers are assumed to be myopic. In the first model, they gave an FPTAS for the optimal pricing strategy in the general case. In the second model, they showed that the revenue maximization problem is inapproximable even in some special case. Their second model is quite similar to ours.

Although there is also a large literature in the field of classical economics studying negative externalities (under various terms, e.g. the Veblen effect, the snob effect, the congestion effect etc.), explicit networks are rarely treated seriously as aforementioned. One of the classical papers in this strand is [19], where the nulclear weapon selling problem was considered from the perspective of network effects. In the more recent computer science literature, compared with positive externalities, network pricing problems with negative externalities are much less investigated. Chen et al. [13] showed that when both positive and negative externalities are allowed in their model, computing any approximate equilibrium is PPAD-hard. However, the complexity status of the problem in the case with only negative externalities is still unknown. The only paper known to us that deals with the network pricing problem with negative externalities is [5] by Bhattacharya et al., although their main focus is on equilibrium computation for given prices rather than pricing. The authors also considered linear externalities, but a combination of single pricing, complete information and strategic consumers. They showed that for any given price, the game that the consumers play is an exact potential game, and provided a set of hardness results. They proved that finding the best equilibrium is NP-hard even for trees, and gave a 2-approximation algorithm for bipartite networks. Along a different line, Alon et al. [2] used the term "negative externality" to mean the harm of discriminative pricing on consumers (because discriminative pricing gives many consumers a feeling of inequality).

All the papers cited above assume that externalities are only exerted between consumers who buy the product. In contrast, for some products or sevices, e.g., public goods, externalities are exerted also from purchasers to nonpurchasers. Our paper is close to [9] in the sense that both papers address strategic substitutes (each player has less incentive to buy when more neighbors purchase), although the network externalities are negative in our settings but positive in their settings of public goods. In the computer science, the public goods pricing problem was also studied by Feldman et al. [16]. Their work differs from ours in two main respects: (i) In our externality model, a consumer's utility is subtractive over the purchases made by this neighbors, whereas in their setting, purchases of neighbors are substitutes. (ii) Technically, they related the pricing problem (where externalities in their model are mathematically expressed in terms of products of neighbors actions) to a single-item auction problem, while we address the pricing problem (where externalities are expressed in terms of sums of neighbors actions) using iterative algorithmic approaches. As noted by the authors [16], their results carry over to a special kind of negative externality, where the valuation of a consumer on the product is positive if and only if the consumer is the only one among her neighbors who possess the product. The aforementioned literature are all on indivisible goods. The network pricing problems for divisible goods with quadratic utilities functions have been studied in [6,10]. Along with [16], a growing number of papers have been addressing the network externality problem from the perspective of mechanism design and auction theory (e.g. [4,14,17]).

3   The Model

Let $G = (V, E)$ be the given undirected network (without self-loops, and possibly associated with a nonnegative integer weight function $w \in \mathbb{Z}_+^{V \times V}$), where $V \equiv [n]$ is the set of n consumers, and E represents the links between pairs of consumers. When the weight function $w \in \mathbb{Z}_+^{V \times V}$ is discussed, it is always assumed that $w_{ij} = w_{ji}$ for all $i, j \in V$ and $w_{ij} = 0$ if and only if $ij \notin E$. Given any consumer $i \in V$ and subset $S \subseteq V$ of consumers, we use $w_i(S) = \sum_{j \in S} w_{ij}$ to denote the sum of weights contributed to consumer i by those in S. Clearly, only i's neighbors can possibly contribute.

We name the model under investigation as PNC (*Pricing with Negative externalities and Complete information*). Let Q, which usually shrinks as the iterative pricing proceeds, denote the set of consumers who do not possess the product.

Each consumer $i \in V$ has an intrinsic value $\nu(i) \in \mathbb{R}_+$, and her *total value* of the product equals $\nu(i) + w_i(Q)$. Initially $Q = V$. The PNC model proceeds as follows.

- *Iterative pricing.* The monopolist seller announces prices $p_1, p_2, \ldots, p_\tau$ sequentially at time $1, 2, \ldots, \tau$.
- *Impatient consumers.* As soon as a price is announced, a consumer in Q buys the product if and only if her current total value is greater than or equal to the current price.
- *Simultaneous moves.* We assume that, for each newly announced price, all consumers in Q make their decisions (buying or not buying) simultaneously.

Note that a consumer in Q who does not purchase at current time t under price p_t may be willing to buy at a later time $t' > t$ under a lower price $p_{t'} < p_t$. For each $t = 1, 2, \ldots, \tau$, let $B(p_t)$ denote the set of consumers who buy the product at price p_t, (i.e., at time t, or in the t-th *round*). We use $\mathrm{R}(\mathbf{p})$ to denote the revenue derived from $\mathbf{p} = (p_1, p_2, \ldots, p_\tau)$, i.e., $\mathrm{R}(\mathbf{p}) = \sum_{t=1}^{\tau} p_t \cdot |B(p_t)|$. In case of $\mathbf{p} = (p_1)$, we often write $\mathrm{R}(\mathbf{p})$ as $\mathrm{R}(p_1)$. The PNC problem is to find a pricing sequence $\mathbf{p} = (p_1, p_2, \ldots, p_\tau)$ such that $\mathrm{R}(\mathbf{p})$ is maximized, where both the length τ and the entries $p_1, p_2, \ldots, p_\tau$ of the sequence are variables to be determined.

4   General Iterative Pricing

In this section, we study the PNC model in the most general setting where no restriction is imposed to the length of the pricing sequence.

Theorem 1. *In the PNC model, computing an optimal pricing sequence is NP-hard, even when the underlying network is an unweighted tree and all the intrinsic values are zero.*

4.1   2-Approximation

As to approximation, we design a very simple greedy algorithm which performs fairly well. For any subnetwork H of G with node set $V(H)$, and any $i \in V(H)$, we use $d_H^w(i) = \sum_{j \in V(H)} w_{ij}$ to denote the weighted degree of i in H.

Theorem 2. *For the PNC model, Algorithm 1 finds a 2-approximate pricing sequence in $O(n^2)$ time.*

Proof. The 2-approximation follows from the observation that the revenue of the iterative pricing output by the algorithm is $\sum_{i \in V} \nu(u) + \sum_{ij \in E} w_{ij}$, while the revenue of the optimal pricing is no more than $\sum_{i \in V} \nu(u) + 2 \sum_{ij \in E} w_{ij}$.   □

Algorithm 1. Iterative Pricing

Input: Network $G = (V, E)$ with weight $w \in \mathbb{Z}_+^{V \times V}$ and intrinsic value $\nu \in \mathbb{R}_+^V$.
Output: Sequence **p** of prices.
1. $G_0 \leftarrow G, \quad t \leftarrow 0$
2. **While** $V(G_t) \neq \emptyset$ **do**
3. $t \leftarrow t + 1$
4. $p_t \leftarrow \max\{\nu(i) + d_{G_{t-1}}^w(i) : i \in V(G_{t-1})\}$
5. $G_t \leftarrow G_{t-1} \setminus B(p_t)$
6. **End-while**
7. Output **p** $\leftarrow (p_1, p_2, \ldots, p_t)$

4.2 Optimal Pricing for Unweighted Split Networks

Network $G = (V, E)$ is a *split network* if its node set V can be partitioned into two sets C and I such that C induces a clique and I is an independent set of G. Clearly, the nodes in I can only have neighbors in C. In case of each node in I adjacent to exactly one node in C, network G is called *core-peripheral*. Core-peripheral networks are widely accepted as good simplifications of many real-world networks and thus have been extensively studied in various environments [8].

We consider the case of uniform intrinsic values, which can be assumed w.l.o.g. to be zeros. Let $d(v) = d_G(v)$ denote the degree of $v \in V$ in G. Suppose that $C = \{v_1, v_2, \ldots, v_k\}$, and $d(v_i) \leq d(v_{i+1})$ for every $i \in [k-1]$. For each $i \in [k]$, note that $v_1, \ldots, v_i$ form a clique set C_i and their neighbors in I form an independent set I_i, and $C_i \cup I_i$ induces a split subnetwork G_i of G with degree sequence

$$d_{G_i}(u_{\ell_i}^i) \leq d_{G_i}(u_{\ell_i-1}^i) \leq \cdots \leq d_{G_i}(u_1^i) \leq d_{G_i}(v_1) \leq \cdots \leq d_{G_i}(v_i),$$

where $I_i = \{u_1^i, u_2^i, \ldots, u_{\ell_i}^i\}$. Apparently, $d_{G_i}(v_h) = d(v_h) - (k - i)$ for every $h \in [i]$. Consider an optimal pricing **p** $= (p_1, \ldots, p_\tau)$ for the PNC problem on G_i, and write the corresponding maximum revenue as $\text{OPT}(G_i)$. One of the following must hold.

- $p_1 = d_{G_i}(v_{h+1})$ for some $h \in [i-1]$, and exactly $(i-h)$ nodes, i.e., $v_{h+1}, \ldots, v_i$, purchase at price p_1, offering revenue $(i-h)p_1 = (i-h)d_{G_i}(v_{h+1})$. It follows that $\tau \geq 2$ and $(p_2, \ldots, p_\tau)$ is an optimal pricing for G_h, giving $\text{OPT}(G_i) = (i-h)d_{G_i}(v_{h+1}) + \text{OPT}(G_h) = (i-h)d(v_{h+1} - k + i) + \text{OPT}(G_h)$.
- $p_1 = d_{G_i}(u_j^i)$ for some $j \in [\ell_i]$ and exactly $(i+j)$ nodes, i.e., $u_j^i, u_{j-1}^i, \ldots, u_1^i$, $v_1, v_2, \ldots, v_i$, purchase at price p_1, offering revenue $(i+j)p_1 = (i+j)d_{G_i}(u_j^i)$. Since the nodes not purchasing at price p_1 are pairwise nonadjacent, it is easy to see that **p** $= (p_1)$ and $\text{OPT}(G_i) = (i+j) \cdot d_{G_i}(u_j^i)$.

For convenience, let $\text{OPT}(G_0)$ stand for real number 0. Then $\text{OPT}(G) = \text{OPT}(G_k)$ can be computed by the following recursive formula (for $i = 1, 2, \ldots, k$):

$$\text{OPT}(G_i)$$
$$= \max \left\{ \max_{h=0}^{i-1} \{\text{OPT}(G_h) + (d(v_{h+1}) - k + i)(i - h)\}, \max_{j=1}^{\ell_i} \{(j + i) \cdot d_{G_i}(u_j^i)\} \right\}.$$

This formula implies the following result.

Theorem 3. *For the PNC model, an optimal pricing sequence for any unweighted split network with uniform intrinsic values can be found in $O(n^2)$ time by dynamic programming.*

5   Approximation by Single Pricing

Finding an optimal single pricing is trivial because it can be chosen from the n valuations of the consumers. Thus it is natural to ask: How does the optimal single pricing work as an approximation to the optimal iterative pricing? We find that the answer is both "good" and "bad", in the sense that single pricing works rather well for many interesting networks with unit weights and uniform intrinsic values, including forests, Erdős-Rényi networks and Barabási-Albert networks; but in general, its worst-case performance, even when restricted to unweighted networks, can be arbitrarily bad. This justifies the importance of the research of iterative pricing, and at the same time poses the interesting question of investigating the relation between single pricing and iterative pricing for more realistic scenarios.

In this section, we restrict our attention to unweighted networks G with uniform intrinsic values, for which we may assume w.l.o.g. that all intrinsic values are zero, and use $\text{OPT}(G)$ to denote the revenue derived from an optimal iterative pricing.

5.1   1.5-Approximation for Forests

We show that the best single price guarantees an approximate ratio of 1.5 for unweighted forests with uniform intrinsic values.

Theorem 4. *For the PNC model, the single pricing p with maximum $p \cdot |B(p)|$ has an approximation ratio of 1.5 for unweighted forests with uniform intrinsic values.*

Remark 1. In Theorem 4, to achieve the approximation ratio 1.5, the single price can be simply chosen between 1 and 2, whichever produces a larger revenue. Moreover, the ratio 1.5 is tight, as shown by the following tree G.

Tree G with $n = 1 + 2k$ nodes is a spider with center of degree k and each leg of length 2 (i.e., the tree obtain from star $K_{1,k}$ by subdividing each link with a node). It is easy to see that the maximum revenue $3k$ is given by pricing sequence $(k, 1)$. However, any single pricing can produce a revenue of at most $\max\{k \cdot 1, 2 \cdot (k+1), 1 \cdot (2k+1)\} = 2k+2$. The tightness follows from $3k/(2k+2) \to 1.5$ ($k \to \infty$).

5.2 Near Optimal Pricing for Erdős-Rényi Networks

For large n, there is a simple algorithm that is "almost optimal" for "almost all" Erdős-Rényi networks $\mathbb{G}(n, \eta(n))$. The network is constructed by connecting n nodes randomly; each link is included in the network with probability $\eta(n)$. This algorithm, which will be referred to as $A(\delta)$, prices only once with price $(1 - \delta)(n-1)\eta(n)$, where $\delta > 0$ is a parameter to be determined by the approximation ratio that we intend to reach.

Theorem 5. *Given arbitrarily small positive number $\epsilon > 0$, set $\delta \in (0,1)$ such that $\frac{1+\delta}{1-\delta} < 1+\epsilon$. Then for the PNC model, Algorithm $A(\delta)$ has an approximation ratio at most $1 + \epsilon$ for asymptotically almost all networks $\mathbb{G}(n, \eta(n))$, as long as*

$$\frac{\eta(n)}{\sqrt{(\ln n)/n}} \to +\infty. \tag{1}$$

To be precise, under condition (1), we have

$$\lim_{n\to\infty} Pr\left(\frac{2|E(\mathbb{G}(n, \eta(n)))|}{r(\mathbb{G}(n, \eta(n)))} \leq 1 + \epsilon\right) = 1,$$

where $E(\mathbb{G}(n, \eta(n))$ is the link set of $\mathbb{G}(n, \eta(n))$, $Pr(\cdot)$ is the probability function, and $r(\mathbb{G}(n), \eta(n))$ is the revenue obtained from the single pricing $(1 - \delta)(n - 1)\eta(n)$.

5.3 $(2 - \epsilon)$-Approximation for Barabási-Albert Networks

The scale-free property (the power-law tail) has been nicely emulated by the multiple-destination preferential attachment growth model introduced by Barabási and Albert [3]. Starting with a small number of nodes (who are originally connected with each other), at each time step a new node enters network $G = (V, E)$, and attaches to β existing nodes. Each of the existing nodes is attached to the new one with a probability that is proportional to its current degree. Such a process is well-known as the *preferential attachment*. Recall that $|V| = n$. Let $\alpha_{n,k}$ be the fraction of nodes with degree k. It is known from [7] that for any fixed $\epsilon > 0$, and any $\beta \leq k \leq n^{1/15}$,

$$\lim_{n\to\infty} Pr\left((1 - \epsilon)\frac{2\beta(\beta + 1)}{k(k + 1)(k + 2)} \leq \alpha_{n,k} \leq (1 + \epsilon)\frac{2\beta(\beta + 1)}{k(k + 1)(k + 2)}\right) = 1. \tag{2}$$

Note by the construction that each node has a degree of at least β. Let Γ be the set of all nodes that have a degree of exactly β. Then

$$\Gamma \text{ is an independent set of } G, \tag{3}$$

because whenever two nodes are connected, the "older" one must have a degree at least $\beta + 1$. Note also that for any fixed $\epsilon > 0$, the inequality $|E| \leq (1 + \epsilon/2)n\beta$ holds for big enough n.

Theorem 6. *Consider the PNC model. For any fixed $\epsilon > 0$, with probability tending to one as $n \to \infty$, the single pricing with price β achieves an approximation ratio of $2 - 2/(2 + \beta) + \epsilon$ for Barabási-Albert network G. To be precise,*

$$\lim_{n \to \infty} Pr\left(\frac{\text{OPT}(G)}{n\beta} \leq 2 - \frac{2}{(2+\beta)} + \epsilon\right) = 1,$$

where $n\beta$ is the revenue obtained by single price β.

In the special case of $\beta = 1$, Barabási-Albert network G is a tree. The approximation ratio $2 - 2/(\beta + 2) = 4/3$ stands in contrast to the ratio 1.5 in Theorem 4 and Remark 1.

5.4 Upper and Lower Bounds for Single Pricing

Having seen the above constant approximations that single pricing achieves, one may ask: can best single pricing always provide good approximations to optimal iterative pricing for unweighted networks with uniform intrinsic values? The following example shows that, in the worst case, the best single price can only guarantee at most a fraction $1/(\ln \ln n)$ of the optimal revenue.

Example 1. The network G with $n = k(k!) + 1$ nodes consists of $\sum_{i=1}^{k} i = k(k + 1)/2$ node-disjoint cliques and one special node which is adjacent to all other nodes, where the number of $(k!/i)$-cliques is i for each $1 \leq i \leq k$.

In the above instance G, there are one node with degree $k(k!)$, which is the special node, and $k!$ nodes with degree $(k!)/i$ for $i = 1, 2, \ldots, k$. Recall that $\text{R}(p)$ denote the revenue under single pricing (p). Note that $\text{R}(k(k!)) = k(k!)$, and $\text{R}((k!)/i) = (i(k!) + 1) \cdot (k!)/i = (k!)^2 + (k!)/i$ for $i = 1, \ldots, k$. Then the best single price is $k!$, which brings a revenue

$$\text{R}(k!) = (k!)^2 + k! = \max_{p \geq 0} \text{R}(p).$$

On the other hand the pricing $\mathbf{p} = (p_1, \ldots, p_{k+1})$ with $p_1 = k(k!), p_{i+1} = (k!)/i$, $i = 1, \ldots, k$, brings revenue $\text{R}(\mathbf{p}) = k(k!) + \sum_{i=1}^{k}(k!)(k!/i - 1) = (k!)^2 \cdot \sum_{i=1}^{k}(1/i)$. When k tends to infinity,

$$\frac{\text{R}(\mathbf{p})}{\text{R}(k!)} = \frac{\sum_{i=1}^{k} \frac{1}{i}}{1 + o(1)} \approx 1 + \ln k = \Theta(\ln \ln n).$$

In complementary to the above example, we show in the following theorem that, with single pricing, one can always assure at least a factor $1/(1 + \ln n)$ of the optimal revenue in unweighed network G with uniform intrinsic values. Let $d_1, d_2, \ldots, d_n$ with $d_1 \geq d_2 \geq \cdots \geq d_n$ be the degree sequence of G.

Theorem 7. $\text{OPT}(G)/\max_{i=1}^{n}\{id_i\} \leq 1 + \ln n.$

Proof. Since $\sum_{i=1}^{n} d_i \geq \text{OPT}(G)$, it suffices to show that

$$\max_{i=1,\cdots,n} \{id_i\} \geq \sum_{i=1}^{n} \frac{d_i}{1 + \ln n}.$$

Suppose on the contrary that $id_i < \frac{\sum_{j=1}^{n} d_j}{1+\ln n}$ for each $1 \leq i \leq n$. Then we have

$$\sum_{i=1}^{n} d_i < \left(\sum_{i=1}^{n} \frac{1}{i}\right) \cdot \frac{\sum_{i=1}^{n} d_i}{1 + \ln n} \implies 1 + \ln n < \sum_{i=1}^{n} \frac{1}{i},$$

which is a contradiction.

6   Conclusion

Our work is an addition to the very limited literature on both pricing with negative network externalities and iterative pricing. The model captures many interesting settings in real-world marketing, and is usually much more challenging than the positive externality counterpart. The hardness result identifies complexity status of a fundamental pricing problem. The algorithms achieve satisfactory performances in general and several important special settings. An interesting direction for future research is to narrow the lower and upper bounds on the approximability of the iterative pricing problem with negative externality. Obtaining more accurate estimations for the optimal pricing is a key to reduce the approximation ratios.

References

1. Akhlaghpour, H., Ghodsi, M., Haghpanah, N., Mirrokni, V.S., Mahini, H., Nikzad, A.: Optimal iterative pricing over social networks (extended abstract). In: Saberi, A. (ed.) WINE 2010. LNCS, vol. 6484, pp. 415–423. Springer, Heidelberg (2010)
2. Alon, N., Mansour, Y., Tenneholtz, M.: Differential pricing with inequity aversion in social networks. In: Proceedings of the Fourteenth ACM Conference on Electronic Commerce, EC 2013, pp. 9–24 (2013)
3. Barabási, A.L., Albert, R.: Emergence of scaling in random networks. Science 286(5439), 509–512 (1999)
4. Bateni, M.H., Haghpanah, N., Sivan, B., Zadimoghaddam, M.: Revenue maximization with nonexcludable goods. In: Chen, Y., Immorlica, N. (eds.) WINE 2013. LNCS, vol. 8289, pp. 40–53. Springer, Heidelberg (2013)
5. Bhattacharya, S., Kulkarni, J., Munagala, K., Xu, X.: On allocations with negative externalities. In: Chen, N., Elkind, E., Koutsoupias, E. (eds.) Internet and Network Economics. LNCS, vol. 7090, pp. 25–36. Springer, Heidelberg (2011)
6. Bloch, F., Quérou, N.: Pricing in social networks. Games and Economic Behavior 80, 243–261 (2013)
7. Bollobás, B., Riordan, O., Spencer, J., Tusnády, G.: The degree sequence of a scale-free random graph process. Random Structures & Algorithms 18(3), 279–290 (2001)

8. Bramoullé, Y.: Anti-coordination and social interactions. Games and Economic Behavior **58**(1), 30–49 (2007)
9. Bramoullé, Y., Kranton, R.: Public goods in networks. Journal of Economic Theory **135**(1), 478–494 (2007)
10. Candogan, O., Bimpikis, K., Ozdaglar, A.: Optimal pricing in networks with externalities. Operations Research **60**(4), 883–905 (2012)
11. Cao, Z., Chen, X., Hu, X., Wang, C.: Pricing in social networks with negative externalities (2015). http://arxiv.org/abs/1410.7263
12. Chen, N.: On the approximability of influence in social networks. SIAM Journal on Discrete Mathematics **23**(3), 1400–1415 (2009)
13. Chen, W., Lu, P., Sun, X., Tang, B., Wang, Y., Zhu, Z.A.: Optimal pricing in social networks with incomplete information. In: Chen, N., Elkind, E., Koutsoupias, E. (eds.) Internet and Network Economics. LNCS, vol. 7090, pp. 49–60. Springer, Heidelberg (2011)
14. Deng, C., Pekeč, S.: Money for nothing: exploiting negative externalities. In: Proceedings of the 12th ACM Conference on Electronic Commerce, EC 2011, pp. 361–370 (2011)
15. Farrell, J., Saloner, G.: Standardization, compatibility, and innovation. The RAND Journal of Economics, 70–83 (1985)
16. Feldman, M., Kempe, D., Lucier, B., Paes Leme, R.: Pricing public goods for private sale. In: Proceedings of the Fourteenth ACM Conference on Electronic Commerce, EC 2013, pp. 417–434 (2013)
17. Haghpanah, N., Immorlica, N., Mirrokni, V., Munagala, K.: Optimal auctions with positive network externalities. ACM Transactions on Economics and Computation **1**(2), 13 (2013)
18. Hartline, J., Mirrokni, V., Sundararajan, M.: Optimal marketing strategies over social networks. In: Proceedings of the 17th International Conference on World Wide Web, WWW 2008, pp. 189–198 (2008)
19. Jehiel, P., Moldovanu, B., Stacchetti, E.: How (not) to sell nuclear weapons. The American Economic Review, 814–829 (1996)
20. Katz, M., Shapiro, C.: Network externalities, competition, and compatibility. The American Economic Review, 424–440 (1985)
21. Kempe, D., Kleinberg, J., Tardos, É.: Maximizing the spread of influence through a social network. In: Proceedings of the Ninth ACM SIGKDD International Conference on Knowledge Discovery and Data Mining, pp. 137–146 (2003)
22. Mossel, E., Roch, S.: On the submodularity of influence in social networks. In: Proceedings of the Thirty-Ninth Annual ACM Symposium on Theory of Computing, STOC 2007, pp. 128–134 (2007)
23. Radner, R., Radunskaya, A., Sundararajan, A.: Dynamic pricing of network goods with boundedly rational consumers. Proceedings of the National Academy of Sciences **111**(1), 99–104 (2014)

Social Relation Based Long-Term Vaccine Distribution Planning to Suppress Pandemic

Donghyun Kim[1], Hao Guo[1], Yuchao Li[2], Wei Wang[2]($\boxtimes$),
Sung-Sik Kwon[1], and Alade O. Tokuta[1]

[1] Department of Mathematics and Physics, North Carolina Central University,
Durham, NC, USA
{donghyun.kim,skwon,atokuta}@nccu.edu, hguo@eagles.nccu.edu
[2] School of Mathematics and Statistics, Xi'an Jiaotong University, Xi'an, China
liyu_chao@126.com, wang_weiw@163.com

Abstract. This paper introduces a new optimization problem which aims to develop a distribution plan of vaccines which will be supplied over time such that an epidemic can be best suppressed until a complete cure for it is invented. We first exploit the concept of temporal graph to capture the projected images of the evolving social relations over time and formally define the social-relation-based vaccine distribution planning problem (SVDP^2) on the temporal graph. Then, we introduce a graph induction technique to merge the subgraphs in the temporal graph into a single directed acyclic graph. Next, we introduce a max-flow algorithm based technique to evaluate the quality of any feasible solution of the problem. Most importantly, we introduce a polynomial time enumeration technique which will be used along with the evaluation technique to produce a **best possible** solution within polynomial time.

1 Introduction

In 2014, the world have witnessed the unprecedented spread of critical disease called Ebloa, which is transmittable from an infected person, who has spent a certain incubation period after the initial infection, to another healthy one via direct contact to bodily fluid from the infected. After the seriousness of the disease was recognized, many efforts were initiated to expedite the development of vaccines to stop the further spread of the disease as well as of a complete cure for it. During last one year, a number of vaccines were tested and several approaches to cure infected ones were tried, which saved several lives. However, the disease is still spreading and many people are dying while the researches for the vaccines and cures are ongoing. Unfortunately, there is no guarantee that Ebloa is the last pandemic on this scale. As a result, the proper preparedness against such calamity is of great urgency to save lives, possibly in the near future.

This work was supported in part by US National Science Foundation (NSF) CREST No. HRD-1345219. This research was jointly supported by National Natural Science Foundation of China under grants 11471005.

M.T. Thai et al. (Eds.): CSoNet 2015, LNCS 9197, pp. 26–34, 2015.
DOI: 10.1007/978-3-319-21786-4_3

It is not difficult to imagine that even a vaccine is once invented against a new pandemic, its near-term availability would be highly restricted. As a result, the development of proper distribution strategy of vaccines over healthy individuals is as important as inventing vaccines and cures of unknown pandemic to reduce the number of the victims of the critical epidemic once happens. Recently, Zhang and Prakash [1] used the information from social relationship to address the issue of selecting those to be vaccinated when the number of currently available vaccines is limited. In this approach, two adjacent nodes with high probability implies there is a great chance of infection from one to the other. This social relation based approach could be promising as the disease usually is transmitted one to another following their (physical) social interaction. However, we found that there is generally a lack of efforts to utilize this approach.

In order to fill this deficiency, this paper investigates the best way to distribute available vaccines which will be supplied over time by exploiting the projected social relations among the members of a society with the objective of minimizing the number of infected people until a complete cure is invented. The list of the contributions of this paper is as follows.

(a) We introduce a new optimization problem, namely social-relation-based vaccine distribution planning problem (SVDP^2), which aims to study the best strategy to distribute regular vaccine supplies over time with the objective of minimizing the number of infected until a complete cure is developed.
(b) We use the concept of temporal graph [10] to capture the projected images of the evolving social relations over time. Then, we introduce a new strategy to reduce the graphs in the temporal graph into a single directed acyclic graph (DAG). Finally, we redefine the proposed optimization problem on this new DAG.
(c) We introduce a new maximum-flow algorithm based strategy to evaluate the performance of any feasible solution of SVDP^2.
(d) We propose a polynomial time exact algorithm for SVDP^2 by exploiting our evaluation strategy.

The rest of this paper is organized as follows. Related work is discussed in Section 2. The formal definition of SVDP^2 is in Section 3. Our main contribution, the polynomial time exact algorithm for SVDP^2 is in Section 4. Finally, we conclude this paper in Section 5.

2   Related Work

We realize that there are several problems which are previously well-studied in the literature. Therefore, we need to explain how our problem is fundamentally different from them. Largely speaking, there are three group of problems related to ours.

Fire-Fighter Problem and Its Variations [4–6]. There are two variations of the the fire-fighter problem [5]. In the context of our problem of interest, the first version, namely MAXSAVE, aims to find a valid vaccine strategy over time to maximize the number of uninfected after a given period. The second version, MINBUDGET, attempts to find a valid vaccine strategy to save the members in a given node subset with a given graph such that the budget for the vaccines (the number of nodes removed) used this purpose is minimized. At a glance, MAXSAVE is similar to our problem of interest, $SVDP^2$. However, $SVDP^2$ is more challenging as MAXSAVE uses a static topology graph while $SVDP^2$ considers a social relation graph which varies over time.

Graph-Cut Problems [2,3]. The main objective of graph-cut problem is that given a graph, to identify a subset of nodes such that after the nodes in the subset are removed from the graph, the resulting graph consists of two connected components in a way that a certain objective function is maximized. One example of such objective is that each of the components should have one designated node s (and t), respectively and the size of the component including t becomes maximized. The main challenges to use a solution for the graph-cut problems for $SVDP^2$ are that (a) the former one assumes we have enough vaccines to contain the epidemic, which is not necessarily true in $SVDP^2$, and (b) the former one also assumes a static graph, which is not necessarily true in our case.

Data-Aware Vaccination Problem [1]. Recently, Zhang and Prakash have investigated the data-aware vaccination problem, the problem of how to best distribute currently available k vaccines over healthy individuals so that the expected number of victims can be minimized with the knowledge of the infection probability from one to another under the assumption that infection of a patient to another happens only one time. In their work, the knowledge of the social network graph which represents the relationship between the people is used to evaluate the likelihood of the disease transmission. Then, a greedy strategy is used to find the best k healthy nodes in the graph such that the average number of patients are minimized. This work is very remote from our work as (a) there is no concept of time-dimension in their work, e.g. an infected individual may infect its neighbor only one time with a probability and the vaccines are only provided at the beginning, and (b) the social network graph is fixed.

Based on our survey, we can conclude that there is no existing work which is directly used to solve $SVDP^2$. In the following section, we provide the formal definition of $SVDP^2$.

3   Problem Definition

This paper uses a temporal graph [10] $\mathcal{G} = \{G_0 = (V_0, E_0), G_1 = (V_1, E_1), \cdots, G_T = (V_T, E_T)\}$, where $G_t \in \mathcal{G}$ captures the social relation among the members of society at the t-th unit moment from the initial moment (0-th moment) to the final moment (T-th moment). After the final moment, it is highly anticipated that a complete cure of the disease will be developed. The time gap between two consecutive moments could range from an hour to weeks. For instance, in case of Ebola, usual incubation time is 2 weeks, and this can be used as a reasonable gap. From the initial relationship among the members in the society, and corresponding graph G_0, the temporal graph $\mathcal{G}$ can be generated by an existing strategy such as [8]. We assume there is a threshold to determine if there should be an edge between a pair of nodes at a moment, which implies that two members at the moment are close enough to infect each other in case that one of them is infected with very high probability. Note that the accuracy of this approach is out of the scope of this paper, and we simple assume that the algorithm used for this purpose is highly precise.

Now, due to the gravity, we list our main assumptions in more detail and corresponding justifications if necessary.

(a) The temporal graph $\mathcal{G} = \{G_0 = (V_0, E_0), G_1 = (V_1, E_1), \cdots, G_T = (V_T, E_T)\}$ representing the social relationship of the members of the society at each moment is known in advance, and is precise. After T unit moments later, a complete cure of the disease will be developed. Any G_i and G_{i+1} may differ in node set or edge set as the relationship can be highly dynamic.
(b) The initial set of infected people I_0 in G_0 is known in advance. I_i will be used to represent the set of infected people in G_i.
(c) After each unit moment from G_i to G_{i+1}, the neighbors of I_i in G_i will be infected in G_{i+1}. We argue that our approach considers the worst-case (in which the infection ratio from two people is 100% if they are related) and thus would be more rigorous to deal with a critical disease like Ebola rather than the probabilistic approach considered by Zhang and Prakash [1].
(d) The initial vaccine supply $\mathcal{Q} = \{Q_0 = (p_0 = 0, q_0), Q_1 = (p_1, q_1), \cdots, Q_l = (p_l, q_l)\}$ are know in advance, where $Q_i = (p_i, q_i) \in \mathcal{Q}$ is the information of i_{th} vaccine supply and p_i is the arrival moment of q_i vaccines.
(c) Shortly after time T, the complete cure for the disease will be developed.

Now, we provide the formal definition of our problem of interest.

Definition 1 (Social-relation-based Vaccine Distribution Planning Problem (SVDP^2)). *Given $\mathcal{G}, \mathcal{Q}, I_0$, and T, the goal of SVDP^2 is to find the best vaccine distribution schedule of the incoming vaccines under the infection model such that the total number of infected people after T unit moments is minimized.*

4   Main Contributions

4.1   Consolidating $\mathcal{G}$ to Integrated Graph $\hat{G}$

Apparently, $\mathcal{G}$ is difficult to deal with as G_i and G_{i+1} in $\mathcal{G}$ may differ in node sets and edge sets for any i. To overcome the difficulty, we introduce a graph consolidation technique to merge the graphs in $\mathcal{G}$ to a new graph $\hat{G} = (\hat{V}, \hat{E})$, and redefine SVDP^2 using $\hat{G}$. This consists of the following steps.

(a) **Node set construction:** Set $\hat{V} \leftarrow \bigcup_{0 \le i \le T} V(G_i)$, where $V(G_i)$ is the set of nodes in G_i. Each node $v_j^{(i)}$ represents the status of node v_j at the i-th moment. In case that there exists a node $v_j^{(i)} \in G_i$ for some i, but $v_j^{(k)} \notin G_k$

for some k, then add a virtual node $w_j^{(k)}$ to $\hat{V}$, e.g. nodes $\{a, b, c, d, e, f, g\}$ in Fig. 1(b).

(b) **Edge set construction:** First, add a directed edge from $v_j^{(i)} \in \hat{V}$ (or alternatively $w_j^{(i)}$) to $v_j^{(i+1)} \in \hat{V}$ (or alternatively $w_j^{(i+1)}$) for each i and j pair: this means an infected node j at i-th moment will stay infected in $i + 1$-th moment (even though the node is outside the area abstracted by the social network). Second, for each $v_j^{(i)} \in G_i$ and its neighbor $v_k^{(i)} \in G_i$, add a direct edge from $v_j^{(i)}$ to $v_k^{(i+1)}$ (or its virtual node $w_k^{(i+1)}$) to $\hat{E}$: this means that a node neighboring to an infected node at i-th moment will keep infected in $i + 1$-th moment.

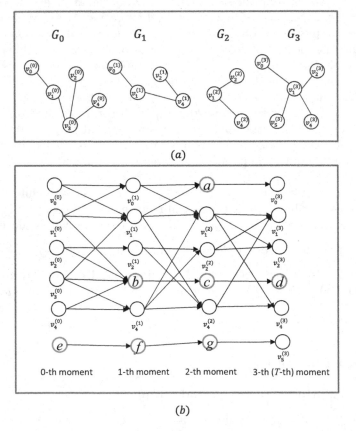

Fig. 1. (Figure (a) illustrates the time temporal graph $\mathcal{G}$ which consists of a series of graphs G_1, G_2, G_3, G_4 representing the social relation at each moment. Figure (b) shows the integrated graph $\hat{G}$ which is induced graph $\mathcal{G}$. In this graph, node a, b, c, d, e, f, g are fake nodes and does not exist. This means that e cannot be infected at the beginning.

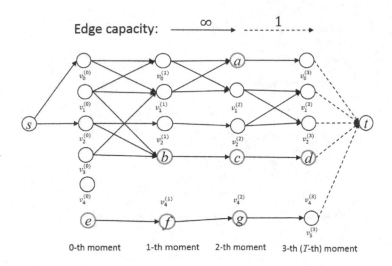

Edge capacity:

0-th moment   1-th moment   2-th moment   3-th (T-th) moment

Fig. 2. In this graph, v_0 and v_2 are initially infected, and an available vaccine is given to v_4 at the initial moment. Then the maximum flow from s to t is equivalent to the number of infected nodes after the T-th moments, which is 4 in this example.

The original infection rule can be applied in a way that when $v_j^{(i)}$ is infected, then its neighbors will be infected, and so on. Due to this, the resulting graph $\hat{G}$ has the following two interesting property.

(a) Once a node $v_j^{(i)}$ is infected, then all $v_j^{(k)}$s such that $k < i$ will be infected.

(b) Once a node $v_j^{(i)}$ is vaccinated, then all $v_j^{(k)}$s such that $k < i$ will be vaccinated.

This means that once we decide a vaccine to $v_j^{(i)}$, then all of the nodes $v_j^{(k)}$ such that $k \geq i$ can be removed from $\hat{G}$. In fact, this is a unique property which distinguishes our problem with the rest of the existing related problems.

4.2 Evaluation of Feasible Solution

In this section, we introduce a max-flow algorithm based strategy to evaluate a feasible solution of SVDP². In detail, given an integrated graph $\hat{G} = (\hat{V}, \hat{E})$, we first add two nodes s, t to $\hat{V}$. Then, add an edge from s to the nodes in $\hat{V}$ which are initially infected. Then, set the edge capacity of all nodes in the current $\hat{G}$ to be ∞. Next, for all nodes at the T-th moment, we add an edge from each of them to t with an edge capacity 1 (see Fig. 3). Suppose the resulting graph is $\hat{G}'$. Then, we prove the following theorem.

Theorem 1. *The maximum $s - t$ flow in $\hat{G}'$ after removing a subset of nodes S which received vaccines, i.e. if a node $v_k^{(i)} \in S$ receives a vaccine, then all nodes*

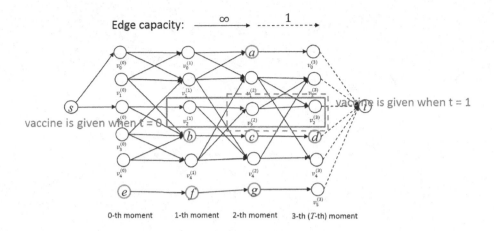

Edge capacity: $\xrightarrow{\infty}$ $\dashrightarrow{1}$

vaccine is given when t = 1

vaccine is given when t = 0

0-th moment   1-th moment   2-th moment   3-th (T-th) moment

Fig. 3. In this graph, v_2 can be vaccinated when $t = 0$, which will make $X_2^{(0)} = \{v_2^{(1)}, v_2^{(2)}, v_2^{(3)}\}$ removed from the graph, or when $t = 1$, which will make $X_2^{(1)} = \{v_2^{(2)}, v_2^{(3)}\}$ removed from the graph

$v_k^{(j)}$ with $j \geq i$ and their corresponding edges are removed from $\hat{G}'$, is equivalent to the number of infected people after the T-th moment.

Proof. Let I_T be the subset of nodes got infected in the final time $t = T$. Let f be a maximum flow with value $|f|$ for the constructed network in Fig. 2. We claim that for the maximum flow f, the flow must be one on the directed edge (v_i^T, t) for any $v_i^T \in I_T$. Otherwise, notice that for each $v_i^T \in I_T$, there is a directed path from s to v_i^T (since v_i^T got infected at moment T). We can take the directed path $(s \to v_i^T \to t)$ as an augmenting path, and increase the flow on every edge of the path by one (note the capacity of each edge from s to v_i^T is infinity, so we can increase the flow as we wish). Then the new flow value would increase by one; contradiction to the maximality of f. Therefore, we have $|f| = |I_T|$. $\square$

4.3 Polynomial Time Exact Algorithm Based on Enumeration

In this section, we discuss how the best possible solution of SVDP^2 can be computed within polynomial time. Our strategy consists of the following steps.

(a) Our key observation on this step is that as we stated in Theorem 1, once a node $v_k^{(i)}$ receives a vaccine, then all nodes $v_k^{(j)}$ with $j \geq i$ and their corresponding edges are removed from $\hat{G}'$. Based on this observation, we first construct a subset $X_k^{(i)} = \{v_k^{(i)}, v_k^{(i+1)}, \cdots, v_k^{(T)}\}$ for each node $v_k^{(i)}$ in $V(\hat{G}') \setminus (\{s, t\} \bigcup V(G_0))$. Note that $X_k^{(i)}$ in fact is the subset of nodes which

should be removed from $\hat{G}'$ once we determined to give a vaccine to $v_k^{(i-1)}$.
This takes polynomial time as the number of such subset $X_k^{(i)}$ is equivalent
to the size of $V(\hat{G}') \setminus \left(\{s,t\} \bigcup V(G_0) \right)$.

(b) Consider $\mathcal{Q} = \{Q_0 = (p_0 = 0, q_0), Q_1 = (p_1, q_1), \cdots, Q_l = (p_l, q_l)\}$. For each
Q_i, we are allowed to pick q_i nodes in $\hat{G}'$ after the p_i-th moments and give
a vaccine to it, which will eliminate all corresponding nodes (i.e. the nodes
in the corresponding $X_k^{(i)}$) from $\hat{G}'$.

Given $\mathcal{Q}$, the number of all possible choices to select nodes to give a vaccine
is bound by $\binom{n}{qT} = O(n^{qT})$, where $q = \max_{1 \le i \le T} q_i$. Then, we just need to
pick the best one among the all possible choices.

As a result, we obtain a polynomial time exact algorithm for SVDP^2. Now, we
show the correctness of this algorithm.

Theorem 2. *Given T and $q = \max_{1 \le i \le T} q_i$ being fixed. The proposed strategy
computes the best possible solution within polynomial time.*

Proof. Note all possible choices strategy of giving of vaccines to nodes in $\hat{G}$
is bounded by $O(n^{qT})$. And According to Theorem 1, each time when can we
use max-flow algorithm to compute the number of infected nodes, which runs
in polynomial time. Thus, the time complexity of our is polynomial under the
given assumptions.

5   Concluding Remarks and Future Works

During the recent decade, we have witnessed several new epidemics which has
threatened the existence of mankind. In most cases, it took a long time to produce sufficient amount of effective vaccines, and it took even long to invent a
complete cure of it. Therefore, it is of great importance to develop an efficient
strategy to minimize the impact of the epidemic while only a limited amount
of vaccines are available. This paper aims to open a discuss on this research
direction, which is relatively not well understood yet. Our approach uses existing social relationship project strategies to capture the images of evolving social
relation which are used to predict the routes of infection of a critical disease.
Then, we develop a polynomial time exact algorithm to establish vaccine distribution plan based on the knowledge of future vaccine production and the exacted
time to discover a complete cure. We believe that this work shows one significant
potential of the information from social network, which is already considered to
be with rich set of information for various applications [11–15]. As a future
work, we plan to further study the problem to introduce a faster algorithm for it
because the running time of our algorithm is very large even though it is polynomial. We are also interested in real data to validate the actual effectiveness of
the proposed approach.

References

1. Zhang, Y., Prakash, B.A.: DAVA: distributing vaccines over networks under prior information. In: SIAM International Conference on Data Mining (SDM 2014), pp. 46–54 (2014)
2. Hayrapetyan, A., Kempe, D., Pál, M., Svitkina, Z.: Unbalanced graph cuts. In: Brodal, G.S., Leonardi, S. (eds.) ESA 2005. LNCS, vol. 3669, pp. 191–202. Springer, Heidelberg (2005)
3. Engelberg, R., Könemann, J., Naor, J.: Cut problems in Graphs with a Budget Constraint. Journal of Discrete Algorithms **5**(2), 262–279 (2007)
4. Finbow, S., MacGillivray, G.: The Firefighter Problem: A Survey of Results, Directions and Questions. Australasian Journal of Combinatorics **43**, 57–77 (2009)
5. Anshelevich, E., Chakrabarty, D., Hate, A., Swamy, C.: Approximation algorithms for the firefighter problem: cuts over time and submodularity. In: Dong, Y., Du, D.-Z., Ibarra, O. (eds.) ISAAC 2009. LNCS, vol. 5878, pp. 974–983. Springer, Heidelberg (2009)
6. Anshelevich, E., Chakrabarty, D., Hate, A., Swamy, C.: Approximability of the Firefighter Problem Computing Cuts over Time. Algorithmica **62**, 520–536 (2012)
7. Wang, N.: Modeling and Analysis of Massive Social Networks, Ph.D. Dissertation (2005)
8. Bringmann, B., Berlingerio, M., Bonchi, F., Gionis, A.: Learning and Predicting the Evolution of Social Networks. IEEE Intelligent Systems **25**(4), 26–35 (2010)
9. Vinterbo, S.A.: Privacy: A Machine Learning View. IEEE Transactions on Knowledge and Data Engineering **16**(8), 939–948 (2004)
10. Kostakos, V.: Temporal graphs. Physica A: Statistical Mechanics and its Applications **388**(6), 1007–1023 (2009)
11. Anifantis, E., Stai, E., Karyotis, V., Papavassiliou, S.: Exploiting Social Features for Improving Cognitive Radio Infrastructures and Social Services via Combined MRF and Back Pressure Cross-layer Resource Allocation. Computational Social Networks **1**(4) (2014)
12. Lu, Z., Fan, L., Wu, W., Thuraisingham, B., Yang, K.: Efficient Influence Spread Estimation for Influence Maximization under the Linear Threshold Model. Computational Social Networks **1**(2) (2014)
13. Kim, H., Beznosov, K., Yoneki, E.: A Study on the Influential Neighbors to Maximize Information Diffusion in Online Social Networks. Computational Social Network **2**(3) (2015)
14. Ai, C., Zhong, W., Yan, M., Gu, F.: A Partner-matching Framework for Social Activity Communities. Computational Social Networks **1**(5) (2014)
15. Ventresca, M., Aleman, D.: Efficiently Identifying Critical Nodes in Large Complex Networks. Computational Social Networks **2**(6) (2015)

Limiting the Spread of Misinformation While Effectively Raising Awareness in Social Networks

Huiyuan Zhang$^{(\boxtimes)}$, Huiling Zhang, Xiang Li, and My T. Thai

Department of Computer and Information Science and Engineering,
University of Florida, Gainesville, FL 32611, USA
{huiyuan,huiling,xixiang,mythai}@cise.ufl.edu

Abstract. In this paper, we study the Misinformation Containment (MC) problem. In particular, taking into account the faster development of misinformation detection techniques, we mainly focus on the limiting the misinformation with known sources case. We prove that under the Competitive Activation Model, the MC problem is NP-hard and show that it cannot be approximated in polynomial time within a ratio of $e/(e-1)$ unless $NP \subseteq DTIME(n^{O(\log \log n)})$. Due to its hardness, we propose an effective algorithm, exploiting the critical nodes and using the greedy approach as well as applying the CELF heuristic to achieve the goal. Comprehensive experiments on real social networks are conducted, and results show that our algorithm can effectively expand the awareness of correct information as well as limit the spread of misinformation.

1 Introduction

With the increasing popularity of online social networks (OSNs), such as Facebook, Twitter and Google+, OSNs have become the most commonly utilized vehicles for information propagation. However, along with genuine and trustworthy information, rumors and misinformation also spread all around the Internet through this convenient and quick dissemination channel, which results in undesirable social effects and even leads to economic losses [1–3]. The rumor of the earthquake in Ghazni province in August 2012 made thousands of people leave their home in panic and be afraid of returning back home [5]. And the rumor about Obama injured originated from Twiiter in June 2011 caused the instability in financial markets. Misinformation about diseases are often observed [6]. For instance, there were many Twitter tweets containing misinformation about swine flu at the outset of the large outbreak in 2009. And the misinformation about vaccinations makes parents withhold immunization from their children [8]. Thus, it is crucial to seek efficient ways to control the inadvertent and intentional spread of misinformation.

Furthermore, once users believe the misinformation they received, they are resistant to change their beliefs, even though there are clear retractions [8]. Thus, rather than making efforts to only eliminate misinformation after it causes users' misunderstandings, negative emotions and further disruptive effects, we

© Springer International Publishing Switzerland 2015
M.T. Thai et al. (Eds.): CSoNet 2015, LNCS 9197, pp. 35–47, 2015.
DOI: 10.1007/978-3-319-21786-4_4

want to disseminate "good information" so as to raise users' awareness, reshape their attitude, and thus reduce their vulnerabilities to misinformation. "Good information" could refer to something for the debunking of misinformation, such as specific recommendations, authorized announcements or valid news.

Related Work. The large size and complex topology of OSNs, and various users' characteristics make this problem more challenging. Some existing works focus on identifying the infected nodes [7], which shed light on how to further design algorithms to prevent the misinformation from disseminating to the whole network. There are some recent attempts on limiting misinformation by initially injecting some good information and letting this good information propagate in the same network to convince other users [1,2,5]. Budak et al. [2] formulated the problem as an optimization problem and gave a greedy solution with approximation guarantees. The β_T^I problem of limiting viral propagation of misinformation is investigated in [1]. Fan et al. [5] studied the containment of rumor originating from a community and obtain the minimum number of needed protectors. But they only aimed to limit the spread misinformation. [4,15] approach this problem in different ways, they want to limit the spread of misinformation by blocking some nodes so that the overall pairwise connectivity can be minimized. However, during the process of limiting the misinformation by using good information, we should also simultaneously propagate this good information to other users who are unaware of this misinformation as soon as possible.

In this paper, we study the problem of minimizing the cost to prevent the spread of misinformation and simultaneously disseminate good information. At first we assume that nodes being active of misinformation are detected. An effective and timely algorithm is proposed to identify the most important disseminators with the minimum total cost to inject correct information into the diffusion. In particular, we should detect a critical round in which we take full protection on them to limit the wide spread of misinformation in advance and also prompt the further propagation of good information. Extensive experiments on real datasets are conducted to evaluate the efficiency and effectiveness of our algorithms and the results show that our algorithms perform significantly well.

Our contributions in this paper are summarized as follows:

- This is the first attempt to limit the misinformation and also maximize the prevalence size of good information. And we introduce Competitive Activation model to represent the competition nature of misinformation and good information.
- For the MC problem, we prove its NP-hardness and show that it cannot be approximated in polynomial time within a ratio of $\frac{e}{e-1}$ unless $NP \subseteq DTIME\ (n^{O(\log \log n)})$.
- The DI algorithm has been developed to find the nodes which can effectively block misinformation and simultaneously expand the influence of good information. And this algorithm is shown to be scalable to large-scale networks and outperforms several other heuristics.

The rest of this paper is organized as follows. Section 1 introduces the competitive activation model. Section 2 and Section 3 give the definition of Misinformation Containment and analysis of its complexity. We propose Dominating Influence algorithm in Section 4, and evaluate the performance of our method in Section 5. Finally, Section 6 concludes this paper.

2   Competitive Activation Model

In this paper, an OSN is modeled as a directed graph $G = (V, E)$, where nodes in V represent users and edges in E represent social ties between each pair of users. The size of a given graph G is $n = |V|$. Starting with a seeding set, information can propagate along edges of the underlying network. It is very possible for a user to be exposed to both misinformation and good information. Negative dominance is used as the tie-breaking rule in Competitive Linear Threshold model [9]. However, considering various characteristics of users, they could make different decisions upon receiving same information. So, we introduce a new parameter *preference* to determine which activation will finally succeed. Our model for the simultaneous spread of misinformation and good information is as follows.

Each node $v \in V$ is associated with two thresholds θ_v^A and $\theta_v^B > 0$, and each edge $(u, v) \in E$ is assigned to two weights $w_{uv}^A, w_{uv}^B \geq 0$ corresponding to misinformation A and good information B. Let I_0^A and I_0^B denote the sets of initial A-active nodes, accepting the misinformation, and B-active nodes, believing good information, respectively. At time t, an inactive node v will become A-active if $\sum_{u \in I_{t-1}^A} w_{uv}^A \geq \theta_v^A$, or will become B-active if $\sum_{u \in I_{t-1}^B} w_{uv}^B \geq \theta_v^B$. When both thresholds have been satisfied, a node will decide to accept which one by its preference, $P_v^i = (\sum_{u \in N_a^{in}(v)} w_{uv}^i)/\theta_v^i$, where $i \in \{A, B\}$ and $N_a^{in}(v)$ is the set of activated in-neighbours of v . It will become A-active if $P_v^A \geq P_v^B$, and vice versa. After accepting one kind of information, a node will stay in this status and not change to accept another one till the end of diffusion process, reflecting the continued influenced effect of information perception.

3   Misinformation Containment and Inapproximability

3.1   Problem Definition

Definition 1. *Misinformation Containment (MC).* *Given misinformation A and good information B spread on a graph $G = (V, E, \theta^A, \theta^B, w^A, w^B)$, where $\theta^i = \{\theta_v^i\}$, $w^i = \{w_{uv}^i\}$ and $i \in \{A, B\}$, while set of I_0^A and k_B are given, this problem aims to find a seeding set for good information I_0^B of size k_B such that we can minimize the number of A-active nodes and simultaneously maximize the number of B-active nodes.*

3.2 Hardness of MC

In this section, we first show the NP-completeness of MC problem by reducing it from the **Maximum Coverage** problem. We further prove the inapproximability of MC which is NP-hard to be approximated within a ratio of $\frac{e}{e-1}$ unless $NP \subseteq DTIME(n^{O(\log \log n)})$.

Theorem 1. *The MC problem is NP-complete.*

Proof. We first consider the decision version of MC problem that asks whether the graph $G = (V, E, w^A, w^B, \theta^A, \theta^B, I_0^A, k_B)$ contains a set of vertices $I_0^B \subset V$ of size k_B such that the number of B-active nodes is at least t_B and the number of A-active nodes is at most t_A where t_A and t_B are positive integers. Given $I_0^B \subset V$, we can easily compute the influence spread of B as well as that of A in polynomial time under the CAM model. This implies MC is in NP.

To prove that MC is NP-hard, we reduce it from the decision version of Maximum Coverage problem defined as follows.

Maximum Coverage. Given a positive integer k, a set of m elements $\mathcal{U} = \{e_1, e_2, \cdots, e_m\}$ and a collection of sets $\mathcal{S} = \{S_1, S_2, \cdots, S_n\}$. The sets may have some elements in common. The *Maximum Coverage* problem asks to find a subset $\mathcal{S}' \subset \mathcal{S}$, such that $|\cup_{S_i \in \mathcal{S}'} S_i|$ is maximized with $|\mathcal{S}'| \leq k$. The decision version of this problem asks whether the input instance contains a subset $\mathcal{S}$ of size k which can cover at least t elements where t is a positive integer.

Reduction. Given an instance $\mathcal{I} = \{\mathcal{U}, \mathcal{S}, k, t\}$ of maximum coverage, we construct an instance $G = (V, E, \theta^A, \theta^B, w^A, w^B, I_0^A, k_B, t_A, t_B)$ of MC problem as follows.

The set of vertices: add one vertex u_i for each subset $S_i \in \mathcal{S}$, one vertex v_j for each element $u_j \in \mathcal{U}$, and a special vertex x.

The set of edges: add an edge (u_i, v_j) for each $e_j \in S_i$ and connect x to each vertex v_j.

Thresholds and weights: assign all vertices the same threshold $\theta^A = \theta^B = \frac{1}{2m}$, and each edges (u_i, v_j) has weight $w_{u_i v_j}^A = 0, w_{u_i v_j}^B = \frac{1}{m}$. In addition, for all edges leaving from x, we assign their weights as $w_{x v_j}^A = \frac{1}{2m}, w_{x v_j}^B = 0$.

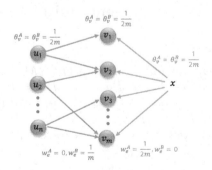

Fig. 1. Reduction from Maximum Coverage to Misinformation Containment

The construction is illustrated in Fig. 1. Finally, set $k_B = k$, $t_B = t + k_B$ and $t_A = m - t + 1$. Let $I_0^A = \{x\}$. We now show the equivalence between two instances.

Suppose that $\mathcal{S}^*$ is a solution to the maximum coverage instance, thus $|\mathcal{S}^*| \leq k$ and it can cover at least t elements in $\mathcal{U}$. By our construction, we can select all the nodes u_i corresponding to subset $S_i \in \mathcal{S}^*$ as a seeding set I_0^B. Thus $|I_B^0| = k = k_B$. Since $\mathcal{S}^*$ can cover at least t elements e_j in $\mathcal{U}$, then I_0^B can influence at least t vertices v_j corresponding to those e_j. Besides, for any v_j, both of A and B's total incoming influence exceed its threshold and $P_{v_j}^A \leq P_{v_j}^B$. Hence, there are at least $t + k_B$ B-active nodes in the MC problem and at most $m - t + 1 = t_A$ A-active nodes.

Conversely, suppose there is a B-seeding set $|I_0^{B*}| = k_B$ such that the number of B-active nodes is at least t_B. For any $v_j \in I_0^{B*}$, we replace it with its adjacent node u_i. This replacement does not reduce the number of B-active nodes. Then the $\mathcal{S}^*$ can be a collection of subset S_i corresponding to those $u_i \in I_0^{B*}$ after the replacement which has exactly size k and the number of elements which it can cover is at least $t_B - k_B = t$.

As MC problem is NP-complete, we further show that the above reduction implies a $\frac{e}{e-1}$-inapproximation factor in the following theorem.

Theorem 2. *The MC problem can not be approximated in polynomial time within a ratio of $\frac{e}{e-1}$ unless $NP \subseteq DTIME(n^{O(\log \log n)})$.*

Proof. We use the above mentioned reduction in the proof of Theorem 1. Suppose that there exists a $\frac{e}{e-1}$-approximation algorithm $\mathcal{H}$ for MC problem. Then $\mathcal{H}$ can return the number of B-active nodes in G with seeding size less than k_B. By our constructed instance, we can obtain the maximum coverage with size t if the the number of B-active nodes in the optimal solution for MC problem is $t + k_B$. Thus algorithm $\mathcal{H}$ can be applied to solve the Maximum Coverage problem in polynomial time. And this contradicts to the NP-hardness of Maximum Coverage problem [10].

4   Dominating Influence Algorithm

In this part, we propose our Dominating Influence (DI) algorithm for MC problem. DI algorithm consists of two sub-algorithms, which are DI-Gateway Nodes Detection and DI-Candidate Selection. DI-Gateway Nodes Detection helps us identify the gateway nodes, which are of significance in enlarging misinformation's influence. Before misinformation's diffusion naturally terminates, we use DI-Candidate Selection to find candidate seeding sets of different searching rounds, which are determined by the set of gateway nodes. Finally, we obtain the best seeds for good information from Dominating Influence algorithm.

4.1   Gateway Nodes Detection

In order to block the spread of misinformation, we should identify which nodes play an important role in its spreading out. In this paper, we use "gateway node" to refer to nodes which help misinformation propagate further. Knowing currently active nodes of misinformation, we can obtain the newly active nodes in each time stamp. Under CAM model, we have the following key observation.

Observation 1. *Each newly activated node in time t must be incident to at least one node that activated in time $t - 1$.*

Algorithm 1. DI-Gateway Nodes Detection

1: **Input:** Two set of nodes I_t^A, I_{t-1}^A
2: **Output:** A set of gateway nodes C_{t-1}
3: $C_{t-1} \leftarrow \emptyset$
4: **for** $i = 1$ to $|I_{t-1}^A|$ **do**
5:   $\delta_{\max} = 0$
6:   **for each** $v \in I_{t-1}^A \setminus C_{t-1}$ **do**
7:     **if** $\delta_v(C_{t-1}) > \delta_{\max}$ **then**
8:       $\delta_{\max} \leftarrow \delta_v(C_{t-1})$
9:     **end if**
10:    **if** $\delta_{\max} = 0$ **then**
11:      Return C_{t-1}
12:    **end if**
13:    $C_{t-1} \leftarrow C_{t-1} \cup \{v\}$
14: **end for**
15: **end for**
16: Return C_{t-1}

According to this observation, we utilize a trace back method to shrink the influence of misinformation step by step. Instead of starting from the inner-most round, we begin with the outer-most round. The reason is to avoid changes from happening in an earlier stage that may result in a cascading behavior to the later round. By simulating the propagation of misinformation, we record the set of nodes I_i^A that activated in round $i, i = 1, 2..., T$. To prevent the further propagation of misinformation to I_t^A, we should deal with nodes in I_{t-1}^A. Rather than targeting all nodes activated in round $t - 1$, we want to find the gateway nodes which contribute to activating the most number of nodes in I_t^A. Thus, we use a greedy approach to sequentially select a node $u \in I_t^A$ maximizing the following marginal gain into set C_{t-1}:

$$\delta_u(C_{t-1}) = f(C_{t-1} + \{u\}) - f(C_{t-1}),$$

where $f(\cdot)$ is the number of newly activated nodes which are incident to the set of selected nodes.

The algorithm terminates and returns the set of gateway nodes C_{t-1} for a given set of A-active nodes I_t^A. The detail of this step is shown in Algorithm 1.

4.2   Candidate Selection

After obtaining the set of gateway nodes, we want to target those nodes and activate them before misinformation reaches. Meanwhile, we desire to enhance users' awareness of good information. To achieve both goals, we present the candidates selection in Algorithm 2, and the core is to iteratively choose a node that maximizes the following marginal gain:

$$\eta_u(I_0^B) = \alpha[\psi(I_0^B + \{u\}) \cap C_{t-1} - \psi(I_0^B) \cap C_{t-1}] +$$
$$\beta[\psi(I_0^B + \{u\}) - \psi(I_0^B)],$$

where $\alpha + \beta = 1$. By adjusting the value of α, and β, we can change the effect on limiting misinformation's influence and expanding the influence of good information.

Algorithm 2. DI-Candidate Selection

1: **Input:** $G = (V, E, w^A, w^B \theta^A, \theta^B)$, C_{t-1} and k_B
2: **Output:** A candidate seed set $I_0^B(t-1)$ of size at most k_B
3: $P \leftarrow \emptyset$, $Q \leftarrow \emptyset$
4: **for** each $v \in C_{t-1}$ **do**
5:     Find node u that is $t-1$-hops away from v
6:      $P \leftarrow P \cup \{u\}$
7: **end for**
8: **for** $u \in P$ **do**
9:     Compute $\eta_u(I_0^B(t-1))$,
10:     Push u into Q
11: **end for**
12: **while** $|I_0^B| \leq k_B$ **do**
13:     **repeat**
14:         $u \leftarrow$ top of Q
15:         Recompute $\eta_u(I_0^B(t-1))$
16:     **until** u stays on top of Q
17:     **if** $\eta_u(I_0^B) \leq 0$ **then**
18:         Return $I_0^B(t-1)$
19:     **end if**
20:     $I_0^B \leftarrow I_0^B + \{u\}$;
21:     Return $I_0^B(t-1)$, $result(A, B, t-1)$
22: **end while**

Since greedy algorithms are always suffering from severe scalability problem, we use two techniques to effectively improve the running time. First, instead of selecting nodes from all over the network, we start from a candidate set P, which consists of nodes that are $t-1$ hops away from the targeted gateway nodes. Second, we employ CELF [11] heuristic to speed up the selection in each iteration. This approach can avoid the exhaustive update, which is extremely time consuming. This algorithm finally returns a candidate seeding set I_0^B as well as the total number of A-active and B-active nodes, respectively.

4.3   DI Algorithm

Incorporating above two algorithms, we obtain the DI algorithm, presented in Algorithm 3. First, we simulate the diffusion of misinformation and obtain termination round T along with the sets of activated nodes $I_t^A, t = 1, \cdots, T$ in each round. Starting with an arbitrary $I_t^A, t \in [1, T]$, by applying Gateway Nodes Detection, we are able to find the set of nodes C_{t-1} that contributed the most to activating nodes in I_t^A. Next, in order to limit the diffusion of misinformation, we should guarantee that the node $v \in C_{t-1}$ should be activated by good information no later than time $t - 1$. This requires us to either let good information reach v earlier than $t - 1$ or activate more of v's neighbors to be B-active nodes in order to make v's preference $P_v^B \geq P_v^A$ at $t - 1$.

Algorithm 3. Dominating Influence Algorithm

1: **Input:** Graph $G = (V, E, w^A, w^B, \theta^A, \theta^B)$, I_0^A and k_B
2: **Output:** A seed set I_0^B of size k_B
3: Simulate A's influence starts with I_0^A
4: Get the termination round T and sets of active nodes $I_i^A, i = 1, ..., T$
5: **for** $t = T$ to 1 **do**
6:      $C_{t-1} \leftarrow$ DI-Gateway Nodes Detection (I_t^A)
7:      $(I_0^B(t), result(A, B, t)) \leftarrow$ DI-Candidate Selection (G, C_{t-1}, k_B)
8: **end for**
9: **for** $t = 1$ to T **do**
10:      Find τ where $argmax_{\tau \in [1,T]}\{B \setminus A | result(A, B, t)\}$
11: **end for**
12: Return $I_0^B(\tau)$

Considering the above time constraint, there will be a trade-off when selecting nodes into the seeding set. If we try to limit the propagation of misinformation at an early stage, the candidate set (which consists of nodes $t - 1$ hops away from C_{t-1}) will be very limited, and thus may lead to decreasing the quality of seeds to disseminate good information. On the contrary, we are able to get a better candidate set by postponing the time to block misinformation, but this may result in increasing the number of A-active nodes dramatically. However, since the termination round of misinformation diffusion is usually a relatively small integer, and by applying the above mentioned enhancements to improve the running time, we are able to go through each C_t where t is from 1 to T searching round in order to find to best seeding set. Eventually, by measuring the difference between number of A-active and B-active nodes for every C_t, we can obtain the best seeding set to contain misinformation and maximally raise users' awareness.

5   Experiment and Evaluation

In this section, we perform various experiments based on the proposed algorithms and heuristics with real-world datasets, and evaluate the performance.

5.1   Dataset Description

We use three real-world networks, which are widely used for information diffusion process analysis, their basic statistics are summarized in Table 1, including:

Gnutella. The snapshot of the Gnutella peer-to-peer file sharing network in August 2002. Nodes represent hosts in the Gnutella network topology and edges represent connections between the Gnutella hosts [13].

Facebook. This dataset contains friendship information among New Orleans regional network on Facebook, spanning from September 2006 to January 2009, where nodes represent users and edges among them are friendship.

Amazon. This network is collected by crawling Amazon webpages. In this graph, an edge (i, j) indicates that product i is frequently co-purchased with product j by customers [14].

Table 1. Basic Information of Investigated Networks

Network	Gnutella	Facebook	Amazon
Nodes	6,301	61,096	262,111
Edges	20,777	1,811,130	1,234,877
Avg. Degree	3.29	29.64	4.71
Type	Directed	Directed	Directed

For graphs we tested on, nodes' thresholds for accepting misinformation and good information are generated uniformly at random in the range $[0, 1]$. To assign the influence weights on each edge, we adopt the method in [12], where we uniformly generate edge weights at random in the range $[0, 1]$, and then normalize the weights of all incoming edges of a node v to let it satisfy that $\sum_{u \in N^{in}(v)} w_{u,v} \leq 1$. Furthermore, for the seeding set of misinformation, we employ the greedy algorithm proposed by Kempe et al. [12], where in each iteration, the node with maximum marginal gain is chosen into the seeds. We are most likely to be able to detect misinformation and take action to contain its spread after it has propagated for a while and leads to undesirable effect [3]. Considering this observation, we introduce a delay d to model the time difference of disseminating good information and misinformation starting out. Compared with random selection, assigning seeds set for misinformation in this way can guarantee the high quality of misinformation initiators, and makes our problem of choosing seeds set for good information so as to limit the influence of misinformation more challenging.

Algorithms Compared. In our experiments, we compare our algorithm with several other heuristics listed as follows:

- Random: Randomly select k_B nodes from $V \setminus I_0^A$ as the seeds for good information in the graph.
- MaxDegree: We choose top k_B nodes from $V \setminus I_0^A$ with highest degree as the seeding set for good information.
- MaxGreedy: The greedy algorithm focuses on maximizing the influence of good information, in which the node with the maximum influence of good information is iteratively picked[16].
- MinGreedy: The greedy algorithm targets on minimizing misinformation propagation; the node with maximum number of A-active nodes blocked is selected in each iteration [2,9].

5.2   Experimental Results

In this part, we first measure the performance of our algorithm, in which we evaluate the number of A-active nodes and B-active nodes as well as their difference across three real world datasets with different number of seeds and rounds. Secondly, we compare the the results from all above mentioned algorithms. Next, we evaluate how time delay impacts the overall performance.

Seeding Set. We first present the spread of misinformation A and good information B achieved by selecting 50 B-seeds at different rounds. We evaluate them based on the number of A-active and B-active nodes, along with the difference between them. Fig. 2 shows two types of information of selecting 50 seeds with initial set $|I_0^A| = 10$ and time delay $d = 2$. The initiators of misinformation are selected by above described method, and before we disseminate good information in the network, misinformation has already activated 83, 205 and 50 nodes in Gnutella, Facebook and Amazon, respectively.

Fig. 2(a), 2(b), 2(c) show that the number of A-active nodes keeps dropping with a larger size of good information seeds. For example, in Gnutella, without adding any B-seeds, the spread of misinformation could reach as many as 851 nodes. However, by adding 50 seeds of good information selected by DI, the active size of misinformation can be limited to only 208 nodes. Conversely, Fig. 2(d), 2(e), 2(f) depict that the amount of B-active nodes increases dramatically with more B-seeds. For the seeds chosen from round 14 in Gnutella, the total number of B-active nodes can be 4749, eventually. Furthermore, we find that the difference between B-active nodes and A-active nodes is steadily increasing with larger budget of the seeding set of good information. It is also fluctuating with different targeting rounds.

Different Methods. Next, we compare the spread of both kinds of information achieved from different heuristics. The comparison is based on the number A-active nodes and B-active nodes and their difference. Fig. 3 shows the spread of misinformation and good information when there are 50 B-seeds and 10 initial A-active nodes, and the time delay $d = 2$ obtained from different heuristics. For limiting the spread of misinformation, MinGreedy is the best among those five methods across three datasets, while Random hardly blocks it. Except for MinGreedy, DI outperforms other heuristics as it effectively prevents the further

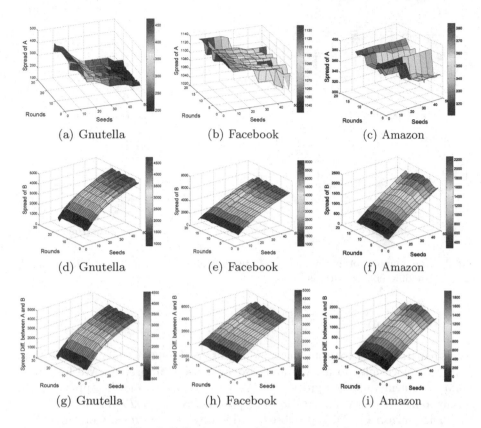

(a) Gnutella   (b) Facebook   (c) Amazon

(d) Gnutella   (e) Facebook   (f) Amazon

(g) Gnutella   (h) Facebook   (i) Amazon

Fig. 2. Influence spread in three networks

propagation of misinformation. As shown in Fig. 3(c), the amount of A-active nodes nodes goes down to 312 finally while it could be 468 without any B-seeds, which means that a 33% of A-active nodes has been reduced.

On the contrary, the number of B-active nodes is increasing sharply for both of the DI and MaxGreedy algorithms. Fig. 3(d) demonstrates that the number of B-active nodes climbs to 4749 and 4608 after selecting 50 nodes by DI and Max-Greedy, while for other three methods, the total number for A-active nodes is less than 1500, similar results can be obtained in Amazon. However, the MaxDegree in Facebook achieves the largest number of nodes accepting misinformation. By digging into the data, we find that there are some super nodes with massive outgoing edges are chosen by MaxDegree, while missed by MaxGreedy. Considering the greedy approach in selecting seeds, some of those super nodes may have less gain than other nodes due to the way we assign edge weights. However, the combination of them could lead to a large cascading influence. Hence, MaxDegree even outperforms MaxGreedy on Facebook. However, seldom nodes accepting misinformation have been reduced compared to our DI.

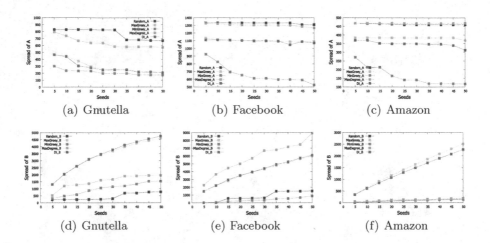

Fig. 3. The number of nodes activated by misinformation and good information achieved by different algorithms

6   Conclusions

In this paper, in order to protect users from potential influence of misinformation, we aim to block misinformation and also raise users' awareness. We formulate the MC problem, and then prove it is NP-complete and cannot be approximated in polynomial time within a ratio of $\frac{e}{e-1}$ unless $NP \subseteq DTIME(n^{O(\log \log n)})$. An efficient algorithm DI is proposed, and extensive experiments on three real-world datasets are conducted. Experiments results show that our algorithm outperforms several other heuristics and well scalable to large-scale social networks.

Acknowledgment. This work is supported in part of NSF Career Award 0953284 and NSF CCF-1422116.

References

1. Nguyen, N.P., Yan, G., Thai, M.T., Eidenbenz, S.: Containment of misinformation spread in online social networks. In: Proceedings of the 4th Annual ACM Web Science Conference (2012)
2. Budak, C., Agrawal, D., El Abbadi, A.: Limiting the spread of misinformation in social networks. In: Proceedings of the 20th International Conference on World Wide Web (2011)
3. Tripathy, R.M., Bagchi, A., Mehta, S.: A study of rumor control strategies on social networks. In: Proceedings of the 19th ACM International Conference on Information and Knowledge Management (2010)
4. Ventresca, M., Aleman, D.: Efficiently identifying critical nodes in large complex networks. Computational Social Networks **2**, 6 (2015)

5. Fan, L., Lu, Z., Wu, W., Thuraisingham, B., Ma, H., Bi, Y.: Least cost rumor blocking in social networks. In: IEEE International Conference on Distributed Computing Systems (ICDCS) (2013)
6. Jin, F., Dougherty, E., Saraf, P., Cao, Y., Ramakrishnan, N.: Epidemiological modeling of news and rumors on twitter. In: Proceedings of the Workshop on Social Network Mining and Analysis (2013)
7. Lim, Y.S., Ribeiro, B., Towsley, D.: Classifying latent infection states in complex networks. Computational Social Networks **2**, 8 (2015)
8. Lewandowsky, S., Ecker, U.K., Seifert, C.M., Schwarz, N., Cook, J.: Misinformation and its correction continued influence and successful debiasing. Psychological Science in the Public Interest **13**(3), 106–131 (2012)
9. He, X., Song, G., Chen, W., Jiang, Q.: Influence blocking maximization in social networks under the competitive linear threshold model. In: SDM (2012)
10. Feige, U.: A threshold of ln n for approximating set cover. Journal of the ACM (JACM) **45**(4), 634–652
11. Leskovec, J., Krause, A., Guestrin, C., Faloutsos, C., VanBriesen, J., Glance, N.: Cost-effective outbreak detection in networks. In: Proceedings of the International Conference on Knowledge Discovery and Data Mining (2007)
12. Kempe, D., Kleinberg, J., Tardos, E.: Maximizing the spread of influence through a social network. In: Proceedings of the Ninth ACM SIGKDD International Conference on Knowledge Discovery and Data Mining (2003)
13. Leskovec, J.: Gnutella peer-to-peer network (2002). http://snap.stanford.edu/data/p2p-Gnutella08.html
14. Leskovec, J.: Amazon co-purchasing network (2003). http://snap.stanford.edu/data/amazon-0302.html
15. Zhang, H., Alim, A.A., Thai, T.M., Nguyen, T.H.: Monitor placement to timely detect misinformation in online social networks. In: Proceedings of the International Conference on Communications (2015)
16. Carnes, T., Nagarajan, C., Wild, S.M., Van Zuylen, A.: Maximizing influence in a competitive social network: a follower's perspective. In: Proceedings of the Ninth International Conference on Electronic Commerce (2007)

Rumor Spreading and Security Monitoring in Complex Networks

Qiyi Han[1], Hong Wen[1(✉)], Jinsong Wu[2], and Mengyin Ren[1]

[1] National Key Laboratory of Science and Technology on Communication,
University of Electronic Science and Technology of China, Chengdu 611731, China
alex_han@163.com, sunlike@uestc.edu.cn, renmy61@hotmail.com
[2] Department of Electrical Engineering, Universidad de Chile, Santiago, Chile
wujs@ieee.org

Abstract. Rumor spreading is a typical phenomenon which poses security threats. Therefore, it is necessary to study the evolution mechanism of rumor spreading in complex networks. Considering the hesitating state in rumor spreading, we supplement fence-sitter group on existing rumor spreading models, then propose a novel SIFR rumor spreading model. The mean-field equations are derived to describe the dynamics of the rumor spreading in complex networks. In order to identify effective rumor control strategies, we further analyze the networks structure and propose an algorithm based on the topology potential theory to search the center nodes and divide the network.

Keywords: Rumor spreading · Monitoring · Complex networks · Topological potential

1 Introduction

Rumors, as an important form of social communications and a typical social phenomenon run through the whole evolutionary history of mankind. If any information circulates without officially publicized confirmation, this phenomenon is called rumor. In other words, rumors are unreliable information which may cause harmful impacts, such as viral marketing, fraud and panic. To understand mechanisms of rumor propagations, it is necessary to study rumor spreading models in complex networks and to investigate the pattern and structure of such a model.

The classical rumor spreading model, named DK model, was proposed by Daley and Kendall [1,2], The development of complex network theory provides the foundation to make the study of rumor spreading step into a new era. Zanette [3,4] first applied complex network theory to study the rumor spreading on small-world networks and proved the critical threshold's existence of the rumor spreading. Moreno et al. [5] established the rumor spreading model on the scale-free network, and found that the uniformity influences the dynamic mechanism of rumor spreading. Nekovee et al. [6] modified the rumor spreading model with consideration of the forgetting mechanism and discussed the thresholds in complex networks. Wang et al. [7] presented a SIRaRu rumor spreading

© Springer International Publishing Switzerland 2015
M.T. Thai et al. (Eds.): CSoNet 2015, LNCS 9197, pp. 48–59, 2015.
DOI: 10.1007/978-3-319-21786-4_5

model and discussed the rumor immunization strategy, and obtained the immunization threshold and spreading thresholds. Zan et al. [8] introduced the status of counterattack and analyzed the self-resistance parameter.

However, most of the previous models have not considered that there are some ignorants who may hear about a rumor and decide to sit on a fence, and not to spread the rumor. After the increasing rumors, the ignorants may be convinced and begin to spread the rumor as a spreader. In addition, rumor control strategy is very important to prevent the rumor from being spread. However, existing research results about rumor control are far less sufficient in the rumor model than those in the infectious disease models. Monitoring is necessary to control the information sharing and spreading channels in order to hinder the spreading of illegal and harmful information [9].

To address the above-mentioned issues, we propose a novel rumor spreading model, called SIFR model, via splitting the infected states with two types of infected states: spreader (S) and fence-sitter (F). The mean-field equations are derived to describe the dynamics of the rumor spreading in complex networks. In order to identify effective rumor control strategies, we combine the complex network theory with the social network analysis to analyze the networks and propose an algorithm based on the topology potential theory to search the center nodes and partition the network.

The rest of the paper is structured as follows. In Section 2, we describe our new rumor model and derive mean-field equation. In Section 3, the parameters in the rumor spreading model are analyzed to develop the rumor control strategy. We introduce the social network analysis theory in Section 4. We describe the security monitoring mechanism and propose the algorithm to partition the networks based on topology potential in Section 6. We conclude in Section 7.

2   SIFR Rumor Spreading Model

Facing the same event, people may have distinct opinions i.e. positive, dubious and negative. In classical SIR rumor spreading model [1,2], the people with positive opinions act as spreaders and the people with negative opinions act as stiflers. However, the people with dubious opinions have fewer descriptions. Therefore, we partition the classical spreader state into two separate states according to whether people provide supports firmly or hesitate indefinitely: spreader and fence-sitter. The fence-sitter state represents those who have heard about the rumor but stopped spreading it temporarily, and would possibly spread the rumor again when convinced.

We consider a closed and mixed group consisting of N individuals as a complex network, where individuals can be represented by vertexes and contacts between people can be represented by edges. We then obtain an undirected graph $G = (V, E)$, where V and E denote the vertexes and the edges, respectively. We assume that the rumor is disseminated through direct contacts of spreaders with the rest of the group, and the rumor spreading process of the SIFR model is shown in Fig. 1.

In the proposed model, we have four states including ignorant (I), spreader (S), fence-sitter (F) and stifler (R). The SIFR rumor spreading rules can be summarized as follows.

(a) When an ignorant meets a spreader, the ignorant may believe the rumor and become a spreader at a rate λ or may not be sure and become a fence-sitter at a rate α or may not believe it and become a stifler at a rate β.
(b) When a fence-sitter meets a spreader, the fence-sitter may be persuaded and become a spreader at a rate μ. If the fence-sitter meets a stifler, he turns into stifler at a rate δ.
(c) When a spreader meets a stifler, the spreader may come to realize the truth and become a stifler at a rate γ or may not be sure and become a fence-sitter sat a rate ω.

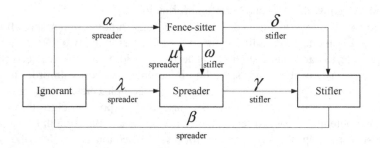

Fig. 1. Structure of SIFR rumor spreading process

Considering the normalization condition:

$$I(t) + S(t) + F(t) + R(t) = 1. \tag{1}$$

The mean-field equations can be described as follows:

$$
\begin{cases}
\dfrac{dI(t)}{dt} = -(\alpha + \lambda + \beta)\bar{k}I(t)S(t), \\[2mm]
\dfrac{dS(t)}{dt} = \lambda\bar{k}I(t)S(t) + \mu\bar{k}F(t)S(t) - \omega\bar{k}S(t)R(t) - \gamma\bar{k}S(t)R(t), \\[2mm]
\dfrac{dF(t)}{dt} = \alpha\bar{k}I(t)S(t) + \omega\bar{k}S(t)R(t) - \mu\bar{k}F(t)S(t) - \delta\bar{k}F(t)R(t), \\[2mm]
\dfrac{dR(t)}{dt} = \delta\bar{k}F(t)R(t) + \gamma\bar{k}S(t)R(t) + \beta\bar{k}I(t)S(t),
\end{cases}
\tag{2}
$$

where $\bar{k}$ denotes the average degree of the network. We assume that there is only one spreader at the beginning of the rumor spreading. The initial conditions for rumor spreading are given as follows:

$$I(0) = \frac{N-1}{N}, S(0) = \frac{1}{N}, F(0) = 0, R(0) = 0. \tag{3}$$

The classical SIR rumor spreading model is a special case of the SFHR model. When $\alpha = 0$, $\beta = 0$, $\delta = 0$, $\mu = 0$, $\omega = 0$, the SIHR model becomes the classical SIR rumor spreading model.

3   Rumor Control Strategy

The parameters which affect the rumor spreading are analyzed. We conclude as follows:

- With the decrease of the infection rate, the peak value of S(t) drops very fast. However, this method is less effective in the decay and life span of rumor;
- With the increase of the immunization rate, the life span of rumor reduces sharply. However, this is less effective in the decay and peak value of S(t);
- With the increase of hesitation rate, the peak value of S(t) decreases sharply. At the same time, the decay and life span are affected a lot;

As the above analysis, the increase of the hesitation rate from spreader to fence-sitter can reduce both the peak value and the life span of rumor spreading. We can adopt different strategies according to the type of rumors, including changing the infection rate, hesitation rate, and immunization rate.

In addition, it has been shown [6,7]   that the average degree $\overline{k}$ of the network is a critical factor which controls the rumor spreading. Especially in target immunization [10,11], the degree distribution and the topology of a network directly affects which nodes and what extent would infect the rumor. Therefore, monitoring the center nodes and partitioning the network are the feasible ways to control the rumor spreading.

4   Social Network Analysis

As an important branch of data mining, social network analysis (SNA) [12] is used to analyze the connection of the units in networks quantificationally. Using SNA, we can identify the leader nodes, monitor the networks, and establish the security mechanism of complex networks effectively.

SNA mainly includes centrality analysis, core-periphery model analysis, and cohesive-subgroups analysis.

4.1   Centrality Analysis

Centrality refers to the power of the network nodes in quantitative analysis, in other words, the evaluation of the importance of the network nodes, which mainly includes three aspects, as showed in Table 1.

- Degree centrality is the direct connections between the actors in the network (point degree in undirected graph and the sum of in-degree and out-degree in directed graph). We determine the status of an actor in the network by the number of the actor's direct contacts to other actors. Larger degree indicates greater power in networks. Degree centrality includes absolute centrality and relative centrality. Absolute centrality refers to the number of the actor's direct contacts to other actors. However, absolute centrality cannot be compared in different scales of networks. Therefore, Freeman [13] proposed relative centrality, which is result of distinguishing absolute centrality with the maximum degree of nodes in the networks

- Betweenness centrality is a measure of a node's ability to control the network's resource, indicating the ability to control the contacts of other nodes. This is equivalent to the number of shortest paths from all vertices to all others that pass through that node. The node plays a role as a bridge in the network with a larger betweenness centrality.
- Closeness centrality is a measure of the ability of not being controlled by other nodes. Larger closeness centrality means higher capacity to against other nodes' control.

Table 1. Centrality analysis

Type	Definition	Description
Degree Centrality	The direct connections of the nodes.	Larger degree centrality means higher probability to be the core of the network.
Betweenness Centrality	The node's ability to control the network's resource.	The node is played a role as a bridge in the network with a larger betweenness centrality.
Closeness Centrality	The ability of not being controlled by other nodes.	Larger closeness centrality means higher capacity to against other nodes' control.

4.2 Core-Periphery Model Analysis

We partition the network into two regions: the core region and the periphery region. The key actors are in the core region. The purpose of core-periphery analysis is to find the actors located in core or periphery area.

Actors or groups may have direct connections with some actors in the network, while they are not directly linked to other actors. Ronald. Burt [14] proposed structural holes to indicate the non-redundant relationship of the actors. There are at least three actors to connect non-redundant actors in a structural hole.

Intermediary plays an intermediary role in the network. There are five categories of intermediary, as showed in Table 2 and Fig. 2.

Table 2. Categories of intermediary

Type	Description
Coordinator	A, B and C are in the same group, B is the coordinator.
Consultant	A and C are in the same group, while B in the other group. B is the consultant.
Gatekeeper	B and C are in the same group, while A in the other group. B is the gatekeeper.
Agent	A and B are in the same group, while C in the other group. B is the agent.
Contact	A, B and C are all in the different groups. B is the contact.

4.3   Cohesive Subgroups Analysis

Cohesive subgroup is a subset in the network, where the actors are contacted closely and actively. General analysis includes four aspects: the characteristics of the contacts between the participants, the connectivity of the internal participants of the subset, contact times between the internal participants of the subset, the closeness of the contacts of the internal participants compared with the external participants. We can describe the network topology effectively based on the cohesive subgroup analysis.

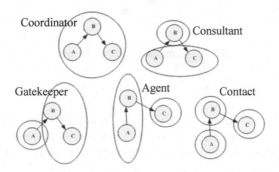

Fig. 2. Five categories of intermediaries

5   Security Monitoring

Aiming at protecting center nodes, security monitoring can control the information sharing and spreading channels to prevent the rumor from spreading extensively.

5.1   Community Structure

As a popular spot in network research, community discovery and division is an important research field of complex networks. With the in-depth study of the network, community structure [15-17], the common nature of many real networks was discovered. The entire network is composed of many communities, which are the collections of the nodes in the network. Nodes contact closely in a collection, while less connections between different collections. A community structure is showed in Fig. 3.

Community structure, generally including many categories, groups, and models, like the connections in the interpersonal networks, is a manifestation of the modularization and heterogeneity of the network.. In-depth study of the community structure is of great importance to analyze the structure and characteristics of the network. Community structure may help improve search performance and accuracy. In this paper, we study the community structure of the network and partition the network based on the data field theory and topology potential algorithm.

Fig. 3. A community structure map

5.2   Data Field

Field is the distribution of an object in space, which expressed by spatial position function. Data field [18], proposed based on the physics field theory, is a virtual space to describe and calculate the effect of the data on the entire number field.

Ω is an n-dimensional space, with the data objects $D = \{x_1, x_2, \ldots x_n\}$, where x_1 is the i-th ($1 \le i \le n$) dimensional coordinate of D representing the similarity of d_i and other data. Ω is a space of number field, and x_i represent the field source in Ω. The interaction effect of d_i formed the data field. Li et al. [19] proposed that every data element can be considered as a particle with unit mass, thus every data element produces an applied field and has applied force for all other data elements in the space, where we determine a data field. The properties of data field are as follows:

- Independence: every data element is not influenced by other elements while radiating energy.
- Superposition: the energy of one object, which is radiated by several elements, is the sum of the independent radiating energy of each element.

Elements of the data field influence each other based on potential function. The closer distances between elements represents stronger impacts. The rules to select potential function are as follows.

For any square integrable function, the following must hold:

$$\iint f(x, y) g(x) g(y) d_x d_y \ge 0 \tag{4}$$

where $f(x, y)$ is a monotone decreasing function of $d(x, y)$, which is a distance function between x and y.

Generally, we choose potential function in the form of the Gaussian potential function: $f(x,y)=e^{-\frac{|d(x,y)|^2}{2\sigma^2}}$, where σ is a factor, and $d(x,y)$ is the Euclidean space distance of x and y.

5.3   Topology Potential

Li et al. [19] proposed topology potential based on data field theory. Suppose that every node in the logical space produces an applied field, each node will be affected by other nodes, while the affection is related to the nodes' logical distance. The topology location of the node, equivalent to the node's potential, which is called topology potential, indicates the ability to influence its neighbor nodes. Topology potential theory has been applied in nodes permutation, community partition, community member discovery, and so on.

The network is expressed as $G=\{V,E\}$, where $V=\{v_1,v_2,...v_n\}$ is the set of nodes, $E\subseteq V\times V$ is the set of edges, and $|E|=m$ is the number of the edges. The interaction effect of the nodes can be indicated as:

$$\varphi(i\rightarrow j)=m_i\times e^{-\frac{d_{ij}}{\sigma}} \tag{5}$$

where $\varphi(i\rightarrow j)$ denotes the impact of v_i to v_j, $m_i\geq 0$ denotes the quality of v_i, d_{ij} denotes the shortest distance between v_i and v_j, and impact factor σ reflects the influence area of each node.

As v_i is influenced by every node in the network, the topology potential can be indicated as:

$$\varphi(v_i)=\sum_{j\in V}(m_j\times e^{-\frac{d_{ij}}{\sigma}}) \tag{6}$$

Regardless of the difference of the inherent property of nodes, we assume $m_i=1(1\leq i\leq n)$, and simplify the formula as:

$$\varphi(v_i)=\sum_{j\in V}e^{-\frac{d_{ij}}{\sigma}} \tag{7}$$

For a given σ, the influence area of every node is approximately limited in $l=\lceil 3\sigma/\sqrt{2}\rceil$ hops. When the distance of nodes is greater than the influence area, the influence between the nodes will rapidly decay to zero.

The topology potential is based on σ's value.

(a) $0 < \sigma < \sqrt{2}/3$: There is no influence between nodes, thus the topology potential is $\varphi(v_i) = m_i = 1$.

(b) $\sqrt{2}/3 < \sigma < 2\sqrt{2}/3$: Every node has impact only on its neighbor nodes, the topology potential is defined as:

$$\varphi(v_i) = 1 + \deg ree \times e^{-\left(\frac{1}{\sigma}\right)^2} \tag{8}$$

(c) $\sqrt{2}l/3 < \sigma < \sqrt{2}(l+1)/3$: The affecting area is l, the topology potential is:

$$\varphi(v_i) = 1 + \sum_{j \in V} n_j(v_i) \times e^{-\left(\frac{j}{\sigma}\right)^2} \tag{9}$$

where $n_j(v_i)$ is the number of nodes located in v_i's l-hop range.

5.4   Community Partition Based on Topological Potential

The network is $G = \{V, E\}$, where $V = \{v_1, v_2, ...v_n\}$ is the set of nodes, $E \subseteq V \times V$ is the set of edges, and $|E| = m$ is the number of the edges. For $\forall v \in V$, if v is not a local maximum potential node, v either belongs to a community, which is called an internal node, or is attracted by several communities simultaneously, named a boundary node.

The steps of community partition algorithm based on topology potential are as follows:

(a) Establish the adjacency matrix A based on the topology structure of the network. Assign a_{ij} as 1 if v_i and v_i is connected or ∞ otherwise.
(b) Calculate each node's degree.
(c) Establish matrix B, where b_{ij} is the number of nodes satisfying that $d_{ik} = j$, $k \in V$, where d_{ik} is the distance between v_i and v_k.
(d) Calculate the optimal impact factor σ when the topology potential entropy is minimal.
(e) Calculate each node's topology potential based on σ and the formula discussed in last section.
(f) Search the local maximum potential nodes and determine the set of community representatives.
(g) Partition the community according to the community representatives.
(h) Output partitioned communities.

5.5   Optimize the Impact Factor σ

We estimate the uncertainty of the network by the characteristic of topology potential entropy H, which is defined as follows.

Given a network $G = \{V, E\}$, where $V = \{v_1, v_2, ...v_n\}$ is the set of nodes, $E \subseteq V \times V$ is the set of edges, and $|E| = m$ is the number of the edges. The topology potential of each node is $\varphi(v_1), \varphi(v_2), ..., \varphi(v_n)$, the topology potential entropy is indicated as:

$$H = -\sum_{i \in V} \frac{\varphi(v_i)}{Z} \log(\frac{\varphi(v_i)}{Z}) \tag{10}$$

where $Z = \sum_{i \in V} \varphi(v_i)$ is the normalization factor.

As H is only related to one unknown parameter σ , we can optimize the impact factor σ by minimize H. Fig. 4 shows a simple network and Fig. 5 shows the relationship between H and σ based on Fig. 4.

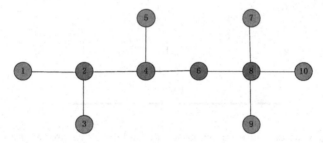

Fig. 4. A simple network

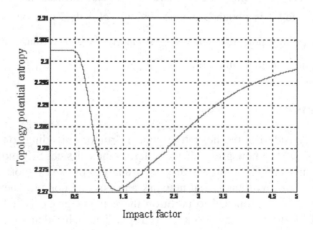

Fig. 5. Relationship between H and σ

Fig. 5 shows that the topology potential entropy approaches the maximum $H_{max} = \lg(n) = \lg 10 \approx 2.3026$ while $\sigma = 0$. With the increase of σ between 0 and 1.3930, H gradually decreases to the minimum $H_{min} = 2.2702$. When $\sigma \geq 1.3930$, , H increase with the increase of σ.

We used the optimal impact factor to calculate each node's topology potential entropy, and the result is showed in Table 3.

Table 3. Topology potential entropy of each node

Node	Topology potential entropy	degree
1	1.8591	1
2	3.0464	2
3	1.8519	1
4	3.1737	3
5	1.8519	1
6	2.831	2
7	1.9791	4
8	3.5165	1
9	1.9791	1
10	1.9791	1

From Table 3, we can observe that node v_4 and node v_8 are two maximum potential nodes, which are two center nodes. Therefore, we can partition this network into two communities. Node v_6 is a boundary node, which is classified in node v_8's community, since the topology potential entropy of node v_8 is greater than that of node v_4. Thus, the two partitioned communities are $\{v_1, v_2, v_3, v_4, v_5\}$ and $\{v_6, v_7, v_8, v_9, v_{10}\}$.

6   Conclusion

In this paper, considering the hesitating state in rumor spreading, our SIFR rumor spreading model adds a new group, fence-sitter, to the classical SIR model. To obtain an effective rumor control strategy, we further analyze the key factors that influence the rumor spreading. With the social network analysis and complex networks theory, an algorithm has been proposed to partition the network group based on the center nodes defined by the topology potential entropy. Through calculating the topology potential entropy for each node in a network, it is easy to find the center nodes in every partitioned network. Therefore the monitoring procedure can be simplified only with the center nodes.

Acknowledgments. This work is supported by the NSFC (Grant No. 61271172), RFDP (Grant No. 20120185110030, 20130185130002 and 20120185110025), SRF for ROCS, SEM and 863 high technology plan (Grant No. 2015AA01A707).

References

1. Daley, D.J., Kendall, D.G.: Epidemics and Rumors. Nature **204**, 11–18 (1964)
2. Daley, D.J., Gani, J.: Epidemic Modelling. Cambridge University Press, Cambridge (2000)
3. Zanette, D.H.: Critical Behavior of Propagation on Small-World Networks. Phys. Rev. E **64**, 050901 (2001)
4. Zanette, D.H.: Dynamics of Rumor Propagation on Small-World Networks. Phys. Rev. E **65**, 041908 (2002)
5. Moreno, Y., Nekovee, M., Pacheco, A.: Dynamics of Rumor Spreading in Complex Networks. Phys. Rev. E **69**, 066130 (2004)
6. Nekovee, M., Moreno, Y., Bianconi, G., Marsili, M.: Theory of Rumour Spreading in Complex Social Networks. Physica A **374**, 457–470 (2007)
7. Wang, J., Zhao, L., Huang, R.: SIRaRu Rumor Spreading Model in Complex Networks. Physica A **398**, 43–55 (2014)
8. Zan, Y., Wu, J., Li, P., Yu, Q.: SICR Rumor Spreading Model in Complex Networks: Counterattack and self-resistance. Physica A **405**, 159–170 (2014)
9. Bao, Y., Huang, W., Yi, C., Jiang, J., Xue, Y., Dong, Y.: Effective deployment of monitoring points on social networks. In: IEEE 2015 International Conference on Computing, Networking and Communications, Communications and Information Security Symposium, pp. 62–66. IEEE Press, Garden Grove (2015)
10. Zanette, D.H., Kuperman, M.: Effects of Immunization in Small-world Epidemics. Physica A **309**, 445–452 (2002)
11. Fu, X., Small, M., Walker, D.M., Zhang, H.: Epidemic Dynamics on Scale-free Networks with Piecewise Linear Infectivity And Immunization. Phys. Rev. E **77**, 036113 (2008)
12. Wasserman, S., Faust, K.: Social Network Analysis: Methods and Applications. Cambridge University Press, Cambridge (1994)
13. Freeman, L.C.: Centrality in Social Networks: Conceptual Clarification. Social Networks **1**, 215–239 (1979)
14. Burt, R.S.: Structural Holes and Good Ideas. American Journal of Sociology **110**, 349–399 (2004)
15. Girvan, M., Newman, M.E.J.: Community Structure in Social And Biological Networks. Proc. of the National Academy of Sciences of the United States of America **99**(12), 7821–7826 (2002)
16. Newman, M.E.J., Girvan, M.: Finding And Evaluating Community Structure in Networks. Phys. Rev. E **69**(2), 026113 (2004)
17. Newman, M.E.J.: Modularity And Community Structure in Networks. Proc. of the National Academy of Sciences of the United States of America **103**(23), 8577–8582 (2006)
18. Wang, L.J., Yang, B.R., Xie, Y.H.: Algorithm of Community Detection Based on Data Fields. Application Research of Computers **28**(11), 4142–4145 (2011)
19. Gan, W.Y., He, N., Li, D.Y., Wang, J.M.: Community Discovery Method in Networks Based on Topological Potential. Journal of Software **20**(8), 2241–2254 (2009)

Preventing and Detecting Infiltration on Online Social Networks

Canh V. Pham[1]([✉]), Huan X. Hoang[1], and Manh M. Vu[2]

[1] University of Engineering and Technology, Hanoi, Vietnam
maicanhki@gmail.com, huanhx@vnu.edu.vn
[2] Faculty of Mathematics and Informatics,
People's Security Academy, Hanoi, Vietnam
vuminhmanh@gmail.com

Abstract. Nowadays, together with the development of internet, online social networks provide lots of benefits for human. People use social networks for different purposes, such as: communicating, information sharing, relations creating or for business purposes, and etc. However, accompanying with benefits of social networks, users also must face the security and privacy risks. These issues have recently paid much attention to. One of these issues is the user can penetrate and steal personal information. The assailants can penetrate through agreeing their friend requests. When confirming this friend request, the user unintentionally discloses personal information. Especially, if the user stolen information is an important person of a specific organization, losses are extremely considerable. Promoted by this phenomenon, in this paper, we propose a new solution to prevent and discover any penetration for a specific user in an organization. First of all, we propose a new model which is called as safety community model in order to protect everybody in the organization. We build a target function orienting to the safety for everybody in the organization. After that, we have designed an effective algorithm to discover the penetration of unsafe factors for specific users in the organization. Tests in social networks are actually implemented, and the result shows that our model can prevent the penetration of outside objects.

Keywords: Infiltration · Social networks · Security · Privacy · Safety community

1 Introduction

Together with the fast development of Online Social Networks (OSNs) in recent years, OSNs makes human lives become more favorable, the user can communicate with each other at any geographic distance. Billions of people in the world use all social networks each day[1]. On average, each person uses 7 hours and 45 minutes each month on Facebook[1][3]; 32 millions of like and comment turns

---

[1] https://www.facebook.com

© Springer International Publishing Switzerland 2015
M.T. Thai et al. (Eds.): CSoNet 2015, LNCS 9197, pp. 60–73, 2015.
DOI: 10.1007/978-3-319-21786-4_6

each day[2]. The user uses social networks for many different purposes: maintaining connection with their friends and families, for business purposes, creating complicated relations or even creating romantic relations. Unfortunately, accompanying with benefits of OSNs, the user must face lots of risks on security and safety on OSNs. One of them is the assailant to penetrate and steal personal information. The user can be stolen information by the assailant, such as: their email address, their friends, their working places, and their organizations. In 2012, Bosman et al. [6] have designed a socialbots which can imitate activities for actual user to collect personal data for defrauding purposes and strategies of spam disseminating in big scale. In 2012, Yashar et al. [7] have introduced an algorithm to collect information of officers in a detailed organization on social network from community and then they can regenerate information on organization. In 2012, Yashar et al. [5] have used Socialbots to acquire information in an organization on social network (OSNs). Socialboot has sent friend requests to users who work in a specific organization in social networks to take valuable information of organization.

Especially, in 2013, Yavial Yashar et al. [4] proposed a method which combines all studies using Socialbot to penetrate into a specific user in a target organization. They have tested on Facebook for three organizations. They penetrate a target user in an organization by make they feel confident through their friends. Firstly, Socialbots finds friends of the target user and send friend request to them. After confirmed by several their friends, Socialbot sends a friend request to the target users. They show that it is easy to penetrate the target user and the success rate of penetration is 50% and 70%, and they also comment that the higher quantity of common friends confirming the friend request is, the higher the penetration ability is. These results show that in fact the information leakage ability of user is very high, the user should select more carefully friends on OSNs. The penetration becomes more and more serious when the user attacked is an important officer of organization, company. They have important information on their organization or company, thus the attacker can use this information to protest against them and cause losses for organizations or companies where they work, these sequences can not be estimated.

Thanks to studies and analysis stated above, we can see that protecting users against penetration is very urgent. Motivated by study of Elyashar in [4], we set a reverse problem: how to detect a penetration for a specific user in an organization or company? In this paper, we propose a solution for this problem, including two tasks: Firstly, we give a new *relationship-measure* to evaluate the relation, and also to identify the importance of relation between two users. Based on this measure, we propose a new model which is called as *Safety Community* (SC) to protect all users in an organization or company. SC is a safe area surrounding all officers in organization, it only includes safe users and reliable relation. The user in organization is recommended to communicate only with users in SC and should be more careful about who do not belong to this model SC. Thereby, we set *Maximize the safety for all users in organization*(MSO)problem on SC model with object function orienting to the safety for everybody in organization.

We also design an effective algorithm for MSO problem and used SC model to prevent the infiltration. Secondly, detecting the penetration of factors which are unsafe for a user in organization. We introduce an solution to detecting the penetration. In this paper, we proposed a solution for preventing and detection infiltration under SC model. Our main contributions are summarized, as follows:

- We introduce a new concept of measure called *relationship- measure* to quantify a relation, and also to identify the importance between two users. We analyze the meaning of the new measure and compare with the previous. Consequently, we propose a SC model to protect all users in an organization. We formulate the *Maximizing for all user in Organization* (MSO) problem on SC model with the object function oriented to the safety of all users in this organization.
- We design an effective algorithm for MSO problem and given a solution to use the SC model for the purpose of preventing and detecting the infiltration.
- We conduct experiments on real social and achieve really impressive result. We show that SC model prevent the infiltration successfully for organizations of different sizes under various OSNs.

The rest of paper is organized as follows. In Section 2, we introduce the graph model and notations which will be user throughout in the paper and also provide a specific description on SC model. Section 3, we formulate MSO problem and give a greedy algorithm for MSO. In section 4, we present a solution for prevent and detecting the infiltration. The experiment results on real-social networks are shown in section 5. Finally, section 6 concludes the paper.

2   Model and Problem Definition

2.1   Network Model

In this paper a social network be represented by a direct and weight graph $G = (V, E, w)$ with V is the set of n network users (or nodes), E is the set of m directed relationships (or edges), w_{uv} is the weight function of edges (u, v) which representing the communication frequency. A set $U \subset V$, $U = \{u_1, u_2, ., u_k\}$ is representing all users in an organization in which we need to protect and k is the cardinality of U. Without loss of generality, we assume that all weights of edges are normalized, i.e $\sum_{(u,v) \in E} w_{uv} = 1$ and $w_{uv} \geq 0$.

2.2   Relationship-Measure

In order to build SC model, firstly we determine a safety relationship between two users. In fact, sometimes we must identify relations (or familiar level) between two users on OSNs.

It is so important, because we can identify their impacts, the interaction between them and also network structure. About identifying the relation between two users, Leskovec et al. [11] comment that your friends in social networks are highly likely to become their friends. Fire et al. [8] introduce a relationship measure which is called as friend-measure. This measure shows that the more connections between neighboring countries of two users, the higher the probability of connecting between two users is. However, this measure only uses in undirected and weighted graph and only identifies relations through mutual friends. In fact, two users can affect each other by word of mouth or peer-pressure through intermediate users between these users.

Therefore, we can evaluate the relation between two users through their intermediate users. Note that, in order to evaluate the relation between two users, we must evaluate this relation in all relations that they have participated. For example, in order to identify the familiar level between u and v, we must consider the interactive frequencies w_{ux}, and also the rate of this relation among all their friends, which means that w_{ux} and w_{vy} with x and y are neighbors of u and v.

By the above analysis, we given the formula to evaluate relationship between two users through t intermediaries users, as follows

$$\varphi(u,v,t) = \frac{\displaystyle\sum_{|P(u,v)|=t} w(P(u,v))}{\displaystyle\sum_{P\in P_{total}(u,t)} w(P)} \qquad (1)$$

where $P(u,v)$ is a path going from v to u (v and u are excluded) of length $|P(u,v)| = t$, $P_{total}(u,t)$ is a path going from u to x (x and u are excluded) of length t, x is any user in social network which from u can come to through t intermediate users, $P_{total}(u,t)$ does not contain a cycle, $w(P(u,v)) = \prod_{(a,b)\in P} w_{ab}$ is the total weight of path $P(u,v)$. However, the formula (1) only reflects the relation correlative to the direction from u to v through t intermediate users, and then to v. To reflect relations between two users in two directions, we define according to the formula, as follows:

$$g(u,v,t) = \varphi(u,v,t) + \varphi(v,u,t) \qquad (2)$$

Finally, we identify $relationship - measure(u,v)$ between two users through T_0 intermediate users, in which T_0 is a predefined parameter:

$$relationship - mesure(u,v,T_0) = \Phi(u,v,T_0) = \sum_{t=0}^{T_0} g(u,v,t) \qquad (3)$$

The meaning of formula (3) allows identifying the relation between users basing on the evaluation of their relation. In large-scaled OSNs, information on edges between users is frequently lacking. Information is usually lost during the process of information retrieval. By using this measure, we can enrich (increase necessary information) of network by calculating the connection between users indirectly. Thanks to this measure, we also broaden the idea of Leskovec and colleagues;

it implies that friends of friends through intermediate users have high probability of becoming friends. The meaning of the formula (3) allows identifying the relationship between users based on their interest together and all the relationships that they have joined. Information is usually lost during the process of retrieval. By using this measure, we can enrich (increase necessary information) of network by calculating the connection between users indirectly. Thanks to this measure, we also broaden the idea of Leskovec et al. [11], it implies that friends of friends through intermediate users have high probability of becoming friends. Now, we use this relationship measure to identify a reliable or safe relationship between two users by using a safe threshold. If $relationship-measure(u, v) \geq \theta$, the relationship between u and v is safe. If not, it is unsafe. Note that, according to this measure, the relation between two users u and v is safe, even when (u, v) is not any edge of graph G. It is safe to evaluate the relationship based on $relationship - measure$, because of reasons, as follows: First of all, this evaluation we can based on a safety threshold for any pairs of nodes u and v in graph G. Secondly, this evaluation is safe by using $relationship - measure(u, v)$ to be against the penetration to a user X in an organization by sending a friend request to mutual friends of X. When we use $relationship - measure(u, v)$ to consider sending a friend request to X with parameter $T_0 \geq 2$, we will restrict the penetration. In this case, we will consider the relationship between S and X through T_0 intermediate people, thus it is possible to avoid making friends through mutual friends of X (or $T_0 = 1$).

2.3 SC Model

In our model, we instruct a Safety Community model based in an important factor which is relationship-measure. Almost all studies on discovering community structure base on maximizing modularity [9] and based on and density function in [10]. However, in this paper, we instruct a community model based on safety relationship between users and orienting to the safety of all users in organization that we should protect. Safety Communication model of set $U = \{u_1, u_2, .., u_k\}$ with $|U| = k$ on the graph $G = (V, E, w)$ denote $G^{SC} = (V^{SC}, E^{SC}, w^{SC}, \theta, T_0)$ where θ, T_0 are predefined parameters. V^{SC}, E^{SC}, w^{SC} be defined by incremental method as follows:

1. *Start*: $V^{SC} = U$. It's mean a initially safety community include all users in the organization.
2. *Incremental users and edges*: From each vertex $u \in V^{SC}$, for each vertex $v \in V \setminus V^{SC}$, if $relationship - measure(u, v) \geq \theta$ then: add v into V^{SC}, edge (u, v) into E^{SC} with $w_{uv}^{SC} = reltionship-measure(u, v)$. Continue this until no more vertex can be add into V^{SC}.

Edges $of G^{SC}$ are safety relationships and it is difficult for Socialbots [4] to achieve it. Note that, edge (u, v) of G^{SC} can belong to G because relationship-measure can be estimated to enrich the network structure. By the incremental method, each user has a safety relationship with user in G^{SC} which will be added.

However, there is a question how to add user with the maximal relationship-measure for all users in G^{SC}. Based on this idea, we build a problem of maximizing safety for all users in organization MSO in the following part.

2.4   Problem Definition

In SC model exits users who have safe relations for a specific user (called as X) in organization and we can not identify their safety for other people in organization. On the other hand, it is a safety which is local, partial and not comprehensive for everybody. Socialbots (S) proposed in [4] can penetrate into X by creating a safe relationship X (through T_0 intermediate users) and then they can penetrate into the safety community. This work can be implemented easily when S has selected people in organization with the fewest number of friends and attack. To evaluate the safety of user v in SC model but this user does not belong to the organization which should be protected (i.e $v \in V_{SC} \setminus U$), we identify the safety of v for U by the following formula:

$$F(v) = \sum_{u_i \in U} \Phi(v, u_i, T_0) \tag{4}$$

On the other hand, a user v is the trust for organization, he must have all trust of everybody in this organization and vice versa. In reality, ensuring the safety of everybody is very important, we introduce a objective function to ensure the safety for all users in set U as follows:

$$H(G_{SC}) = \sum_{v_k \in V^{SC} \setminus U} F(v_i) = \sum_{v_k \in V^{SC} \setminus U} \sum_{u_k \in U} \Phi(v_i, u_k, T_0) \tag{5}$$

Now, we should maximize the safety for all users in $V^{SC} \setminus U$ for set U, which means finding the maximum function H(G). In SC model, the number of edges increases when we use the incremental method of edges and vertices into G^{SC} based on the safety relationship for G^{SC}, indirectly into set U. Therefore, we only have a number of user enough to create a safety relationship and we provide a user threshold to restrict the number of vertices of G^{SC} required.

Problem 1 (MSO Problem)
Given a direct and weight graph $G = (V, E, w)$ underlying an OSN. A set users $U = \{u_1, u_2, .., u_k\}$ is representing all users in organization which we need protect. The MSO problem construct a Safety Community (SC) model $G^{SC} = (V^{SC}, E^{SC}, w^{SC}, \theta, T_0)$ of U on G with the goal of:

$$\text{maximum: } H(G_{SC}) = \sum_{v_k \in V^{SC} \setminus U} F(v_i) = \sum_{v_i \in V^{SC} \setminus U} \sum_{u_k \in U} \Phi(v_i, u_k, T_0)$$

subject to: $|V^{SC}| \leq \Delta$

In the next section, we analysis the complexity of MSO and give a solution to MSO.

3   Solution to MSO Problem

3.1   An Algorithm Calculate Relationship-Measure

By using the ideas of Rippling algorithm, we can calculate $P_{total}(u, t+1)$ by using $P_{total}(u, t)$. Called Q_{old} set is all vertices which can came from u through t intermediaries users. Now, we go through $t + 1$ intermediate users be call set Q_{new} by using Q_{old}. $P(u, v)$ is selected from $P_{total}(u, t)$ which have end vertex is v, similarly for $P(u, v, t+1)$, doing so until $t = T_0$. Base on this idea, we proposed an algorithm to calculate $relationship - measure$ for each pair of vertices (u, v).

Algorithm 1. Relationship-measure

Data: $G = (V, E, w), T_0$.

Result: $\Phi(u, v), \forall u, v \in V, (u \neq v)$

begin

  $P(u, v) \leftarrow 0$;

  $Q_{old} \leftarrow \emptyset; Q_{new} \leftarrow \emptyset;$

  for $u \in V$ **do**

    $t \leftarrow 0;$

    $Q_{new} \leftarrow u;$

    while $t \leq T_0$ **do**

      $P_{total}(u, t) \leftarrow 0; P(u, v) \leftarrow 0$;

      $Q_{old} \leftarrow Q_{new}; Q_{new} \leftarrow \emptyset;$

      for $x \in Q_{old}$ **do**

        \\ Find all vertices can come to and calculate $P_{toal}(u, t), P(u, v)$

        for $v \in V, w_{xv} \neq 0$ **do**

          $Q_{new} \leftarrow Q_{new} \cup \{v\};$

          if $P(u, x) = 0$ **then**

            $P(u, v) \leftarrow w_{xv}$;

          else

            $P(u, v) \leftarrow P(u, x).w_{xv}$;

          $P_{total}(u, t) \leftarrow P_{total}(u, t) + P(u, v)$;

      \\ calculate $\varphi(u, v, t)$ of vertices which came to.

      for $v \in Q_{new}$ **do**

        $\varphi(u, v, t) = \dfrac{P(u, v)}{P_{total}(u, t)};$

        \\ When calculated both directions

        if $\varphi(v, u, t) \neq 0$ **then**

          $g(u, v, t) = \varphi(u, v, t) + \varphi(v, u, t);$

          $\Phi(u, v) \leftarrow \Phi(u, v) + g(u, v, t);$

    $t \leftarrow t + 1$;

Lema 1. *The time of complexity for calculate the relationship − measure for each pair vertices by Algorithm 1 is $O(M_0 n^2)$.*

3.2 Greedy Algorithm for MSO Problem

We introduce a simple greedy algorithm in Algorithm 2, this algorithm for MSO consists of two step.

Algorithm 2. Greedy Algorithm

Data: $G = (V, E, w), \Phi(u, v), \theta, U = \{u_1, u_2, \ldots, u_k\}$.
Result: $G^{SC} = (V^{SC}, E^{SC}, w^{SC})$.
begin
$\quad V^{SC} \leftarrow U; C \leftarrow \emptyset$;
$\quad$ **Initialize Candidate set:**
$\quad$ **for** $v \in V^{SC}$ **do**
$\quad\quad$ **for** $x \in V \setminus V^{SC}$ **do**
$\quad\quad\quad$ **if** $\Phi(v, x) \geq \theta$ **then**
$\quad\quad\quad\quad$ $C \leftarrow C \cup \{x\}$;

$\quad$ **Update Candidate set:**
$\quad$ **while** $|V^{SC}| \leq \Delta$ or $C = \emptyset$ **do**
$\quad\quad$ Chose $u \in C$ such that $F(u) = \max_{x \in C} F(x)$;
$\quad\quad$ $C \leftarrow C \setminus \{u\}$;
$\quad\quad$ $V^{SC} \leftarrow V^{SC} \cup \{u\}$;
$\quad\quad$ Add the edge (v, u) into E^{SC} which from v can go to u in previous
$\quad\quad$ $w_{vu}^{SC} = \Phi(v, u)$;
$\quad\quad$ **for** $x \in V \setminus \{C \cup V^{SC}\}$ **do**
$\quad\quad\quad$ **if** $\Phi(u, x) \geq \theta$ **then**
$\quad\quad\quad\quad$ $C \leftarrow C \cup \{u\}$;

$\quad$ Return G^{SC};

Step 1. Initialize Candidate set The candidate C is a set of all users with safety relationship with any user in U. And then, we select users with maximal safety for all users in U. In this step, the set V^{SC} of model SC including U, it is obvious because U set of users who need be protected in SC model and C will be updated in step 2.

Step 2. Update Candidate set: In this step, set C be updated, user in C with maximizing safely for all users in U be calculated by (4) will be add into V^{SC} and remove from C and the vertices has safety relationship with each vertex in V^{SC} be add into C, this work is continued until satisfying: $|V^{SC}| = \Delta$.

Theorem 1. *The Greedy algorithm has a time complexity* $O((k + \Delta)n)$, *where* n *is number of vertices of* G *and* $k = |U|$.

4 Solutions to Prevent and Detect Infiltration

In this section we use the solution of the problem MSO to perform the prevention and detection of the infiltration of factors unsafe outside the SC model.

Prevention: We use SC model to prevent from unsafe risks, especially Social-bots proposed in [4]. When any user who does not belong to SC model sends friend request to users belonging to organization, this request will be deleted. It will prevent from penetration to steal information. Socialboot in [4] only penetrates when confirming the friend request of specific number of mutual friends for target user, thus using relationship-measure will restrict this penetration. Prevention also restrictions the subjective element of the user when Socialboot deceived by the trust through numbers mutual friend.

Detecting: In case u is an important user in organization, we must consider whether confirming friend requests more strictly. Suppose the set $X = \{u_1, u_2, \ldots, u_n\}$ is a set of important users in U and the threshold of making friends for them are correlative to: $\theta_1, \theta_2, \ldots, \theta_n$. When a user in SC model sends a friend request to user $u_i \in X$. We will calculate $relationship-measure(S, u_i)$, if $relationship-measure(S, u_i) \geq \theta_i$, the friend request is confirmed, if not, the friend request is deleted. Although S belongs to SC, it is unfamiliar with u. In this case, S wants to attach X through t ($t \geq 1$) intermediate users, the complication of this case is $O(n^t)$, in which n represents mutual friends of X and intermediate users. It is difficult to implement because the complication is very big, thus S sends too much friend requests.

5   Experimental

5.1   Datasets

We examine the performance of our proposed algorithm on difference real-world OSNs, including: Facebook, Slashdot, Epinions with different size in Table 1. For Facebook, we use the dataset includes users in New Orleans spanning from September 2006 to January 2009 [12]. The data includes more than 63,000 users connected by more than 1.5 million friendship links represents the interaction frequency between the users and average node degree of 23.5. We calculate and normalized the weight of each edge is proportional to the frequency of interaction on the whole network. For Slashdot and Epinions, we used the dataset by the approach of Leskovec [13] and Richardson [14]. For each Dataset: Slashdot, Epinions we randomly assign weight of each edge respectively and normalized them.

Table 1. Dataset

Datasets	Nodes	Edges	Avg. Degree
Facebook	60,000	1,500,000	23.5
Slashdot	82,168	948,464	11.7
Epinions	75,879	508,837	6.7

5.2   Experiment Results

In our experiment, firstly for each OSN, we select an organization which is a group of users to have an intimate relation, in which the number of users is k. And then, we use the method of Socialbot (S) [4] to implement attacks to specific users in each organization. Secondly, we construct and optimize SC model for organizations with purpose of preventing attacks. If S does not belong to SC, deleting any friend request of S to all users in organization, thus we should only check whether S belongs to SC or not? We call organizations in Facebook, Slanshdot, Epinions social networks correlatively to: U_1, U_2, U_3, they are organizations which should be protected against penetration. In order to evaluate results more complicatedly, we select organizations with different dimensions k = 50, k = 100, k = 200 and k = 500. Selecting organizations with different dimensions to evaluate impacts of users in organization through function F(x) defined by (4), which affect results of building SC model.

We can see in experiments of Alyshar [4], their socialboot can penetrate successfully the target user when they achieve a specific number of mutual friends. Therefore, we suppose that each user in organization U has a threshold of friend request confirming, it is defined as follows:

$$\theta_i = \frac{\text{The number mutual friend need for acceptance of } u_i}{\text{the number of friends of } u_i}$$

In detecting and optimizing SC model, we select parameters for this experiment, they are: Number of intermediate users, limited number of users in SC model correlatively to: $T_0 = 5$, $\Delta = 10k$. We use different safety thresholds θ for each organization and organization size, because we believe that it depends on structure and properties of each network.

Simulation of Socialbots' attack: We implement to simulate attacks of Socialbot S according to the method proposed in on datasets. In each organization, we select at random 10 users to implement attacks and we create a Socialbot to attack to them. To attack target users, S finds all of their friends to send friend requests, and then S sends friend requests to target users. We choose the size of the organization respectively $k = 50, 100, 200, 500$ to perform experiments. Figure 1 reports results that we have reproduced this attack of S for each organization with different size on data set. These results are also correlative to results of Elyashar [4]. In general, the more S has mutual friends with target user, the higher the probability of confirming friend request of target user. Our results show that it is very easy to infiltrate to target user with high successfully rate. Next, we suppose S has not penetrated to the target user, and we implement to construct SC model by algorithm 2 for each different organization, and then we check whether S can belong to SC model for each organization or not? If SC does not contain S, S is isolated and the friend request will be deleted. Table 2 reports the results of SC model and protection capabilities of SC model respectively with $k = 50, 100, 200, 500$. We tested in organizations of various sizes from 50 to 500 to evaluate comprehensively when k change.

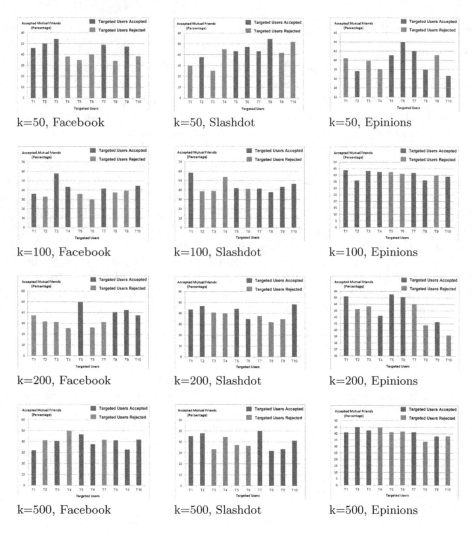

Fig. 1. Results of Socialboot attack for each organization and in different sizes

SC model: SC model in each circumstance has different size (number of vertices), it depends on the safety threshold. We introduce different safety thresholds for each different OSNs and different size. When selecting organizations in Facebook, its numbers of members are 50, 100 and 200, V^{SC} always reaches the maximal level $\Delta = 10k$, because we select the threshold θ. However, it only affects size of G^{SC} not affect the protective and functionality of prevention of G^{SC} because G is optimized by algorithm 2. In addition, we see that with same size of organizations, organizations in Facebook have G^{SC} higher than other networks, although we select the safety threshold θ always higher then other networks. It shows the difference in size of SC model depends on structure of

Table 2. The results construct SC model and prevention

Size of Org	Organization	θ	Size of G^{SC}	S belong G^{SC}
	U_1-Facebook	3.0	500	No
k=50	U_2-Slashdot	2.7	306	No
	U_3-Epinions	2.5	245	No
	U_1-Facebook	2.98	1,000	No
k=100	U_2-Slashdot	2.67	504	No
	U_3-Epinions	2.43	297	No
	U_1-Facebook	2.87	2,000	No
k=200	U_2-Slashdot	2.52	1,210	No
	U_3-Epinions	2.21	949	No
	U_1-Facebook	2.3	3,266	No
k=500	U_2-Slashdot	1.92	2,508	No
	U_3-Epinions	1.81	2,968	No

OSNs. The index which represents a network structure is average level of each OSNs higher than size of correlative SC model.

The prevention of SC model: We see that S can not penetrate into G^{SC} in any circumstance, and our purpose and expectation are totally achieved. For organizations with small scale of 50 to organizations which scales are 10 times as high as. Although Socialbot sends friend requests to target users, but they can penetrate into G^{SC}. Especially, when size of this organization is $k = 500$ and SC model has the number of members up to 3,266 in Facebook, 2508 in Slashdot and 2.968 in Epinions, S can not become a member in this model. It proves that relationship-measure introduced by us can evaluate the familiar level, and so that it is possible to restrict the penetration of Socialboot.

6   Conclusion

In this paper, we study a solution to detect and prevent the penetration of Socialbot which is proposed in [4]. We design a new measure to quantify the familiar and belief level between two people. Especially, in order to prevent from penetration, we build SC models and propose MSO problem to optimize SC models. Our experiment on actual data sets show that SC model is able to prevent impressively. In the future, we have a plan of researching this issue in social networks with more suitable model? Whether we can find suitable solutions or not?

Appendix

The proof of Lema 1

Proof. The first loop for contribute a factor n to the time complexity. For the while loop, when $t = 0$, $Q_{old} = u$. $t = 1$, Q_{old} contain all friends of u, $|Q_{old}| = M$, where M representing number of friend of each user. $t = 2$, Q_{old} contain friend of friend of u thus, $|Q_{old}| = M^2$. Similarly, $t = T_0$, Q_{old} contain friend of all T_0-th intermediate users. Thus, this work implement lost $1 + M + M^2 + ... + M^{T_0}$ of time complexity, where M^t is presenting number of friend of t-th intermediate users. We note that, the number of friend less than the number of vertices, thus $M_0 = 1 + M + M^2 + ... + M^{T_0} << n$. The two last for loop can be done in $n + M^t$ step, because calculating of Q_{new} must visit all the vertices in order to find the appropriate vertices and now $|Q_{new}| = M^t$. The time of complexity is $O(n + M^t) = O(n)$, due to $M^t << n$. The total of time complexity is $O((1 + M + M^2 + ... + M^{T_0})n.n) = O(M_0 n^2)$. $\Box$

The proof of Theorem 1

Proof. In the Initialize Candidate step, it is easy to see that this step contribute a factor kn to the time complexity, where $k_1 = |U|$. In the Update Candidate step, the while loop lost Δ of time complexity. The selection of user $u \in C$ has $H(u)$ maximizing implement $|C|$ of time complexity. The for loop implement lost $|V \setminus C|$ so the total of this step is $O(\Delta(|C| + |V \setminus C|)) = O(\Delta.|V|) = O(\Delta.n)$ time complexity. Hence, the complexity of algorithm is $O(kn) + O(\Delta.n) = O((k + \Delta)n)$. $\Box$

References

1. O'Cass, A., Fenech, T.: Webretailing adoption: exploring the nature of internet users Webretailing behaviour. Journal of Retailing and Consumer Services **10**, 81–94 (2003)
2. social media and internet statistics. http://thesocialskinny.com/216-social-media-and-internet-statistics-september-2012
3. new social media stats for 2012. http://thesocialskinny.com/99-new-social-media-stats-for-2012/
4. Elyashar, A., Fire, M., Kagan, D., Elovici, Y.: Homing Socialbots: Intrusion on a specific organizations employee using Socialbots. In: IEEE/ACM International Conference on Advances in Social Networks Analysis and Mining (2013)
5. Elyashar, A., Fire, M., Kagan, D., Elovici, Y.: Organizational intrusion: organization mining using socialbots. In: ASE International Conference On Cyber Security, Washington D.C., USA (2012)
6. Boshmaf, Y., Muslukhov, I., Beznosov, K., Ripeanu, M.: Design and Analysis of a Social Botnet, July 9 (2012)
7. Fire, M., Puzis, R., Elovici, Y.: Organization Mining Using Online Social Networks. ACM Transactions on Embedded Computing Systems 9(4), Article 39 (2012)
8. Fire, M., Tenenboim, L., Lesser, O., Puzis, R., Rokach, L., Elovici, Y.: Link prediction in social networks using computationally efficient topological features. In: SocialCom/PASSAT, pp. 73–80. IEEE (2011)
9. Fortunato, S.: Community detection in graphs. Physics Reports 486(3–5), 75–174 (2010)
10. Fortunato, S., Castellano, C.: Community structure in graphs. eprint arXiv 0712.2716 (2007)

11. Leskovec, J., Huttenlocher, D., Kleinberg, J.: Predicting positive and negative links in online social networks. In: Proceedings of the 19th International Conference on World Wide Web, WWW 2010, pp. 641–650. ACM, New York (2010)
12. Viswanath, B., Mislove, A., Cha, M., Gummadi, K.P.: On the evolution of user interaction in facebook. In: 2nd ACM SIGCOMM Workshop on Social Networks (2009)
13. Leskovec, J., Lang, K., Dasgupta, A., Mahoney, M.: Community Structure in Large Networks: Natural Cluster Sizes and the Absence of Large Well-Defined Clusters. Internet Mathematics 6(1), 29–123 (2009)
14. Richardson, M., Agrawal, R., Domingos, P.: Trust Management for the Semantic Web. In: Fensel, D., Sycara, K., Mylopoulos, J. (eds.) ISWC 2003. LNCS, vol. 2870, pp. 351–368. Springer, Heidelberg (2003)

Consensus on Social Influence Network Model

Zhenpeng Li[1] and Xijin Tang[2]([⊠])

[1] Department of Applied Statistics, Dali University, Dali 671003, China
lizhenpeng@amss.ac.cn
[2] Academy of Mathematics and Systems Sciences,
Chinese Academy of Sciences, Beijing 100190, China
xjtang@amss.ac.cn

Abstract. Many studies show that opinions formation displays multiple patterns, from consensus to polarization. Under the framework of the social influence network by Friedkin and Johnsen (1999) and based on random walk on graph, we rigorously prove that for a social group influence system, with static social influence structure, the group consensus is almost a quite certain result. In addition, we prove the lower bounds on the convergence time m for random walk P^m to be close to its final average consensus (wisdom group decision making) state, given an arbitrary initial opinions profile vector and one small positive error ϵ. Although our explanations are purely based on mathematic deduction, it shows that the latent social influence structure is the key factor for the persistence of disagreement and formation of opinions convergence or consensus in the real world social group.

Keywords: Social influence network theory · Random graph · Opinions dynamics

1 Introduction

Originally from decentralized decision making, consensus problems have an old history, such as the models introduced in DeGroot(1974)[1], Friedkin and Johnsen (1999)[2] and Friedkin (2011)[3]. From social psychological point of view, this line of research began with French's formal theory of social power[4], a simple model of collective opinion formation in a network of interpersonal influencing social group. As a step forward, Friedkin presented the social influence network theory, which considered both cognitive and structural aspects, and focused on the contributions of networks of interpersonal influence to the formation of interpersonal agreements and group consensus.

Over the past few years, models of the convergence of opinion or consensus problem in social systems have been the subjects of a considerable amount of recent attention in the fields such as motion coordination of autonomous agents [5,6], distributed computation in control theory [5,7,8], randomized consensus algorithms [9,10], and sensor networks about data fusion problems [11]-[15].

© Springer International Publishing Switzerland 2015
M.T. Thai et al. (Eds.): CSoNet 2015, LNCS 9197, pp. 74–83, 2015.
DOI: 10.1007/978-3-319-21786-4_7

Most of the growing interests in consensus problems (both algorithms and practical applications) are based on probabilistic settings. This might be due to the unpredictability of the environment where the communication between agents occurs [9], and the random characteristics of influences or interactions among agents in systems (man made or social systems).

Recently, the study of opinion dynamics has started to attract the attention of the control community, who with the bulk of motivation have developed about methods to approximate and stabilize consensus, synchronization, and other coherent states. However, comparing with many man-made or engineering systems, social systems do not typically exhibit a consensus of opinions, but rather a persistence of disagreement, i.e. polarization patterns. The ubiquitous group polarization phenomena can be observed from political election to carbon dioxide emissions debate[16]. In a social system, the difficulty in arriving at a collective consensus state roots in the fact that the process of opinion formation can rarely be reduced to accepting or rejecting the consensus of others, as exemplified by Arrow's dilemma of social choice [17]. On the contrary, in most cases individuals construct their options in a complex interpersonal environment or with their prior identities (e.g. prior beliefs, prejudices and social identities etc.), their views are often in a state of disagreement or not easily changed, due to opinion-dependent limitations in the network connectivity and obstinacy of the agents as pointed in Ref. [18]. This phenomenon shows the complexities of social control in social economic systems.[1]

Consensus as one of the important and regular group opinions dynamic pattern is generally observed in a relative smaller group discussion and barging process. Friedkin and Johnsen's social influence network theory emphasizes that the interpersonal influence social structure (or social influence matrix) is the underling precondition for the group consensus or opinion convergence. In that model, the initial social influence structure of group of actors is assumed to be fixed during the entire process of opinion formation. However, with the evolution of time stamp, considering both stubborn and susceptible effects, the interpersonal influence structure can be regarded as a dynamic recursive process. For this reason, the interpersonal influence structure in their model is also dynamic, as described in Section 2.

From social influence matrix point of view, in large scale group opinions dynamic processes, the group belief is difficult to reach convergence, let alone consensus state, since social influence structures are generally unconnected, not to mention the social impact relations can be positive, negative or neutral. For example, on-line highlighted discussion, political or social hot spot debates often display polarization patterns [19].

In this paper, our interests concentrate on the precondition for consensus formation in a social group based on Friedkin's model. From interpersonal network structure point of view, our investigation presents the conditions for the

---

[1] *In classical sociological field, social control refers to the occurrence and effectiveness of ongoing efforts in a group to formulate, agree upon and implement collective courses of action.*

formation of group opinions convergence and consensus. We investigate the opinions convergence phenomenon over a group of N individuals with a random walk social influence structure, and for any given initial opinions distribution, i.e. the opinions evolution problem with a (time-variant) linear dynamic model driven by random matrices. Our analytic proof provides strict mathematic explanations for the deterministic characterization of the ergodicity, which can be used for studying the consensus over random graphs and the formation of opinion parties. The proof procedures are self-contained and based on ergodic theorem of Markov chain and eigenvalues of random graph, as introduced in Ref.[20].

2   Problem Formulation and Terminology

Social influence network theory presents a mathematical formalization of the social process of opinions changes that unfold in a social network of interpersonal influences. The spread of influence among individuals in a social network can be naturally modeled under a probabilistic framework, here, we briefly describe the classical Friedkin and Johnsen's model to illustrate how the opinion dynamics arise in the context of social networks.

Let $W = [w_{ij}]$ is a $N \times N$ row random matrix of interpersonal influence, i.e. for each i, w_{ij} denotes for the individual j's social influence to i, $\sum_j w_{ij} = 1$. $A = diag(a_1, a_2, ..., a_N)$ is a $N \times N$ diagonal matrix of individuals susceptibilities to interpersonal influence on the opinion. In a group of N persons, with the initial $N \times 1$ opinions vector $y^{(1)}$, the updating opinions vector $y^{(t)}$ in the interpersonal opinions influence system is described by Equ.(1),

$$y^{(t+1)} = AWy^{(t)} + (I - A)y^{(1)} \tag{1}$$

Definition 1. *The system (1) reaches the convergence state if, for any initial opinions vector $y^{(1)}$, it holds that $\lim_{t \to \infty} y^{(t)} = y^*$.*

Definition 2. *The system (1) reaches consensus state if, for any initial opinions vector $y^{(1)}$, and each $1 \le i, j \le N$, it holds that $\lim_{t \to \infty} |y_i^{(t)} - y_j^{(t)}| = 0$, where $|.|$ is the symbol of the absolute value. This means that, as a result of the social influence process, in the limit they have the same belief on the subject.*

As a consequence of system (1), the opinion profile at time $t \in Z \ge 0$ is equal to

$$y^{(t+1)} = \widehat{W}^t y^{(1)}, \tag{2}$$

where $\widehat{W}^l = (AW)^t + (\Sigma_{k=0}^{t-1}(AW)^k)(I - A)$ is the reduced-form coefficients matrix, discribing the total or net interpersonal effects that transform the initial opinions into equilibrium opinions, and for any entry $\widehat{w}_{ij}^t$ in $\widehat{W}^t$, satisfies $0 \le \widehat{w}_{ij}^t \le 1$, $\sum_j \widehat{w}_{ij}^t = 1$. According to Def.1, under suitable conditions, when

$t \to +\infty$ if $I - AW$ is nonsingular, the system (1) arrives at convergence equilibrium opinions profile y^*, where $y^* = \lim\limits_{t\to\infty} y(t) = (I - AW)^{-1}(I - A)y^{(1)}$. When $t \to +\infty$, we have

$$\lim_{t\to\infty} \widehat{W}^t = \lim_{t\to\infty} \{(AW)^t + \sum_{k=0}^{t-1}(AW)^k(I - A)\} = (I - AW)^{-1}(I - A) = V. \quad (3)$$

Given large enough time stamp t, and a sufficiently small positive real number ε, V can be approximated by $\widehat{W}^t$. Furthermore, according to the approximation error $||\widehat{W}^t - V|| \le \varepsilon$ (where $||.||$ denotes the matrix norm), we can obtain the time stamp's upper bound and lower bound as $ln(||V|| - \varepsilon)/ln(||\widehat{W}||) \le t \le ln(||V|| + \varepsilon)/ln(||\widehat{W}||)$, where $||\widehat{W}|| = ||AW + I - A||$.

Followed the same lines of the convergence results by Ishii and Tempo (2010) [21], and Golub and Jackson (2010) [22], by showing the ergodicity property, Frasca, et al.(2013) proved the convergence result of system (1)[18]. Touri and Nedic (2011) studied the ergodicity and consensus problem with a linear discrete-time dynamic model driven by stochastic matrices [23].

It should be noted according to Def.1, that equilibrium opinions may settle on the mean of group members' initial opinions, a compromise opinion that differs from the initial ones, or altered opinions that do not form a consensus. When consensus is formed in system (1), i.e. as $t \to +\infty$, $\widehat{W}^t$ will have the form of a stratification of individual contributions as following,

$$\widehat{W}^t = \begin{bmatrix} \widehat{w}^t_{11} & \widehat{w}^t_{22} & \dots & \widehat{w}^t_{NN} \\ \widehat{w}^t_{11} & \widehat{w}^t_{22} & \dots & \widehat{w}^t_{NN} \\ \vdots & \vdots & \vdots & \vdots \\ \widehat{w}^t_{11} & \widehat{w}^t_{22} & \dots & \widehat{w}^t_{NN} \end{bmatrix},$$

which suggests that the initial opinion of each individual makes a particular relative contribution to the emergent consensus.

3   Random Walk on Weighted Graph

In this section, without the lose of the generality of system (1), we firstly introduce the weighted adjacency random matrix, the weighted Laplacian and the transition matrix of the random walk, then we present the conditions for a group opinions consensus. Here we use the canonical graph symbol $G(V, E)$ in which V and E denote vertexes and edges respectively.

A weighted undirected graph G is defined as $w : V \times V \longrightarrow R$ such that $w_{ij} = w_{ji}$, if $\{i, j\} \notin E(G)$ then $w_{ij} = 0$. In the context, the weighted degree d_i of a vertex i is defined as $d_i = \sum_j w_{ij}$, $vol(G) = \sum_i d_i$ denotes the volume of the graph G. For a general weighted undirected graph G, the corresponding random walk is determined by transition probabilities $p_{ij} = Pr(x_{t+1} = j|x_t = i) = w_{ij}/d_i$, which are independent of i. Clearly, for each vertex i satisfies $0 \leq p_{ij} \leq 1, \sum_i p_{ij} = 1$, in other words, transition matrix P is row stochastic matrix. In addition if for any $j \in V(G)$ satisfying $\sum_j p_{ij} = 1$, then transition matrix P is named double stochastic matrix.

In this study, based on random walk on a graph, with the aim to prove the Friedkin and Johnsen's social influence system conclusions rigorously, we define transition matrix P on graph $\widehat{W}^t$ with entries $p_{ij} = Pr(x_{t+1} = j|x_t = i) = \widehat{w}_{ij}^t/\widehat{d}_i^t$, where $\widehat{d}_i^t = \sum_j \widehat{w}_{ij}^t$, and matrix L as follows:

$$L_{ij} = \begin{cases} \widehat{d}_i^t - \widehat{w}_{ii}^t & \text{if } i = j, \\ -\widehat{w}_{ij}^t & \text{if } i \text{ and } j \text{ are adjacent,} \\ 0 & \text{otherwise.} \end{cases} \tag{4}$$

where $\widehat{w}_{ij}^t \in \widehat{W}^t$ is defined in Equations (2) and (3). Let T denote the diagonal matrix with the (i, i)-th entry having value $\widehat{d}_i^t$ as following

$$T = \begin{bmatrix} \widehat{d}_1^t & \cdots & \cdots & 0 \\ 0 & \widehat{d}_2^t & \cdots & 0 \\ \vdots & \vdots & \vdots & \vdots \\ 0 & \cdots & \cdots & \widehat{d}_N^t \end{bmatrix}, \tag{5}$$

we set $T^{-1}(i, i) = 0$ for $\widehat{d}_i^t = 0$, and if $\widehat{d}_i^t = 0$ we say i is an isolated vertex. Then the graph $\widehat{W}^t$'s Laplacian matrix ζ is defined to be the form $\zeta = T^{-1/2}LT^{-1/2}$, and each entry in ζ is listed as following,

$$\zeta_{ij} = \begin{cases} 1 - \dfrac{\widehat{w}_{ii}^t}{\widehat{d}_i^t} & \text{if } i = j, \text{and } \widehat{d}_i^t \neq 0, \\ -\dfrac{\widehat{w}_{ij}^t}{\sqrt{\widehat{d}_i^t \widehat{d}_j^t}} & \text{if } i \text{ and } j \text{ are adjacent,} \\ 0 & \text{otherwise.} \end{cases} \tag{6}$$

Since ζ is symmetric, its eigenvalues are all real and non-negative. Let the eigenvalues of ζ be $\{\lambda_i | i = 0 : N - 1\}$ in increasing order of λ_i, such that $0 = \lambda_0 \leq \lambda_1 \leq \cdots \leq \lambda_{N-1}$. Furthermore, it is easy to check that transition matrix P satisfies $P = T^{-1/2}(I - \zeta)T^{1/2}$, and $\mathbf{1}TP = \mathbf{1}T$, where $\mathbf{1}$ is unit vector.

Definition 3. *The random walk P^m is said to be irreducibility if for any $i, j \in V$, there exists some t such that $p_{ij}^t > 0$. Definition 3 ensures the graph P^m is strongly connected.*

Definition 4. *The random walk P^m is aperiodic if the greatest common divisor of the lengths of its simple cycles is 1, i.e., $\gcd\{m : p_{ii}^m > 0\} = 1$ for any state i.*

Definition 5. *The random matrix P is said to be ergodic if there is an unique $n \times 1$ stationary distribution vector π satisfying $\lim_{m \to \infty} P^m (y^{(1)})' = \pi$, where $'$ is the transpose operation.*

Definition 6. *The random matrix P is convergent if $\lim_{m \to \infty} P^m (y^{(1)})'$ exists, for any initial vectors beliefs $y^{(1)}$.*

The social influence exchange among the N agents may be represented by a graph $G(V, E_m)$ with the set E_m of edges given by $E_m = \{(i,j)|p_{ij}^m > 0\}$. But this condition is not sufficient to guarantee consensus of dynamic system (1) as stated in Ref.[24]. This motivates the following stronger version Definition 7, as addressed in Refs.[25,26].

Definition 7. *(Bounded interconnectivity times). There is some $B \geq 1$ such that for each nodes pairs $(i,j) \in E_\infty$, agent j sends his/her social impact to neighbor i at least once at every B consecutive time slots, i.e. the graph $(G(P), E_m \bigcup ... \bigcup E_{(m+B-1)})$ is strongly connected. This condition is equivalent to the requirement that there exists $B \geq 1$ such that $(i,j) \in E_m \bigcup ... \bigcup E_{m+B-1}$ for all $(i,j) \in E_\infty$ and $m \geq 0$.*

Definition 5 is the well-known result that aperiodicity is necessary and sufficient for convergence in the case where P is strongly connected. In other words, the necessary conditions for the ergodicity of P are (i) *irreducibility*, (ii) *aperiodicity*, i.e., Def.5 is equivalent to Defs. 3 and 4. If Def.5 holds, Def.6 satisfies.

If a Markov chain is irreducible and aperiodic, i.e. Def.3 (or Def.3's stronger version Def.7) and Def.4 are both satisfied, or equivalently Def.5 holds, then P converges to its corresponding steady distribution. This conclusion is fairly easily verified by adapting theorems on steady-state distributions of Markov chains, such as the proof provided in Ref.[27]. From another alternative, we will prove this result by spectrum graph theorem in the following section.

For above Defs.3-7, we summarize the associated results in the following Theorem 1, then we emphasize on consensus result proof and converge time derivation.

Theorem 1. *If P is a random matrix, the following are equivalent:*
(i)P is aperiodic and irreducible.
(ii)P is ergodic.
(iii) P is convergent, there is a unique left eigenvector p_s of P corresponding to eigenvalue 1 whose entries sum to 1 such that, for every $y^{(1)}$,
$\left(\lim_{m \to \infty} P^m (y^{(1)})' \right)_i = \pi(i), where \quad \pi(i) = (p_s)'(y^{(1)})'$ for every i.

Both (i) and (ii) in Theorem 1 are the well-known results. Next we focus on the proof of (iii) based on spectral graph theory. Theorem 1 presents the conditions for the formation of opinions convergence.

4   The Convergence of Opinions Profile on Random Graph

In this section, with the above Defs. 3,4 or 7, we prove that the convergence of group opinions over general weighted and undirected random graph are almost surely. In addition, we prove the lower bounds on the convergence time t for random walk P^t to be close to its stationary distribution, given an arbitrary initial distribution and small positive error ϵ. We note that this proof is based on spectrum graph theorem, which is different with Markov chains methods, such as in [9,10,11,18].

Proof. In a random walk associated with a weighted connected graph G, the transition matrix P satisfies $1TP = 1T$, where 1 is the vector with all elements are scalar 1. Therefore the stationary distribution is exactly $\pi = 1T/vol(G)$. We show that for any initial opinions profile distribution $y^{(1)}$, when m is large enough, $P^m y^{(1)}$ converges to the stationary distribution π in the sense of L_2 or Euclidean norm. We write $y^{(1)}T^{-1/2} = \sum_i a_i e_i$, where e_i denotes the orthonormal eigenfunction associated with λ_i. Because $e_0 = 1T^{1/2}/\sqrt{vol(G)}$ and $< y^{(1)}, 1 > = 1$, $||.||$ represents the L^2 norm, we have $a_0 = \frac{<y^{(1)}T^{-1/2}, 1T^{1/2}>}{||1T^{1/2}||} = \frac{1}{\sqrt{vol(G)}}$. We then have

$$||y^{(1)}P^m - \pi|| = ||y^{(1)}P^m - 1T/vol(G)|| = ||y^{(1)}P^m - a_0 e_0 T^{1/2}||$$

$$= ||y^{(1)}T^{-1/2}(I - \zeta)^m T^{1/2} - a_0 e_0 T^{1/2}|| = ||\sum_{i \neq 0}(1 - \lambda_i)^m a_i e_i T^{1/2}||$$

$$\leq (1 - \lambda')^m \frac{max_j \sqrt{\widehat{d_j^t}}}{min_j \sqrt{\widehat{d_j^t}}} \leq e^{-m\lambda'} \frac{max_j \sqrt{\widehat{d_j^t}}}{min_j \sqrt{\widehat{d_j^t}}} \tag{7}$$

where

$$\lambda' = \begin{cases} \lambda_1, & \text{if } 1 - \lambda_1 \geq \lambda_{N-1} - 1 \\ 2 - \lambda_{N-1}, & \text{else.} \end{cases}$$

Given any $\epsilon > 0$, for Equ.(7) we have

$$e^{-m\lambda'} \frac{max_j \sqrt{\widehat{d_j^t}}}{min_j \sqrt{\widehat{d_j^t}}} \leq \epsilon, \tag{8}$$

then we have $\frac{max_j \sqrt{\widehat{d_j^t}}}{\epsilon \, min_j \sqrt{\widehat{d_j^t}}} \leq e^{m\lambda'}$, so $m \geq \frac{1}{\lambda'} log(\frac{max_j \sqrt{\widehat{d_j^t}}}{\epsilon min_j \sqrt{\widehat{d_j^t}}})$.

With the symmetry of transition probability P^m, we easily check that $||y^{(1)}P^m - \pi'|| = ||(y^{(1)}P^m - \pi')'|| = ||(y^{(1)}P^m)' - \pi'|| = ||(P^m)'(y^{(1)})' - \pi'|| = ||P^m(y^{(1)})' - \pi||$.

With this we conclude that after $m \geq [\frac{1}{\lambda} log(\frac{max_j \sqrt{d_j}}{\epsilon min_j \sqrt{d_j}})]$ steps, the L_2 distance between $P^m(y^{(1)})'$ and its stationary distribution π' is at most ϵ. Thus, P^m converges to a matrix with all of whose rows are equal to the positive vector $\pi' = (\pi_1, \pi_2, ..., \pi_N)'$, when a consensus is formed in Friedkin and Johnsen's model. Accordingly, we have $(\lim_{t\to\infty} y^{(t)})_i = \sum_{i=1}^{N} \pi_i y_i^{(1)}$ almost surely with ε approximating error corresponding to t updating steps.

In the herding example, there is consensus (of sorts), while which could lead to the wrong outcome or misunderstandings (misdirections) for the whole social group, such the "Mob phenomenon" of French revolution described by *Gustave LeBon*. In this case, group consensus is equivalent to the unwisdom of crowds. If group consensus to be emerged at certain slot t^*, such that $y^{(t^*)} = \frac{1}{N} \sum_{i=1}^{N} y_j^{(1)}$, for each j in a social group, we say that the society is wise, i.e. each individual arrives the group average initial opinions profile.

One special case of the above theorem is when P is a double random matrix. With this condition, the matrix has vector $\mathbf{1}$ as their common left eigenvector at all times, and therefore all the entries of the state vector converge to $(1/N)(\mathbf{1}^T y^{(1)})\mathbf{1} = (1/N)\sum_{j=1}^{N} y_j^{(1)} \mathbf{1}$, in other words, the mean of the initial N individual's opinion profile, with probability 1. This special case is addressed in Ref.[28], we say this group is a wise social group, as introduced in Ref.[22].

5   Conclusions

In this study, from random walk aspects, we investigate the well-known Friedkin and Johnsen's model. We define a weighted random walk P based on the social influence matrix. If P satisfies ergodicity, i.e. aperiodic and irreducible, Friedkin and Johnsen's model converges to the average consensus of the initial group opinions profile (the wise group decision making steady state) is almost surely. Furthermore, we prove the lower bounds on the convergence time m for random walk P^m to be close to its average consensus, given an arbitrary initial distribution and a small positive error ϵ.

Acknowledgments. This research was supported by National Natural Science Foundation of China under Grant Nos.71171187 and 61473284, and the Scientific Research Foundation of Yunnan Province.

References

1. DeGroot, M.: Reaching a consensus. Journal of the American Statistical Association **69**(345), 118–121 (1974)
2. Friedkin, N.E., Johnsen, E.C.: Social influence networks and opinion change. In: Lawler, E.J., Macy, M.W. (eds.) Advances in Group Processes vol. 16, pp. 1–29 (1999)
3. Friedkin, N.E.: A formal theory of reflected appraisals in the evolution of power. Administrative Science Quarterly **56**(4), 501–529 (2011)

4. French, J.: A formal theory of social power. Psychological Review **63**, 181–194 (1956)
5. Jadbabaie, A., Lin, J., Morse, A.S.: Coordination of groups of mobile autonomous agents using nearest neighbor rules. IEEE Transactions on Automatic Control **48**(6), 988–1001 (2003)
6. Blondel, V., Hendrickx, J.M., Olshevsky, A., et al.: Convergence in multiagent coordination, consensus, and flocking. In: The 44th IEEE Conference on Decision and Control and the European Control Conference 2005, pp. 2996–3000 (2005)
7. Olfati-Saber, R., Murray, R.M.: Consensus problems in networks of agents with switching topology and time-delays. IEEE Transactions on Automatic Control **49**(9), 1520–1533 (2004)
8. Moreau, L.: Stability of multiagent systems with time-dependent communication links. IEEE Transactions on Automatic Control **50**(2), 169–182 (2005)
9. Fagnani, F., Zampieri, S.: Randomized consensus algorithms over large scale networks. IEEE Journal on Selected Areas in Communications **26**(4), 634–649 (2008)
10. Tsitsiklis, J.N.: Problems in decentralized decision making and computation. Ph.D. Dissertation, Massachusetts Institute of Technology (1985)
11. Kempe, D., Dobra, A., Gehrke, J.: Gossip-based computation of aggregate information. In: Proceedings of the 44th Annual IEEE Symposium on Foundations of Computer Science, pp. 482–491. IEEE (2003)
12. Intanagonwiwat, C., Govindan, R., Estrin, D.: Directed diffusion: a scalable and robust communication paradigm for sensor networks. In: Proceedings of the 6th Annual International Conference on Mobile Computing and Networking, pp. 56–67. ACM (2000)
13. Zhao, J., Govindan, R., Estrin, D.: Computing aggregates for monitoring wireless sensor networks. In: Proceedings of the First IEEE International Workshop on Sensor Network Protocols and Applications, pp. 139–148. IEEE (2003)
14. Xiao, L., Boyd, S., Lall, S.: A scheme for robust distributed sensor fusion based on average consensus. In: The 4th International Symposium on Information Processing in Sensor Networks, 63–70. IEEE (2005)
15. Dimakis, A.G., Sarwate, A.D., Wainwright, M.J.: Geographic gossip: efficient aggregation for sensor networks. In: Proceedings of the 5th International Conference on Information Processing in Sensor Networks, pp. 69–76. ACM (2006)
16. Galam, S.: From 2000 bush-gore to 2006 italian elections: voting at fifty-fifty and the contrarian effect. Quality and Quantity Journal **41**, 579–589 (2007)
17. Arrow, K.J.: Social Choice and Individual Values. Wiley, New York (1951)
18. Frasca, P., Ravazzi, C., Tempo, R., et al.: Gossips and prejudices: Ergodic randomized dynamics in social networks. arXiv:1304.2268 (2013)
19. Yardi, S., Boyd, D.: Dynamic Debates: An Analysis of Group Polarization Over Time on Twitter. Bulletin of Science, Technology & Society **30**(5), 316–327 (2010)
20. Chung, F.R.K.: Spectral graph theory. American Mathematical Soc. 92 (1997)
21. Ishii, H., Tempo, R.: Distributed randomized algorithms for the PageRank computation. IEEE Transactions on Automatic Control **55**(9), 1987–2002 (2010)
22. Golub, B., Jackson, M.O.: Naive learning in social networks and the wisdom of crowds. American Economic Journal: Microeconomics, 112–149 (2010)
23. Touri, B., Nedic, A.: On ergodicity, infinite flow, and consensus in random models. IEEE Transactions on Automatic Control **56**(7), 1593–1605 (2011)
24. Bertsekas, D., Tsitsiklis, J.N.: Parallel and Distributed Computation: Numerical Methods. Prentice Hall (1989)

25. Nedic, A., Ozdaglar, A.: Distributed subgradient methods for multi-agent optimization. IEEE Transactions on Automatic Control **54**(1), 48–61 (2009)
26. Olshevsky, A., John, N.: Tsitsiklis. Convergence Speed in Distributed Consensus and Averaging. SIAM Journal on Control and Optimization **48**(1), 33–55 (2009)
27. Kemeny, J.G., Snell, J.L.: Finite Markov Chains. Springer (1960)
28. Boyd, S., et al.: Randomized gossip algorithms. IEEE Trans. Information Theory **52**(6), 2508–2530 (2006)

Social Influence Spectrum with Guarantees: Computing More in Less Time

Thang Dinh[1]([⊠]), Hung Nguyen[1], Preetam Ghosh[1], and Michael Mayo[2]

[1] Department of Computer Science, Virginia Commonwealth University,
Richmond, VA 23284, USA
{tndinh,hungnt,pghosh}@vcu.edu
[2] US Army Engineer Research and Development Center,
3909 Halls Ferry Road, Vicksburg, MS 39180, USA
Michael.L.Mayo@usace.army.mil

Abstract. Given a social network, the Influence maximization (InfMax) problem seeks a seed set of k people that maximize the expected influence for a viral marketing campaign. However, a solution for a particular seed size k is often not enough to make informed choice regarding budget and cost-effectiveness.

In this paper, we propose the computation of *influence spectrum* (InfSpec), the maximum influence at **each** possible seed set' sizes (i.e. $k = 1..n$), thus provide optimal decision making for any range of budget or influence requirements. As none of the existing methods for InfMax are efficient enough for the task in large networks, we propose LISA, the first linear time algorithm for InfSpec (and also InfMax). LISA runs in an expected time $O(\epsilon^{-2}(m + n))$ and returns a $(1 - 1/e - \epsilon)$-approximate influence spectrum with high probability. Using statistical decision theory, LISA has an asymptotic optimal running time (in addition to its optimal approximation guarantee). In practice, LISA also surpasses the state-of-the-art InfMax methods, taking less than 5 minutes to process a network of 41.7 million nodes and 1.5 billions edges.

Keywords: Influence spectrum · Influence maximization · Approximation algorithm · Linear-time algorithm

1 Introduction

The Influence maximization (InfMax) problem seeks to find a seed set of k *influential* individuals in a social network that can (directly and indirectly) influence the maximum number of people. It stands a fundamental problem in computational social networks with many applications in viral marketing, controlling epidemic disease, virus/worm propagation, and so on. Kempe et al. [1] was the first to formulate InfMax as a combinatorial optimization problem on the two pioneering diffusion models, namely, Independent Cascade (IC) and Linear Threshold (LT). Since InfMax is NP-hard, they provide a natural greedy algorithm that yields $(1 - 1/e - \epsilon)$-approximate solutions for any $\epsilon > 0$. This celebrated work

© Springer International Publishing Switzerland 2015
M.T. Thai et al. (Eds.): CSoNet 2015, LNCS 9197, pp. 84–103, 2015.
DOI: 10.1007/978-3-319-21786-4_8

has motivated a vast amount of work on InfMax in the past decade [2–8]. Nevertheless, the proposed methods cannot find satisfactory solutions in billion-edge networks. They either scale poorly[2,4,5] or have no approximation guarantee [3]. Even the state-of-the-art method in [8] scales poorly with the number of seeds.

On top of the challenge to solve the InfMax problem in billion-size network, we often need to compute seed sets for multiple sizes k in order to make informed choice regarding budget and cost-effectiveness. Moreover, a viral campaign marketing might go through multiple phases. The planning of the expenses for each phase cannot be done properly without knowing the influence for multiple ranges of the number of seeds. Going beyond InfMax, the authors in [2] optimize the size-influence ratio (the expected number of influenced individuals per seed node) or finding the min-seed set that can influence a large fraction of networks [9]. However, these approaches still give only one solution, that may not suite the multi-objective nature of decision making processes.

In this paper, we propose the computation of *influence spectrum* (InfSpec), the maximum influence (and the corresponding seed sets) for all possible seed sizes from $k = 1$ up to n. The influence spectrum gives better insights for decision making and resource planning in viral marketing campaigns. Given the influence spectrum, we can find the solutions for not only InfMax but also cost-effective seed set [2] and min-seed set selection [9] problems (with the best approximation guarantees). As useful as it is, no one has ever considered computing InfSpec due to the perception that it seems extremely computational expensive. The fact is computing InfSpec implies solving of n InfMax instances with seed size as large as n. Unfortunately, InfMax methods either do not scale well with large seed sets [1,2,4,5] or resort to heuristics [3], i.e., obtained results could be arbitrarily worse than the optimal ones. One might look into adapting some greedy methods for InfMax for the task, e.g., solving InfMax with $k = n$ and output the solutions for all intermediate values of $k = 1..(n - 1)$. When attempting to adapt the state-of-the-art method for InfMax in [8] for InfSpec, we obtain a prohibitive high time complexity $O(n(m + n) \log n\epsilon^{-2})$, rendering the algorithm unsuitable for the task.

We introduce LISA the first linear-time approximation algorithm to compute InfSpec in billion-size networks. Given arbitrarily small $\epsilon, \delta > 0$, our algorithm has an expected running time $O(\epsilon^{-2} \ln \frac{2}{\delta}(m + n))$ and output $(1 - \frac{1}{e} - \epsilon)$-approximate influence spectrum with high probability. Also, LISA requires only an additional $O(n)$ space. The proposed algorithm has both optimal approximation guarantees and optimal time-complexity (up to a constant factor) and outperforms the state-of-the-art methods for InfMax in practice. In particular, when $\epsilon = 0.2$ and $\delta = 1/n$, it takes about 5 minutes on a network with 41.7 million nodes and 1.5 billion edges. In comparison, it is 10 times faster than TIM+ [8], the fastest known method with approximation guarantee for InfMax, when $k = 1000$ and is several magnitudes of order faster than TIM+ for larger k. Moreover, LISA provides significantly higher quality seed sets than TIM+'s for large network. Our contributions are summarized as follows.

- We introduce the problem of computing InfSpec that enables full spectrum analysis of influence in networks.
- We propose LISA a linear-time approximation algorithm for InfSpec with a factor $(1-1/e-\epsilon)$. LISA is also the *first linear-time approximation algorithm* for InfMax and Min-Seed selection problems (with best possible approximation factors). Moreover, it outputs InfMax solutions for all seed set sizes at the same time, effectively *"killing all birds with one stone"*.
- We provide theoretical analysis to show the superiority of LISA in terms of time-complexity over existing methods (with approximation guarantees) for InfMax.
- Using sequential analysis theory, we derive the optimal number of samples needed to approximate influence of a queried seed set with a given level of accuracy. Our method potentially sets new standards on simulating diffusion processes in networks.
- We perform experiments on large networks up to billions of edges. As LISA is *easy to implement* and requires no complicated parameter estimation or tuning, it runs in *orders of magnitude faster* than the state-of-the-art methods for InfMax, including those without guarantees. Also, its *solution quality* is very close to that of the (slow) greedy method and outperforms TIM+'s for large-scale networks. Finally, in moderate and large networks, LISA uses the *least amount of memory* among all the methods.

Table 1. Time complexity of InfMax methods (with approximation guarantee $1-1/e-\epsilon$ and probabilistic guarantee $1-\delta$)

Name	Time complexity	Note
Greedy (Kempe et al.[1])	$O(k^3 mn\log n/\epsilon^2)$	$\delta = 1/n$
RIS (Borgs et al. [10])	$O(k(m+n)\log^2 n/\epsilon^3)$	$\delta = 1/n$
TIM/TIM+ (Tang et al. [8])	$O(k(m+n)\log n/\epsilon^2)$	$\delta = 1/n$
LISA (this paper)	$O((m+n)\log n/\epsilon^2)$	$\delta = 1/n$
LISA (this paper)	$O((m+n)\epsilon^{-2}\ln\frac{2}{\delta})$	for any $\delta > 0$

Related Works. Kempe et al. [1] formulated the influence maximization problem as an optimization problem. They show the problem to be NP-complete and devise an $(1-1/e-\epsilon)$ approximation algorithm. Since InfMax encodes MaxCoverage problem as a special case, InfMax cannot be approximated within a factor $(1 - \frac{1}{e} + \epsilon)$ [11] under a typical complexity assumption. Later, computing the exact influence is shown to be #P-hard [3]. Leskovec et al. [2] study the influence propagation in a different perspective in which they aim to find a set of nodes in networks to detect the spread of virus as soon as possible. They improve the simple greedy method with the lazy-forward heuristic (CELF), which is originally proposed to optimize submodular functions in [12], obtaining an (up to) 700-fold speed up.

Several heuristics are developed to derive solutions in large networks. While those heuristics are often faster in practice, they fail to retain the $(1 - 1/e - \epsilon)$-approximation guarantee and produce lower quality seed sets. Chen et al. [13] obtain a speed up by using an influence estimation for the IC model. For the LT model, Chen et al. [3] propose to use local directed acyclic graphs (LDAG) to approximate the influence regions of nodes. In a complement direction, there are recent works on learning the parameters of influence propagation models [14,15]. The influence maximization is also studied in other diffusion models including the majority threshold model [16] or when both positive and negative influence are considered [17,18] and when the propagation terminates after a predefined time [16,19]. Recently, InfMax across multiple OSNs have been studied in [20,21].

Recently, Borgs et al. [10] makes a theoretical breakthrough and presents an $O(kl^2(m + n) \log^2 n/\epsilon^3)$ time algorithm for InfMax under IC model. Their algorithm (RIS) returns a $(1-1/e-\epsilon)$-approximate solution with probability at least $1 - n^{-l}$. In practice, the proposed algorithm is, however, less than satisfactory due to the rather large hidden constants. In a sequential work, Tang et al. [8] reduce the running time to $O((k + l)(m + n) \log n/\epsilon^2)$ and show that their algorithm is also very efficient in large networks with billions of edges. Nevertheless, Tang's algorithm scales poorly with the number of seeds k and the estimation of the number of sampling times is both complicated and far from optimal.

For comparison, we summarize time-complexity of all algorithms that produce $(1 - 1/e - \epsilon)$-approximation solution with probability $1 - 1/n$ in Table 1. Our proposed algorithm LISA has the best time-complexity while compute all n different seed set at the same time. Note that we follow [10] and [8] to set $\delta = 1/n^l$. However, we argue that it is sufficient and better to set δ to be a small constant, say $\delta = 10^{-6}$, in practice.

2   Model and Problem Definition

In this section, we formally define the InfSpec problem and present an overview of Borgs et al. and Tang et al's methods [8,10]. For simplicity, we focus on Linear Threshold (LT) model, however, our solution can be extended easily to Independent cascade (IC) model.

2.1   Problem Definition

We abstract a social network using a weighted graph $\mathcal{G} = (V, E, w)$ with $|V| = n$ nodes and $|E| = m$ directed edges. Each edge $(u, v) \in E$ is associated with a weight $w(u, v) \in [0, 1]$ and $\sum_{u \in V} w(u, v) \leq 1$.

Linear Threshold (LT) model. Given a seed set $S \subseteq V$, the influence cascades in G happen in rounds. At round 0, all nodes in S are activated and the others are not activated. Each node v selects a random threshold λ_v uniformly at random in range $[0, 1]$. In a round $t \geq 1$, an inactivated node v becomes activated if $\sum_{\text{activated neighbor } u} w(u, v) \geq \lambda_v$. Once node v gets activated, it will remain activated til the end. The process stops when no more nodes get activated.

Let $\mathbb{I}(S)$ denote the expected number of activated nodes given the seed set S, where the expectation is taken over all λ_v values from their uniform distribution. We call $\mathbb{I}(S)$ the *influence spread* of S under the LT model.

The LT model is equivalent to the reachability in the *live-edge* graphs, defined in [1]: Given a graph $\mathcal{G} = (V, E, w)$, for every $v \in V$, select at most one of its incoming edges at random, such that the edge (u, v) is selected with probability $w(u, v)$, and no edge is selected with probability $1 - \sum_u w(u, v)$. Each live-graph G generated from $\mathcal{G}$ is also called a sample graph. The influence spread of a seed set S is same as the expected number of nodes reachable from S over all possible sample graphs.

Definition 1 (Influence Maximization (InfMax)). *Given $k \leq n$, find a seed set of size k that maximizes $\mathbb{I}(S)$.*

Definition 2 (Influence Spectrum (InfSpec)). *For all $k = 1, 2, \ldots, n$, find $\hat{S}_k$ of size k that maximizes $\mathbb{I}(\hat{S}_k)$.*

Since InfMax is an NP-hard problem, it follows that InfSpec is also an NP-hard problem. Also, since we cannot approximate InfMax with a factor $1 - 1/e + \epsilon$ unless $NP \subseteq DTIME(n^{\log \log n})$ [1].

Greedy approach. The Greedy approach in [1], referred to as the Greedy, starts with an empty seed set $S = \emptyset$, and iteratively adds to S a node u that leads to the largest increase in the objective, i.e.,

$$u = \arg \max_{v \notin S}(\mathbb{I}(S \cup \{v\}) - \mathbb{I}(S))$$

The main bottle-neck is that we have to repeatedly compute $\mathbb{I}(S)$ to a reasonable accuracy using Monte-Carlo method. To estimate $\mathbb{I}(S)$, we first generate a sample graph g of $\mathcal{G}$ using the live-edge model: select for each node $v \in \mathcal{G}$ at most one of its incoming edges at random, such that the edge (u, v) is selected with probability $w(u, v)$, and no edge is selected with probability $1 - \sum_u w(u, v)$. We then measure the number of nodes reachable from S in g, say $R_g(S)$. After generating enough sample graphs g (typically $ns = 10,000$ samples [1]), we can take the average of $R_g(S)$ as an estimation of $\mathbb{I}(S)$.

To a select a node u, we may have to perform up to n estimations of $\mathbb{I}(.)$ that require generating ns samples each. Thus, Greedy with its $O(k \times ns \times mn)$ time complexity is computationally prohibitive for networks with millions of nodes. Even the recently improved heuristics CELF and CELF++ [5] do not scale well for large networks.

In next section, we will discuss a recent sampling strategy introduced by Borgs et al. [10] for InfMaxand several challenges in using that approach to obtain a linear-time algorithm for InfMax.

3 Estimating Influence via Reverse Influence Sampling

Borgs et al. [10] introduce a novel approach, called Reverse Influence Sampling (RIS), to estimate the influence in Independent Cascade model. In summary, for

Table 2. Table of Symbols

Notation	Description
n, m	#nodes, #links of graph $\mathcal{G} = (V, E)$, respectively
$\mathbb{I}(S)$	Influence Spread of seed set $S \subseteq V$. For $v \in V$, $\mathbb{I}(v) = \mathbb{I}(\{v\})$
OPT_k	The maximum $\mathbb{I}(S)$ for any size-k node set S
S_k^*	An optimal size-k seed node, i.e., $\mathbb{I}(S_k^*) = OPT_k$
$m_{\mathcal{H}}$	#hyperedges in hypergraph $\mathcal{H}$
$deg_{\mathcal{H}}(S), S \subseteq V$	#hyperedges incident at some node in S. Also, $deg_{\mathcal{H}}(v)$ for $v \in V$
c	Sampling constant $c = 2(e - 2) \approx \sqrt{2}$
λ_L	Complexity factor of LISA, $\lambda_L = \frac{8c}{e}(1 - \frac{1}{e})(e - 1 - \frac{c}{2})\epsilon^{-2} < 4.6\epsilon^{-2}$

each node u uniformly chosen at random, instead of generating enough sample graphs to estimate u's influence, they travel in a *reversed graph* to infer which nodes can influence u. Repeating that process multiple time will provide us with information on the influence landscape of the network.

In this section, we present an adapted version of RIS for LT model. The RIS procedure to generate a random hyperedge $\mathcal{E}_j \subseteq V$ in LT model is summarized in Algorithm 1. After choosing a starting node u randomly, we attempt to select an *in-neighbor* v of u, i.e. (v, u) is an edge of $\mathcal{G}$, according to the edge weights. Then we "move" to v and repeat, i.e. to continue the process with v replaced by u. The procedure stops when we encounter a previously visited vertex or no edge is selected. The hyperedge is then returned as the set of nodes visited along the process.

Algorithm 1. Reverse Influence Sampling in LT model (RIS-LT)

Input: Weighted graph $\mathcal{G} = (V, E, w)$
Output: A random hyperedge $\mathcal{E}_j \subseteq V$.
$\mathcal{E}_j \leftarrow \emptyset$
Pick a node v uniformly at random.
Repeat
    Add v to $\mathcal{E}_j$
    Attempt to select an edge (u, v) using live-edge model
    if edge (u, v) is selected **then** Set $v \leftarrow u$.
Until $(v \in \mathcal{E}_j)$ OR (no edge is selected)
Return $\mathcal{E}_j$

The key insight into why random hyperedges generated via RIS can capture the influence landscape is stated in the following lemma.

Lemma 1. *Given a fixed seed set $S \subseteq V$, for a random hyperedge $\mathcal{E}_j$,*

$$\Pr[\mathcal{E}_j \cap S \neq \emptyset] = \frac{\mathbb{I}(S)}{n}$$

The proof is similar to that for IC model in [10] and is omitted.

Thus we can apply Monte-Carlo method to estimate the influence of a given seed set S, i.e., to generate enough hyperedges (aka samples) and compute the frequency that the hyperedges intersect with S. Even better, we only need to generate the hyperedges once, and can reuse the hyperedges to approximate the influence of as many seed sets as we want. This is a huge advantage comparing to the traditional greedy [1], in which we have to perform an excessive number of BFS to estimate nodes' influence. All we need to figure out is the number of sample times (i.e. number of hyperedges) needed to estimate nodes' influence at a desired level of accuracy.

3.1   Number of Samples (Hyperedges)

This section focuses on the number of samples (hyperedges) needed to achieve a pre-determined performance guarantee. As the number of samples directly decide the running time, it is critical to minimize the number of samples (keeping the same performance guarantees). For example, Borgs et al.'s method requires at least $48\frac{m+n\log n}{\epsilon^3 OPT_k}$ hyperedges to find a $(1 - 1/e - \epsilon)$-approximate of InfMax with probability at least $1 - 1/n^l$, while Tang et al.'s [8] needs only $(8 + \epsilon)\frac{k(m+n)}{\epsilon^2 OPT_k}$ hyperedges to provide the same guarantees. Here $OPT_k = \max_{|S|=k, S \subseteq V}\{\mathbb{I}(S)\}$, the maximum influence of any size-k seed set. Hence, the Tang et al.'s is asymptotically $\frac{1}{\epsilon}\log n$ times faster than the Borgs et al's.

Let Z be a random variable distributed in $[0, 1]$ with mean $\mathbb{E}[Z] = \mu$ and variance σ_Z^2. Let $Z_1, Z_2, \ldots, Z_T$ be independently and identically distributed (i.i.d.) realizations of Z. A *Monte Carlo estimator* for μ_Z is

$$\tilde{\mu} = \frac{1}{T}\sum_{i=1}^{T} Z_i.$$

$\tilde{m}u$ is said to be an (ϵ, δ)-approximation of μ, for $0 < \epsilon, \delta \leq 1$, if

$$\Pr[(1 - \epsilon)\mu \leq \tilde{\mu} \leq (1 + \epsilon)\mu] \geq 1 - \delta.$$

Let $\rho(\epsilon) = \max\{\sigma^2, \epsilon\mu\}$. The Generalized Zero-One Estimator Theorem in [22] prove that if

$$T = 2c\ln\frac{2}{\delta}\frac{\rho(\epsilon)}{\epsilon^2\mu^2} \tag{1}$$

then $\tilde{\mu} = \frac{1}{T}\sum_{i=1}^{T} Z_i$ is an (ϵ, δ)-approximation of μ. Moreover, the number of sampling time is (asymptotic) optimal (by a constant factor) [22].

In this paper, we are interested in the random variable Z with realizations

$$Z_j = \min\{|S \cap \mathcal{E}_j|, 1\},$$

where S is a fixed seed set and $\mathcal{E}_j$ is a random hyperedge generated by Algorithm 1. From Lemma 1, Z is Bernouli random variable with mean $\mu_Z = \mathbb{I}((S)/n$ and variance $\sigma_Z^2 = (1 - \mu_Z)\mu_Z$.

A major obstacle in using Eq. (1) to derive the optimal number of samples is that we do not know σ_Z^2 and μ_Z, the quantity we're trying to estimate. Let $S_k^* = \arg\max_{|S|=k, S \subseteq V}\{\mathbb{I}(S)\}$, and $OPT_k = \mathbb{I}(S_k^*)$. If we can come up with a close bound on OPT_k, we will know the necessary number of hyperedges to capture the influence landscape. After that, InfMax and InfSpec can be reduced to the classic MaxCoverage problem [23] as shown in [8,10].

Thus, the key to the efficiency of the two previous studies in [8,10] are the methods to probe and estimate the value of OPT_k. With the better probing and estimating techniques, TIM and TIM+ in [8] reduce the time-complexity in [8] by a factor $O(1/\epsilon \log n)$, making the first scalable method for InfMax in billion-size networks. However, the number of sample times in [8] is still far from optimal, especially for large seed sets. As a consequence, the two algorithms scale poorly with large number of seeds.

4   Linear-Time Approximation Algorithm

In this section, we propose the first linear-time [1] that returns a $(1 - 1/e - \epsilon)$-approximate InfSpec with probability at least $1 - \delta$ for any constant $\epsilon \in (0, 1/2]$ and $\delta \in (0, 1)$.

Our algorithm, named LISA, is presented in Algorithm 2. It consists of 2 phases: 1) Phase-A: Generate a sufficient number of hyperedges using RIS (Algorithm 1) and 2) Phase-B: Solve an instance of MaxCoverage using a greedy approach. While the overall architecture is similar to those in [8,10], the key difference is in how we determine the number of necessary hyperedges.

Borgs et al. [10] generates hyperedges until a pre-defined number of edges explored by the algorithm and only provide a low successful probability 2/3. While the authors suggest that their algorithm can be repeated multiple times to boost up the success probability, this approach leads to very inefficient implementation. Tang et al. [8] estimates OPT_k via the average cost of RIS, called EPT. However, their approach still require generating as many as k times more hyperedges than necessary. Differently, we propose a novel stopping rule: *we stop generating hyperedges once the maximum degree in the hypergraph reaches a constant* $\Upsilon_L(\epsilon, \delta)$. Here, the degree of a node in the hypergraph is the number of hyperedges that contain the node. Later we show our stopping rule guarantees a 'rich' enough hypergraph to estimate nodes' influence and small enough hyperedges to make the algorithm run in linear-time.

---

[1] In [8,10], $\delta = 1/n^l$. In that case, our algorithm has an $O((m + n) \log n \epsilon^{-2} \ln \frac{2}{\delta})$ time-complexity and is no longer linear. Nevertheless, it remains the lowest known time-complexity approximation algorithm for InfMax.

Algorithm 2. Linear-time Influence Spectrum Algorithm (LISA)

Input: Precision parameters $\epsilon \in (0, 1/2]$ and $\delta \in (0, 1)$, weighted graph $\mathcal{G}$
Output: Influence spectrum $\hat{I} = \{\hat{i}_1, \hat{i}_2, \ldots, \hat{i}_n\}$.
1: $\Upsilon_L \leftarrow 1 + \lambda_L \ln \frac{2}{\delta}$
2: $\mathcal{H} \leftarrow (V, \mathcal{E} = \emptyset)$.

Phase A: Max-Degree Sampling
3: **repeat**
4:    Generate $\mathcal{E}_j \leftarrow RIS - LT(\mathcal{G})$
5:    Add $\mathcal{E}_j$ to $\mathcal{E}$.
6: **until** $\max\limits_{v \in V} deg_{\mathcal{H}}(v) \geq \Upsilon_L$.

Phase B: Computing Influence Spectrum:
7: $\hat{i}_0 \leftarrow 0$
8: $m_{\mathcal{H}} = |\mathcal{E}|$ ▷ Fixed parameter for the rest of the algorithm
9: **for** $i = 1$ **to** n **do**
10:    $\hat{v}_i \leftarrow \arg\max\limits_{v} deg_{\mathcal{H}}(v)$
11:    $\hat{i}_i \leftarrow \hat{i}_{i-1} + \frac{deg_{\mathcal{H}}(v)}{m_{\mathcal{H}}} n$
12:    Remove $\hat{v}_i$ and its incident hyperedges from $\mathcal{H}$
13: **return** $\hat{I} = \{\hat{i}_1, \hat{i}_2, \ldots, \hat{i}_n\}$ and $\hat{S} = \{\hat{v}_1, \hat{v}_2, \ldots, \hat{v}_n\}$

Our algorithm is easy to implement and requires no parameters rather than ϵ and δ. In practice, it scales very well with billion-size networks and large seed sets. It proves to be the fastest algorithm known for InfMax while maintaining superior solution quality at the same time.

4.1 Approximation Guarantees

We use the following version of Chernoff-Hoeffding's inequality with the proof presented in the appendix.

Lemma 2. *For any fixed $T > 0$,*

$$\Pr[\hat{\mu} \geq (1 + \epsilon)\mu] \leq e^{\frac{-T\mu\epsilon^2}{2c}}$$

and

$$\Pr[\hat{\mu} \leq (1 - \epsilon)\mu] \leq e^{\frac{-T\mu\epsilon^2}{2c}}.$$

Theorem 1. *Let $\hat{S}_k = \{\hat{v}_1, \hat{v}_2, \ldots, \hat{v}_k\}$, the set of the first k nodes selected by LISA. For all $k = 1..n$,*

$$\mathbb{I}(\hat{S}_k) \geq (1 - \frac{1}{e} - \epsilon)OPT_k \tag{2}$$

with probability at least $1 - \delta$. Also, for $\epsilon \leq 1/4$,

$$\Pr[(1 - \epsilon)\mathbb{I}(\hat{S}_k) \leq \hat{i}_k \leq (1 + \epsilon)\mathbb{I}(\hat{S}_k)] > 1 - \delta. \tag{3}$$

Proof. Proof of (2). To compute the influence spectrum, LISA employs the greedy algorithm for the **Max-Coverage** problem in [23], selecting in each step the node that is incident with the most number of hyperedges. Let S_k^* be an optimal size-k seed set, i.e., $\mathbb{I}(S_k^*) = OPT_k$. Since the greedy algorithm has an approximation ratio $(1 - 1/e)$ [23] We have

$$\deg_{\mathcal{H}}(S_k) \geq (1 - 1/e)\, \mathsf{MaxCover}(k) \geq (1 - 1/e)\deg_{\mathcal{H}}(S_k^*) \qquad (4)$$

where $\mathsf{MaxCover}(k) = \max_{|S|=k, S \subseteq V} deg_{\mathcal{H}}(S)$. Let $\hat{o}_k = n \times deg_{\mathcal{H}}(S_k^*)/m_{\mathcal{H}}$, an unbiased estimator of $\mathbb{I}(S_k^*) = OPT_k$. By definition, we have

$$\hat{i}_k = n \times \deg_{\mathcal{H}}(S_k)/m_{\mathcal{H}} \geq (1 - \frac{1}{e})n \times \deg_{\mathcal{H}}(S_k^*)/m_{\mathcal{H}} = (1 - \frac{1}{e})\hat{o}_k \qquad (5)$$

Let $\eta = \frac{\epsilon e}{c + 2 - \epsilon e} > 0$, we will show in the rest of the proof that

$$Pr[\hat{i}_k \geq (1 + \eta)\mathbb{I}(\hat{S}_k)] + Pr[\hat{o}_k \leq (1 - \eta)OPT_k] < \delta$$

or, equivalently,

$$Pr[\mathbb{I}(\hat{S}_k)] \geq 1/(1 + \eta)\hat{i}_k \text{ AND } \hat{o}_k \geq (1 - \eta)OPT_k] > 1 - \delta.$$

This will lead to

$$\mathbb{I}(\hat{S}_k) \geq 1/(1 + \eta)\hat{i}_k \geq (1 - \frac{1}{e})(1 - \eta)\hat{o}_k \geq \frac{1 - \eta}{1 + \eta}(1 - \frac{1}{e})OPT_k \qquad (6)$$

$$= (1 - \frac{1}{e} - (1 - \frac{1}{e})\frac{2\eta}{1 + \eta})OPT_k = (1 - \frac{1}{e} - \epsilon)OPT_k \qquad (7)$$

with probability at least $1 - \delta$.

We now show that $Pr[\hat{i}_k \geq (1 + \eta)\mathbb{I}(\hat{S}_k)] < \delta/2$. For each random hyperedge $\mathcal{E}_j$ generated using reverse influence sampling, let $X_j = \min\{|\mathcal{E}_j \cap \hat{S}_k|, 1\}$ be Bernoulli random variables with mean $\mu_X = \mathbb{I}(\hat{S}_k)/n$. Define $T(\eta, \delta) = \frac{\deg_{\mathcal{H}}(\hat{S}_k)}{(1+\eta)\mu_X}$, we have

$$Pr[\hat{i}_k \geq (1 + \eta)\mathbb{I}(\hat{S}_k)] = Pr[\frac{\sum_{j=1}^{m_{\mathcal{H}}} X_j}{m_{\mathcal{H}}} \geq (1 + \eta)\mu_X] = Pr[T(\eta, \delta) \geq m_{\mathcal{H}}] \qquad (8)$$

$$= Pr[\sum_{j=1}^{T(\eta, \delta)} X_j \geq \deg_{\mathcal{H}}(\hat{S}_k)] = Pr[\frac{\sum_{j=1}^{T(\eta, \delta)} X_j}{T(\eta, \epsilon)} \geq (1 + \eta)\mu_X] \qquad (9)$$

Since $deg_{\mathcal{H}}(\hat{S}_k) \geq deg_{\mathcal{H}}(\hat{v}_1) = \max_v deg_{\mathcal{H}}(v) = \Upsilon_L$, we have $T(\eta, \delta) \geq \frac{\Upsilon_L}{(1+\eta)\mu_X} > 2c \ln \frac{2}{\delta} \eta^{-2} \mu_X^{-1}$. Apply Lemma 2 with $T(\eta, \delta)$, we have

$$Pr[\hat{i}_k \geq (1 + \eta)\mathbb{I}(\hat{S}_k)] = Pr[\frac{\sum_{j=1}^{T(\eta, \delta)} X_j}{T(\eta, \delta)} \geq (1 + \eta)\mu_X] \qquad (10)$$

$$\leq e^{\frac{-T(\eta, \delta)\mu_X \eta^2}{2c}} < e^{\frac{-2c \ln \frac{2}{\delta} \eta^{-2} \mu_X^{-1} \mu_X \eta^2}{2c}} = \frac{\delta}{2} \qquad (11)$$

Similarly, we can show $Pr[\hat{o}_k \le (1-\eta)OPT_k] < \delta/2$ using the second half of Lemma 2. This completes the proof of (2).

Proof of (3). From (10), $Pr[\hat{i}_k \ge (1+\eta)\mathbb{I}(\hat{S}_k)] < \delta/2$. A similar proof to (10) also yields $Pr[\hat{i}_k \le (1-\eta)\mathbb{I}(\hat{S}_k)] < \delta/2$. These two together lead to

$$\Pr[\hat{i}_k \ge (1+\eta)\mathbb{I}(\hat{S}_k)] + \Pr[\hat{i}_k \le (1-\eta)OPT_k] < \delta/2 + \delta/2 = \delta$$

For $\epsilon \le 1/4$, $\eta = \frac{\epsilon e}{c+2-\epsilon e} < \epsilon$. Therefore

$$Pr[\hat{i}_k \ge (1+\epsilon)\mathbb{I}(\hat{S}_k)] + \Pr[\hat{i}_k \le (1-\epsilon)OPT_k] \tag{12}$$

$$< \Pr[\hat{i}_k \ge (1+\eta)\mathbb{I}(\hat{S}_k)] + \Pr[\hat{i}_k \le (1-\eta)OPT_k] < \delta. \tag{13}$$

This yields the second part in (3).   □

Time Complexity. Phase B of LISA can be implemented in a linear-time in terms of the total size of the hyperedges. As we shall show later in the space complexity section, the expected total size of the hyperedges is $O(\Upsilon_L n)$. Thus Phase has an expected time complexity $O(\Upsilon_L n)$.

In detail, we store nodes in a list D of size $\max_{v \in V} deg_{\mathcal{H}}(v) \le m_{\mathcal{H}}$ in which the D_t contains all nodes of degree t in $\mathcal{H}$. Updating the degree of a node is as simple as moving the node to the position in D corresponding with the new degree. Thus updating node degree can be done in $O(1)$. To find maximum degree node, we maintain a variable Δ initialized as $\max_{v \in V} deg_{\mathcal{H}}(v)$. Whenever D_Δ is empty, we keep decreasing Δ until reaching a non-empty D_Δ or reaching $\Delta = 0$. Since the nodes' degrees never increase, in the worst-case Δ will decrease to 0 and maintaining Δ takes a $\max_{v \in V} deg_{\mathcal{H}}(v) \le m_{\mathcal{H}}$ time. Therefore, the time complexity of LISA depends mostly on the time complexity of Phase A.

We shall bound the time-complexity of Phase-A via the number of edges examined. Keeping track of the maximum degree in the hypergraph is relatively easy and can be done with little additional cost.

Lemma 3. *The expected number of edges examined by* LISA *is at most* $4.6 \ln \frac{2}{\delta} \epsilon^{-2} m$.

Proof. The proof consists of two parts 1) bound the expected number of hyperedges $m_{\mathcal{H}}$ and 2) estimate the mean number of edges visited per reverse influence sampling.

Number of hyperedes: Let $v^* = \arg\max_{v \in V} \mathbb{I}(v)$, the most influential node. Note that v^* is not necessary the same with $\hat{v}_1$, selected by LISA. Define $Y_j = |\{v^*\} \cap \mathcal{E}_j\}|$—, a random variable with mean $\mu_Y = \mathbb{I}(v^*)/n$.

Denote by $T_{max}(\Upsilon_L)$ and $T^*(\Upsilon_L)$ the random variables that correspond to the numbers of sampled hyperedges until $\max deg_{\mathcal{H}}(v) = \Upsilon_L$ and $deg_{\mathcal{H}}(v^*) = \Upsilon_L$, respectively. Clearly, $T_{max}(\Upsilon_L) = m_{\mathcal{H}} \le T^*(\Upsilon_L)$, hence,

$$\mathbb{E}[T_{max}(\Upsilon_L)] \le \mathbb{E}[T^*(\Upsilon_L)].$$

Using Wald's equation [24], and that $\mathbb{E}[T^*(\Upsilon_L)] < \infty$ we have

$$\mathbb{E}[T^*(\Upsilon_L)]\mu_Y = \Upsilon_L$$

Therefore,

$$\mathbb{E}[m_{\mathcal{H}}] = \mathbb{E}[T_{max}(\Upsilon_L)] \leq \mathbb{E}[T^*(\Upsilon_L)] = \frac{\Upsilon_L}{\mu_Y}.$$

Average number of edges visited per reverse influence sampling: The reverse influence sampling procedure picks a source vertex u uniformly at random. Then for each vertex v, it will examine all in-neighbors of v with a probability $\mathbb{I}(v, u)$, the probability that v can reach to u over all sample graphs of $\mathcal{G}$ (aka the probability that v influences u). Thus the mean number of edges examined by the procedure is

$$\frac{1}{n} \sum_{u \in V} (\sum_{v \in V} \mathbb{I}(v, u) d^-(v)) = \frac{1}{n} \sum_{v \in V} d^-(v) \sum_{u \in V} I(v, u) \tag{14}$$

$$= \frac{1}{n} \sum_{v \in V} d^-(v) \mathbb{I}(v) \leq \frac{1}{n} \sum_{v \in V} d^-(v) \mathbb{I}(v^*) = \frac{m}{n} \mathbb{I}(v^*) \tag{15}$$

Therefore, the expected number of edges examined by LISA is at most

$$\frac{m}{n} \mathbb{I}(v^*) \frac{\Upsilon_L}{\mu_Y} = m \mu_Y \frac{\Upsilon_L}{\mu_Y} = m(1 + \lambda_L \ln \frac{2}{\delta}) \approx 4.6 \epsilon^{-2} \ln \frac{2}{\delta} m \tag{16}$$

This yields the proof. $\qquad\square$

Theorem 2. *LISA has an expected running time $O(\ln \frac{2}{\delta} \epsilon^{-2}(m + n))$.*

Proof. Since Phase A has a time complexity $O(\Upsilon_L n)$ and Phase B has an expected runntime $O(\Upsilon_L m)$, it follows that the expected time complexity of LISA is $O(\Upsilon_L(m + n)) = O(\ln \frac{2}{\delta} \epsilon^{-2}(m + n))$. $\qquad\square$

Space Complexity. Besides an $O(m + n)$ space to hold $\mathcal{G}$, we show that on average only an additional $O(n)$ space is sufficient to hold the hyperedges. Thus, LISA has an expected linear space complexity $O(m + n)$. [2]

Lemma 4. *The expected additional space to store all the hyperedges is $O(\Upsilon_L n)$.*

Proof. From the proof of Lemma 3, the expected number of hyperedges is at most Υ_L/μ_Y with $\mu_Y = \max_{v \in V} \mathbb{I}(v)/n$. The mean size of a hyperedge can be computed as

$$1/n \sum_{u \in V} \sum_{v \in V} \mathbb{I}(v, u) = 1/n \sum_{v \in V} \mathbb{I}(v) \leq n \mu_Y$$

Therefore, the expected value of the total sizes of all hyperedges is at most

$$\frac{\Upsilon_L}{\mu_Y} \times n \mu_Y = \Upsilon_L n.$$

This completes the proof. $\qquad\square$

---

[2] Indeed there exists $\beta > 0$ so that the algorithm takes an $O(m + bn)$ space with exponentially small probability $e^{-\beta b}$.

4.2 InfMax

The InfMax problem can be solved by first running LISA and returning $\hat{S}_k = \{\hat{i}_1, \hat{i}_2, \ldots, \hat{i}_k\}$. The approximation and running time follow directly.

Theorem 3. *There exists a randomized algorithm that returns a $(1 - 1/e - \epsilon)$-approximate of the InfMax problem in an expected time $O(\epsilon^{-2} \ln \frac{2}{\delta}(m + n))$.*

5 Efficient InfSpec Validation

For very large networks, many algorithms for InfMax do not use an adequate number of samples, resulting in the inaccurate estimation of their seeds' influence. For example, the algorithms in [1,3] use a fixed number of samples (typically 10,000 or 20,000) which is not sufficient for large networks or ones with small edge weights. In [8], TIM and TIM+ opt for high error rate $\epsilon = 0.2$ (in Twitter) to reduce computational time, unfortunately this may lead to estimation of seeds' influence with up to 20% error rate.

In this section, we provide a fast and memory-efficient algorithm, called EIVA, to estimate the influence of all n seed sets $\hat{S}_1, \hat{S}_2, \ldots, \hat{S}_n$ where $\hat{S}_k = \{\hat{v}_1, \hat{v}_2, \ldots, \hat{v}_k\}$. Here we assume $\hat{v}_1, \hat{v}_2, \ldots, \hat{v}_n$ are given as the output of LISA or other algorithms for InfMax. While having the same time-complexity with LISA. Moreover, EIVA do not store hyperedges but only a single array to store the values of $\hat{i}_k$, thus its space-complexity does not depend on the accuracy parameters ϵ and δ.

Algorithm 3. Efficient Influence Spectrum Validation Algorithm (EIVA)

Input: Weighted graph $\mathcal{G}$, seed nodes $\hat{S} = \{\hat{v}_1, \hat{v}_2, \ldots, \hat{v}_n\}$ and $\epsilon, \delta \in (0, 1)$
Output: (ϵ, δ)-approximation of $\mathbb{I}(\hat{S}_k), k = 1..n$ where $\hat{S}_k = \{\hat{v}_1, \hat{v}_2, \ldots, \hat{v}_k\}$.
1: $\Upsilon_L \leftarrow 1 + \lambda_L \ln \frac{2}{\delta}$
2: $T \leftarrow 0, \hat{i}_t \leftarrow 0, \forall t = 0..n$
3: **repeat**
4:     Generate random hyperedge $\mathcal{E}_j \leftarrow RIS - LT(\mathcal{G})$.
5:     $t_{min} = \arg \min_t \{\hat{v}_t \in \mathcal{E}_j\}$
6:     $\hat{i}_{t_{min}} \leftarrow \hat{i}_{t_{min}} + 1$
7:     $T \leftarrow T + 1$
8: **until** $\hat{i}_1 \geq \Upsilon_L$.
9: **for** $t = 1$ **to** n **do**
10:     $\hat{i}_t \leftarrow \hat{i}_{t-1} + \hat{i}_t \times n/T$
11:**return** $\hat{I} = \{\hat{i}_1, \hat{i}_2, \ldots, \hat{i}_n\}$

Specifically, EIVA, shown in Algorithm 3, repeatedly generates a hyperedge $\mathcal{E}_j$ in each step. It then looks for the smallest index t_{min} that $\hat{v}_t \in \mathcal{E}_j$. Observe that all seed sets $\hat{S}_k, k \geq t_{min}$ will cover hyperedge $\mathcal{E}_j$. Instead of increasing the value of all $\hat{i}_k, k \geq t_{min}$, EIVA only increases $\hat{i}_k$ by one. Finally, the values of $\hat{i}_k$ will be aggregated at the end, lines 9 and 10. This smart update strategy reduces the worst-case time-complexity per hyperedge from $O(n)$ to $O(1)$. Hence, we'll be able to compute all the influence of the seed sets much faster.

Lemma 5. *EIVA (Algorithm 3) computes (ϵ, δ)-approximate for the influence of all seed sets in time $O(\epsilon^{-2} \ln \frac{2}{\delta}(m+n))$ and **only an** $\theta(n)$ **additional space** (exluding the space to store the graph).*

6 Experiments

In this section, we evaluate the performance of LISA in four real-world social networks. Notably, our algorithm can solve the problem in Twitter with 1.5 billion edges in only few minutes.

6.1 Experiment Settings

Table 3. Datasets' Statistics

Datasets	NetHEPT	NetPHY	DBLP	Twitter
Nodes	15K	37K	655K	41.7M
Edges	59K	181K	2M	1.5G
Type	undirected	undirected	undirected	directed
Avg. degree	4.1	4.87	6.1	70.5

Datasets. We perform our experiments in four datasets: NetHEPT, NetPHY, DBLP, and Twitter. The basic statistics of these networks are summarized in Table 3. *NetHEPT, NetPHY* and *DBLP* are collaboration networks taken from the "High Energy Physics - Theory", "Physics" sections of arXiv.org and "Computer Science Bibliography". These undirected networks were frequently used in previous works [3,25,26]. In the networks, nodes and edges represent authors and co-authorship, respectively. Specially, the largest network is a large portion of *Twitter*, crawled in July 2009 with 41.7 million nodes and 1.5 billion edges [27]. s

Algorithms. We compare our LISA algorithm with with 3 state-of-the-art methods: 1) CELF++ [26], the fastest implementation of the greedy algorithm, 2) Simpath [25], a high-quality solution that improves the greedy solution using look-ahead technique, and 3) LDAG [3], a scalable heuristics (i.e., no approximation guarantee).

Metrics. For each algorithm, we measure 1) the *spread of influence*, i.e., the expected number of influenced nodes eventually, 2) the running time, and 3) the

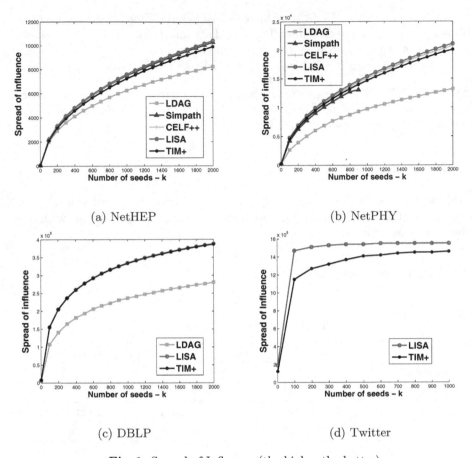

(a) NetHEP

(b) NetPHY

(c) DBLP

(d) Twitter

Fig. 1. Spread of Influence (the higher the better)

peak memory consumption. Note that we only need to run LISA once to get the metrics for all different $k = 1..n$, in contrast, we have to run the other algorithms for each value of k. We terminate any algorithms that take more than 24 hours to finish.

Parameters. We set $\epsilon = 0.1$ and $\delta = 1/n$ for LISA and TIM+, unless otherwise mentioned. For CELF++, we use the pruning threshold μ of 10^{-3}. For LDAG, we use the influence parameter $\theta = 1/320$ to control the size of the local DAG constructed for each node as recommended by the authors. For Simpath, we also set the pruning threshold μ to 10^{-3} and look-ahead value l to 4 as suggested in [25]. Finally, we revalidate the spread of influence of the outputed seed sets using EIVA with very high accuracy level: $\epsilon = 0.1$ and $\delta = 1/n^3$.

Weights Settings. We adopt the methods in [1] to calculate the influence weights on edges. More precisely, we assign the weight on an edge (u, v) as

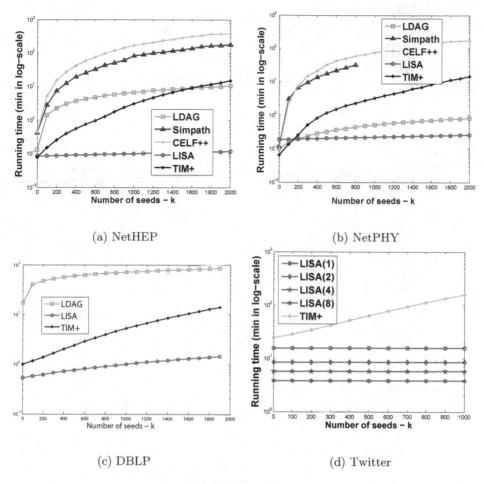

(a) NetHEP

(b) NetPHY

(c) DBLP

(d) Twitter

Fig. 2. Running time

$b_{uv} = \frac{A(u,v)}{D(v)}$ where $A(u,v)$ is the number of actions both u and v perform, and $D(v)$ is the in-degree of node v, i.e., $N(v) = \sum_{u \in N^i n(v)} A(u,v)$.

Enviroment. Our code is written with C++ and compiled with GCC 4.7. All our experiments are carried out using a Linux machine with a 3.4GHz Intel Xeon CPU and 32GB memory of RAM.

6.2   Results

Solution Quality. The quality of the algorithms, measured as the expected number of influenced nodes eventually and termed *spread of influence* are shown in Figure 1. Our algorithm *LISA* shows the best quality solutions in all datasets. It is better than both CELF++ and TIM+. Especially, only LISA and TIM+ can run the largest dataset Twitter and in that case LISA outperfoms TIM+ by

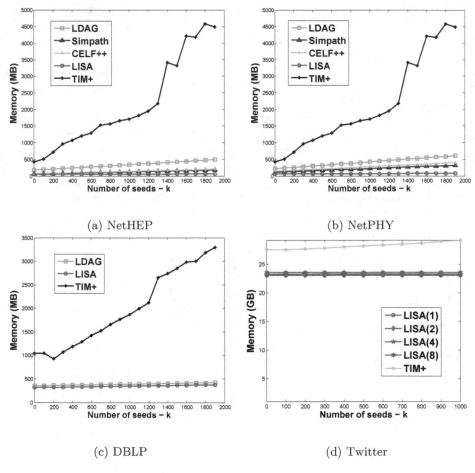

(a) NetHEP

(b) NetPHY

(c) DBLP

(d) Twitter

Fig. 3. Memory usage

a large margin. LDAG, being a fast heuristic without performance guarantee, perform poorly in comparison to the others.

Running Time. Fig. 2 shows the time taken by various algorithms against the size of the seed set on the four datasets. Obviously, *LISA, marked with red circles, is the fastest of all.* Across all the datasets, it outperforms the LDAG heuristic that provides no performance guarantee and is up to 10 times faster than the runner-up TIM+ with the same theoretical guarantees. For the Twitter dataset which contains billions of edges, only TIM+ and LISA are scalable enough for the task. LISA takes only 15 minutes to find all size-k seed sets and its parallel version with 8 threads, denoted by LISA(8), does so in less than 5 minutes.

Memory Consumption. We show the memory usages of all the algorithms in Fig. 3. Again, *LISA consistently has the smallest footprint* in temrs of the memory. It uses even less memory than the LDAG heuristic. In comparison with

TIM+, LISA uses up to ten times less memory TIM+. For the largest dataset Twitter, only 23GB memory is consumed depsite that we did not try to optimize the code for memory consumption.

Parallelization. The metrics for the (simple) parallelized versions of LISA are shown in Figs. 2 and 3 under the names LISA(2), LISA(4), LISA(8). Here the number in the brackets is the number of threads used. The parallel utility decreases when the number of threads increases. Nevertheless, we achive at least 50% parallel utility in all cases. Also, we observe almost no increase in the memory consumption with more threads. We anticipate the parallel utility will be much greater if we carefully tweak LISA for parallelization.

7 Conclusion

We propose the computation of *influence spectrum* (InfSpec) to give better insights for decision making and resource planning in viral marketing campaigns. To compute InfSpec, we design LISA, the first linear time algorithm for InfSpec. LISA runs in an expected time $O(\epsilon^{-2}(m + n))$ and returns a $(1 - 1/e - \epsilon)$-approximate influence spectrum with high probability. In practice, LISA also surpasses the state-of-the-art InfMax methods, taking less than 5 minutes to process a network of 40.6 million nodes and 1.5 billions edges. While the analysis of LISA is based on LT model, all the results also hold for IC model and the generalized model that combine both LT and IC in [28]. In the future, we will attempt to push the limit further to develop *sublinear time approximation algorithms* for InfSpec and InfMax problems.

Appendix

Lemma 2. For any fixed $T > 0$,

$$\left.\begin{array}{l} \Pr[\hat{\mu} \geq (1 + \epsilon)\mu] \\ \Pr[\hat{\mu} \leq (1 - \epsilon)\mu] \end{array}\right\} \leq e^{\frac{-T\mu\epsilon^2}{2c}}.$$

Proof. Let Z be a distribution on points in the interval $[0, 1]$ with mean μ_Z and variance σ_Z and $\rho_Z = \max\{\sigma_Z^2, \epsilon\mu_Z\}$. Define $\xi_k = \sum_{i=1}^k (Z_i - \mu_Z)$. According to the lemma 4.6 in [22], for any $\beta \leq c\rho_Z$, we have

$$\left.\begin{array}{l} \Pr[\xi_N/N \geq \beta] \\ \Pr[\xi_N/N \leq -\beta] \end{array}\right\} \leq e^{\frac{-N\beta^2}{2c\rho_Z}} \leq e^{\frac{-N(\epsilon\mu_Z)^2}{2c\rho_Z}} \leq e^{\frac{-N\epsilon^2\mu_Z}{2c}}.$$

The last two steps hold since $\epsilon\mu_Z \leq \beta \leq c\rho_Z$ and $\rho_Z \leq \mu_Z$. $\qquad\square$

References

1. Kempe, D., Kleinberg, J., Tardos, É.: Maximizing the spread of influence through a social network. In: KDD 2003, pp. 137–146. ACM, New York (2003)
2. Leskovec, J., Krause, A., Guestrin, C., Faloutsos, C., VanBriesen, J., Glance, N.: Cost-effective outbreak detection in networks. In: ACM KDD 2007, pp. 420–429. ACM, New York (2007)
3. Chen, W., Wang, C., Wang, Y.: Scalable influence maximization for prevalent viral marketing in large-scale social networks. In: ACM KDD 2010, pp. 1029–1038. ACM, New York (2010)
4. Goyal, A., Lu, W., Lakshmanan, L.: Simpath: An efficient algorithm for influence maximization under the linear threshold model. In: 2011 IEEE 11th International Conference on Data Mining (ICDM), pp. 211–220. IEEE (2011)
5. Goyal, A., Lu, W., Lakshmanan, L.: Celf++: optimizing the greedy algorithm for influence maximization in social networks. In: Proceedings of the 20th International Conference Companion on World Wide Web, pp. 47–48. ACM (2011)
6. Cohen, E., Delling, D., Pajor, T., Werneck, R.F.: Sketch-based influence maximization and computation: Scaling up with guarantees. In: Proceedings of the 23rd ACM International Conference on Conference on Information and Knowledge Management, pp. 629–638. ACM (2014)
7. Ohsaka, N., Akiba, T., Yoshida, Y., Kawarabayashi, K.i.: Fast and accurate influence maximization on large networks with pruned monte-carlo simulations. In: Twenty-Eighth AAAI Conference on Artificial Intelligence (2014)
8. Tang, Y., Xiao, X., Shi, Y.: Influence maximization: Near-optimal time complexity meets practical efficiency. In: Proceedings of the 2014 ACM SIGMOD International Conference on Management of Data, pp. 75–86. ACM (2014)
9. Long, C., Wong, R.C.: Minimizing seed set for viral marketing. In: Proceedings of the 2011 IEEE 11th International Conference on Data Mining, ICDM 2011, pp. 427–436. IEEE Computer Society, Washington, DC (2011)
10. Borgs, C., Brautbar, M., Chayes, J., Lucier, B.: Maximizing social influence in nearly optimal time. In: Proceedings of the Twenty-Fifth Annual ACM-SIAM Symposium on Discrete Algorithms, SODA 2014, pp. 946–957. SIAM (2014)
11. Feige, U.: A threshold of ln n for approximating set cover. Journal of ACM 45(4), 634–652 (1998)
12. Minoux, M.: Accelerated greedy algorithms for maximizing submodular set functions. In: Stoer, J. (ed.) Optimization Techniques. LNCIS, vol. 7., 234–243. Springer (1978)
13. Chen, N.: On the approximability of influence in social networks. SIAM Journal of Discrete Mathematics 23(3), 1400–1415 (2009)
14. Goyal, A., Bonchi, F., Lakshmanan, L.: Learning influence probabilities in social networks. In: Proceedings of the third ACM International Conference on Web Search and Data Mining, pp. 241–250. ACM (2010)
15. Kutzkov, K., Bifet, A., Bonchi, F., Gionis, A.: Strip: stream learning of influence probabilities. In: Proceedings of the 19th ACM SIGKDD International Conference on Knowledge Discovery and Data Mining, pp. 275–283. ACM (2013)
16. Dinh, T., Zhang, H., Nguyen, D., Thai, M.: Cost-effective viral marketing for time-critical campaigns in large-scale social networks. IEEE/ACM Transactions on Networking (2014)
17. Zhang, H., Dinh, T., Thai, M.: Maximizing the spread of positive influence in online social networks. In: 2013 IEEE 33rd International Conference on Distributed Computing Systems (ICDCS), pp. 317–326 (July 2013)

18. Li, Y., Chen, W., Wang, Y., Zhang, Z.: Influence diffusion dynamics and influence maximization in social networks with friend and foe relationships. In: Proceedings of the Sixth ACM International Conference on Web Search and Data Mining, pp. 657–666. ACM (2013)

19. Chen, W., Lu, W., Zhang, N.: Time-critical influence maximization in social networks with time-delayed diffusion process. In: Twenty-Sixth AAAI Conference on Artificial Intelligence (2012)

20. Shen, Y., Dinh, T.N., Zhang, H., Thai, M.T.: Interest-matching information propagation in multiple online social networks. In: Proceedings of the 21st ACM International Conference on Information and Knowledge Management, CIKM 2012, pp. 1824–1828. ACM, New York (2012)

21. Nguyen, D., Zhang, H., Das, S., Thai, M., Dinh, T.: Least cost influence in multiplex social networks: Model representation and analysis. In: 2013 IEEE 13th International Conference on Data Mining (ICDM), pp. 567–576 (December 2013)

22. Dagum, P., Karp, R., Luby, M., Ross, S.: An optimal algorithm for monte carlo estimation. SIAM J. Comput. **29**(5), 1484–1496 (2000)

23. Vazirani, V.: Approximation Algorithms. Springer (2001)

24. Wald, A.: Sequential Analysis. John Wiley and Sons (1947)

25. Goyal, A., Lu, W., Lakshmanan, L.V.: Simpath: An efficient algorithm for influence maximization under the linear threshold model. In: 2011 IEEE 11th International Conference on Data Mining (ICDM), pp. 211–220. IEEE (2011)

26. Goyal, A., Lu, W., Lakshmanan, L.V.: Celf++: optimizing the greedy algorithm for influence maximization in social networks. In: Proceedings of the 20th International Conference Companion on World Wide Web, pp. 47–48. ACM (2011)

27. Kwak, H., Lee, C., Park, H., Moon, S.: What is Twitter, a social network or a news media? In: WWW 2010: Proceedings of the 19th International Conference on World Wide Web, pp. 591–600. ACM, New York (2010)

28. Kempe, D., Kleinberg, J.M., Tardos, É.: Influential Nodes in a Diffusion Model for Social Networks. In: Caires, L., Italiano, G.F., Monteiro, L., Palamidessi, C., Yung, M. (eds.) ICALP 2005. LNCS, vol. 3580, pp. 1127–1138. Springer, Heidelberg (2005)

Improve Network Clustering via Diversified Ranking

Bing-Jie Sun, Hua-Wei Shen$^{(\boxtimes)}$, and Xue-Qi Cheng

CAS Key Laboratory of Network Data Science and Technology,
Institute of Computing Technology, Chinese Academy of Sciences,
Beijing 100190, China
shenhuawei@ict.ac.cn

Abstract. Clustering is one fundamental task in network analysis. A widely-used clustering method is k-means clustering, where clustering is iteratively refined by minimizing the distance between each data point and its cluster center. For k-means clustering, one key issue is initialization, which heavily affects its accuracy and computational cost. This issue is particularly critical when applying k-means clustering to graph data where nodes are not embedded in a metric space. In this paper, we propose to use diversified ranking method to initialize k-means clustering, i.e., finding a set of seed nodes. In diversified ranking, seed nodes are figured out by considering their centrality and diversity in a unified manner. With seed nodes as starting points, k-means clustering is used to cluster nodes into groups. We apply the proposed method to detect communities in synthetic network and real-world network. Results indicate that the proposed method exhibits high effectiveness and efficiency.

Keywords: Clustering · Seed node · Diversified ranking · K-means

1 Introduction

In recent years we witnessed an explosive growth of interests on complex network analysis [1–5]. Clustering is a fundamental task of network analysis, which assigns nodes to communities, groups of nodes with relatively denser connections within groups but sparser connections between them [6]. A widely-used clustering method is k-means clustering, which iteratively refines clustering starting with a random initialization. Unfortunately, random initialization often results in two problems, low clustering accuracy and high computational complexity, especially for graph data without an explicit metric space. When being applied to graph data, the initialization of k-means is generally completed via finding a set of seed nodes [7,8].

Several efforts have been made to develop efficient methods that could deal with the problem of initializing k-means clustering [9]. K-means++ [10] aims to spread out seed nodes in the whole network without considering the centrality of nodes. PageRank [11] selects the nodes with high centrality, but without considering whether these nodes are densely connected or not. To overcome these

© Springer International Publishing Switzerland 2015
M.T. Thai et al. (Eds.): CSoNet 2015, LNCS 9197, pp. 104–115, 2015.
DOI: 10.1007/978-3-319-21786-4_9

problems, diversified ranking [12–18] is proposed to select a set of seed nodes with both high centrality and high diversity. In addition, several methods integrated ranking and clustering method to detect communities in several specific networks [15,16]. However, few methods are effective at finding a high quality seed set to reduce the number of iterations and to improve the clustering performance.

In order to improve the performance of k-means clustering on graph data, we need to find a high-quality set of seed nodes, which has both high centrality and diversity. In this paper, we propose a two-stage framework by integrating diversified ranking method with clustering method. In the first stage, we use diversified ranking method to find a high-quality set of seed nodes. In the second stage, we start from the seed set to find communities through iterative node adjustment using k-means method. We implemented the proposed framework by employing an improved GRASSHOPPER diversified ranking method in the first stage and a state-of-the-art clustering method (i.e., k-means clustering) in the second stage. This two-stage framework exhibits two potential benefits: 1) it circumvents the shortage of random selection of seed nodes, balancing efficiency and accuracy via efficient heuristic in the first stage and a principled optimization in the second stage; 2) the two stages of the framework itself are independent, i.e., each stage could be optimized independently. Thus, we can choose any diversified ranking method to help our second stage clustering method improve its performance. Taken together, this paper opens the door of community detection on graph to integrate diversified ranking with clustering methods based on seed nodes.

2   Related Work

Diversified ranking on graph is proposed to find an explicit global order in a graph. We use V to represent the node set of a graph G, W to represent the adjacent matrix. Each element w_{ij} of W is the weight of the edge between node i and j. For unweighed network, $w_{ij} = 1$ when there is an edge between node i and j; otherwise $w_{ij} = 0$. Diversified ranking aims to find a node set $S \subseteq V$, so that nodes in S have both high centrality and diversity. Existing diversified ranking methods on graph could be roughly classified into three categories [19]: 1) diversified ranking based on maximizing marginal benefit; 2) diversified ranking based on random walk with competition; 3) diversified ranking based on mutual reinforcement between ranking and clustering.

Marginal Benefit Maximization. Zhai *et al.* proposed the Subtopic [20,21] model to cover as much as possible the subtopics of a query. Diversified ranking is used to exclude the similar subtopics from each other. For efficiency, Lin *et al.* studied the property of the objective function of diversified ranking, e.g., Submodular [22]. Generally, the objective function satisfies the following three requirements: submodularity, non-decreasing and $f(\emptyset) = 0$. With these requirements, the optimization result $f(S)$ using greedy strategy is no poorer than $(1 - \frac{1}{e})$ of the optimal result $f(S^*)$, i.e., $f(S) \geq (1 - \frac{1}{e})f(S^*)$. Li *et al.* argued that the above diversified ranking methods use little topological information of the graph. Hence, they proposed the Expansion [18] method to include the node whose neighbours cover the largest number of additional nodes of the network into the seed set. Generally, diversified ranking is used in the methods to select the important and dissimilar nodes as the top-k ranked nodes. Different similar measurements correlate to different diversified ranking methods.

Random Walk With Competition. This line of methods has two typical examples. The first is random walk based on node self-reinforcement, introducing the mechanism of "rich-get-richer" to random walk. The transition probability between nodes is dynamically changed during the process of random walk. The more frequently the node is visited, the larger probability it will be visited in the next iteration of random walk [13]. Eventually, the k richest nodes in the network are taken as the top-k ranked nodes. The second is random walk with absorbing state. This method turns the node with the largest expected number of visits into sink state each time [17,23,24]. Eventually, the k sink nodes in the network are the top-k ranked nodes. These methods can discover relatively good results, but neither of them can explain the diversity of the top-k nodes.

Mutual Reinforcement Between Ranking and Clustering. Sun *et al.* claimed that they integrated ranking and clustering together firstly in their work RankClus [15] and NetClus [16]. They start from a random k-part community partition and mapped each node to a k-dimensional mixture vector according to the membership probability. Then, they use PageRank to identify the local central nodes in each community and update node assignment. Ranking and clustering are iteratively optimized. The target of these two methods is not to find an optimal seed nodes set to improve the clustering method, and can only be used on specific information networks [15]. Other methods such as LTR [25] and LeaderRank [26] are also proposed to improve the quality of ranking algorithms.

To improve the performance of clustering on graph data, there are two main problems: 1) node mapping; 2) seed nodes selection. Spectral clustering techniques [27] make use of the spectrum (eigenvalues) of the similarity matrix of the data to perform dimensionality reduction before clustering in fewer dimensions. Generally, k eigenvectors for some k, are computed, and then another algorithm (e.g. k-means clustering) is used to cluster points by their respective k components in these eigenvectors. The node mapping problem can be solved elegantly in this way. However, when faced large graphs spectral clustering have ill-conditioned problem [28] and convergence problem. Liu *et al.* proposed a community detection method using a dissimilarity-index-based k-means and a diffusion-distance-based k-means [29]. However, the two methods only measure the diversity of the nodes without considering their centrality. Other ranking methods like RankClus [15] and NetClus [16] are proposed to integrate ranking with clustering. The two methods can mutually reinforce both the ranking and clustering performance through the full use of community information and clustering-central information.

3   The Proposed Method

In this paper we propose a two-stage framework as shown in Fig. 1. This framework could be briefly described as follows:

1) The first stage considers both the centrality and diversity of the network nodes to select a high-quality set of seed nodes.

2) The second stage employs a state-of-the-art clustering method, e.g., k-means, starting from the seed nodes to assign each node to clusters.

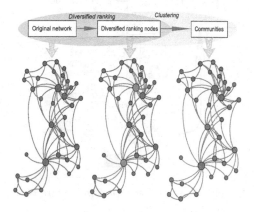

Fig. 1. (Color online) Two-stage framework for detecting communities. We take the karate club network as an example to illustrate the process of the two-stage framework. The seed nodes identified in the first stage is coloured as green. Communities are differentiated from each other by color.

To implement the framework into an efficient method for detecting communities, two principles or rules are critical: 1) in the first stage, the diversified ranking method should consider both the centrality and diversity of the nodes, and find a high-quality set of seed nodes; 2) in the second stage, the clustering methods should start from the seed set and iteratively cluster the nodes in k groups. The first stage guarantees the effectiveness and efficiency of the second stage. The second stage refines the clustering to an optimal solution.

The calculation framework of our method is shown in Fig. 2. We select k nodes in the first stage as the initial cluster centers of the network. We map all nodes to an n-dimensional vector using **Google matrix** used by Google's PageRank algorithm, where each dimension j of the vector $\overrightarrow{v_i}$ represents the possibility that node i can visit the corresponding node j in random walk, following

$$v_{ij} = \lambda * A_{ij} + (1 - \lambda) r_j, \tag{1}$$

where A represents the state transition matrix and r represents the PageRank value of each node. In this case, each vector represents the link preference of the corresponding node to the other nodes in the network. We can measure the similarity between the nodes and cluster centers and assign the nodes to the closest cluster represented by one seed node. The cluster centers will be updated iteratively until convergence. To guarantee the convergence of our method, we set the damping factor $\lambda = 0.85$ to guarantee the detailed balance of the Markov process in order to obtain a definite seed set. K-means clustering is employed in the second stage to find the final clustering, using the seed nodes for initialization.

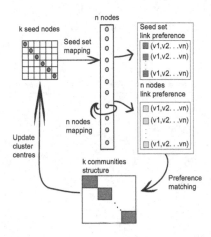

Fig. 2. (Color online) Implementation procedure of our method

3.1   The Improved **GRASSHOPPER**

GRASSHOPPER method is proposed by Zhu *et al.* [23]. In GRASSHOPPER, a highly ranked item is a representative of a local group in this set. Once the node is turned into absorbing state, it covers its neighbour nodes. GRASSHOPPER requires three inputs: a graph W, a probability distribution r that encodes the prior ranking, and a weight $\lambda \in [0,1]$. The distribution r is the PageRank values under random walk. At each step the random walker moves to a neighbour node with probability λ and $(1 - \lambda)$ to a random state. P is the state transition matrix, $P = \lambda * P + (1 - \lambda) * 1r^{\top}$. The first node of the seed set is the largest PageRank node found under the stationary distribution of random walk. GRASSHOPPER turns the ranked items into absorbing states. Once a node is turned into absorbing state, random walk will be absorbed and stay there. The matrix P can written as

$$P = \begin{bmatrix} I_S & 0 \\ R & Q \end{bmatrix}, \tag{2}$$

where S is the set of nodes ranked so far. The selection of the next node is different. They select the state with the largest expected number of visits as the next item in GRASSHOPPER ranking.

$$v = \frac{N^{\top}1}{n - |S|}, \tag{3}$$

where N is the fundamental matrix that gives the expected number of visits in absorbing random walk, $|S|$ is the size of seed set S, and n is the number of nodes in the whole network. However, the expected number of visit cannot explain why the seed set is diversified and sometimes include some redundant nodes. In this paper, we consider both the expected number of visit and the expansion of a node together. We introduce a parameter δ to balance the two parts, so that the expected score matrix v' is

$$
\begin{aligned}
v'_j &= \delta \frac{\sum_i N_{ij}^\top 1}{n - |S|} + (1 - \delta) \frac{|Nb(S \bigcup \{j\})| - |Nb(S)|}{n} \\
&= \delta \frac{\sum_i N_{ij}^\top 1}{n - |S|} + (1 - \delta) \frac{|Nb(\{j\})| - |Nb(S)|}{n},
\end{aligned}
\tag{4}
$$

where $Nb(j)$ is the neighbour nodes of node j. In this paper, we set $\delta = 0.3$. So the final objective function is $s_{|S|+1} = \arg\max_{i=|S|+1}^{n} v_i$, where, $s_{|S|+1}$ is the score of the $(|S|+1)$th node to be ranked. We demonstrate the difference between the original GRASSHOPPER and our modified GRASSHOPPER in Fig. 3. As shown in Fig. 3, the seed nodes discovered by our method is better in both centrality and diversity.

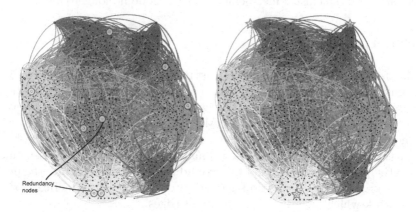

Fig. 3. (Color online) Seed nodes identified by the original GRASSHOPPER (left) and the modified one (right). Seed nodes are highlighted by circles or stars respectively. Communities, represented by different colors, are obtained through modularity optimization, offering us a reference to validate the effectiveness of seed nodes.

3.2 K-means Clustering Method

In the second stage of our framework we use k-means method as the clustering method. Actually, any clustering method that could start with a predefined seed set can be used here. For a graph G with n nodes, if we suppose to use k-means, we must embed these nodes to a metric space. In this paper, we map each node i to an n-dimensional vector $\overrightarrow{v_i}$. Each dimension j in this vector $\overrightarrow{v_{ij}}$ represents the probability that node i visit j under random walk, as defined in Eq. (1). So, we have $\sum_{j=1}^{n} \overrightarrow{v_{ij}} = 1$. We allow self-edges in graph. If two nodes have similar connection pattern with all n nodes in the graph, we can say they are similar and the distance between the two nodes is close. The distance between node i and j is

$$d_{ij} = \sqrt{\sum_{t=1}^{n} (\overrightarrow{v_{it}} - \overrightarrow{v_{jt}})^2}. \tag{5}$$

Initially, we map each node to an n-dimensional vector and calculate the distance between each node and the k seed nodes $S = \{s_1, s_2, \cdots s_k\}$. We assign each node to the closest cluster. c_i' represents the cluster which node i belongs to. So we get k clusters $C = \{c_1, c_2, \cdots c_k\}$. We update the cluster center

$$s_i = \frac{\sum_j^n \overrightarrow{v_j} \delta(c_j', c_i)}{|c_i|}. \tag{6}$$

The new cluster centers may be virtual nodes embedded in the n-dimensional space. We iteratively optimise the cluster centers and the nodes assignment until convergence i.e., the cluster membership of each node no longer changes.

4   Experiments

To validate our method, we apply it to synthetic benchmark networks, where community structure is known a prior, and a real-world network. We use "benchmark" tool [30] to generate synthetic networks. For accuracy evaluation, we use the Normalized Mutual Information (NMI) [31] to measure the extent to which each method could accurately identify the communities planted in the benchmark networks. Higher NMI represents higher similarity of node assignments between the proposed method and the ground truth. To demonstrate the performance of our method on real-world network, we test it on the American football teem network which have 115 nodes and 613 edges.

4.1   Evaluation on Synthetic Networks

We generate the synthetic networks with the number of nodes $N = 1000$ and with the power exponents of the degree distribution and the community size distribution being $(\alpha, \beta) = (2, 1)$. We set the average degree of nodes as $\frac{N}{50}$ and the maximum degree as $\frac{N}{10}$ and the range of community size as $[\frac{N}{20}, \frac{3*N}{20}]$. The mixing parameter u varies from 0.0 to 0.8 with the interval 0.1.

Firstly, we demonstrate the performance of our method compared with the random seed set selection method. Our method uses a high quality set of seed nodes obtained by the improved GRASSHOPER. The comparison method uses 50 random seed sets for each of the u value. As shown in Fig. 4, our method is more accurate and keeps stable compared with the one using random seed set. To

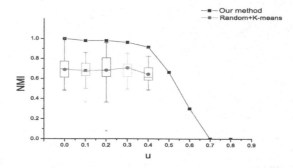

Fig. 4. (Color online) Comparison between our method and the standard k-means clustering with randomly-selected seed nodes. For each network, we run the k-means clustering for 50 times, depicting the mean and variance caused by randomness of seed nodes. When $u > 0.4$, limited by the huge computational cost, we don't give the results. Our method give a definite high quality seed set each time, the performance of k-means keep the same good in repeated experiments.

further illustrate how well our method performs, we compare our method with the random seed set method with 100 different random initializations. We select the benchmark network when $u = 0.2$. We examine the number of iterations required before convergence and the final value of the NMI. As shown in Fig. 5, our method requires the smallest number of iterations with respect to all the 100 implementations. Meanwhile, our method always outperforms these implementations of random seed set method at obtaining high-quality community structure, reflected by the highest NMI value.

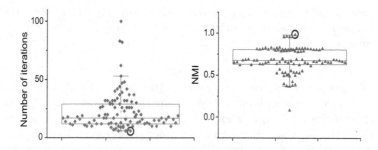

Fig. 5. (Color online) The performance of our method compared with 100 randomly-selected seed sets on benchmark network. The box-plot illustrates the distribution of the results from the 100 implementations of random-selected seed sets, while our method is marked by a circle. Here we scatter all points only for clarity.

Secondly, we compare the proposed method with several state-of-the-art methods to demonstrate the effectiveness of our method, as shown in Fig. 6.

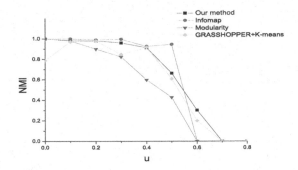

Fig. 6. (Color online) Comparison between our method and several state-of-the-art community detection methods

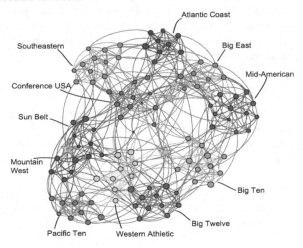

Fig. 7. (Color online) Communities detected by our method in American football game network. Labels are the name of conference to which these teams belong. Five independent teams are marked by star.

Louvain [32] proposed by Blondel *et al.* is the most widely used community detection method for its high efficiency. The performance of our method is better than that of the Louvain method for almost all the u values. Infomap [33] is claimed one of the most accurate non-overlapping community detection methods recently. The accuracy of our method is almost as well as Infomap when $u < 0.5$ which is the usual scene of real-world networks. To demonstrate the improvement of our method compared with the traditional GRASSHOPPER, we also compared the difference between the clustering accuracy of our improved method and the clustering method combined with GRASSHOPPER. As shown in Fig. 6, we can see our method is better than that of the "GRASSHOPPER+k-means" method.

4.2   Test on Real-World Networks

To further validate our method and offer some intuition about why it works well, we apply it on a real-world network, the American football game network [34] which has 115 nodes and 613 edges. Nodes in the network represent football teams, and edges represent games between these teams. As shown in Fig. 7, we can see our method identified the conferences the football teams belong to, except the independent teams. Actually, any link-based algorithm, e.g., modularity optimization method, cannot recognise these independent nodes.

5   Conclusion

In this paper, we proposed a two-stage framework to circumvent the problem of seed selection suffered by k-means clustering in network clustering. This framework is motivated by two key insights: 1) high quality seed set can improve the performance of clustering method in graph; 2) high quality seed set are the nodes set that have both high centrality and low diversity. We implement the two-stage framework by integrating an improved diversified ranking method, GRASSHOPPER, and a state-of-the-art clustering method, e.g., k-means. Results on synthetic networks and real-world networks demonstrate that the proposed method is accurate and efficient at detecting communities of networks.

Acknowledgments. This work was funded by the National Basic Research Program of China (the 973 program) under grant numbers 2014CB340405, 2013CB329602, and the National Natural Science Foundation of China with Nos 61232010, 61472400, 61202215. This work is also funded by the National Information Security Program of China (the 242 program) with grant number 2014A137.

References

1. Albert, R., Barabási, A.L.: Statistical mechanics of complex networks. J. Reviews of Modern Physics **74**(1), 47 (2002)
2. Newman, M.E.J., Girvan, M.: Finding and evaluating community structure in networks. J. Physical Review E **69**(2), 026113 (2004)
3. Shen, H.W., Cheng, X.Q., Guo, J.F.: Exploring the structural regularities in networks. J. Physical Review E **84**(5), 056111 (2011)
4. Gopalan, P.K., Blei, D.M.: Efficient discovery of overlapping communities in massive networks. J. Proceedings of the National Academy if Sciences **110**(36), 14534–14539 (2013)
5. Sun, B.J., Shen, H.W., Cheng, X.Q.: Detecting overlapping communities in massive networks. J. EPL **108**(6), 68001 (2014)
6. Girvan, M., Newman, M.E.J.: Community structure in social and biological networks. J. Proceedings of the National Academy if Sciences **99**(12), 7821–7826 (2002)
7. McDaid, A., Hurley, N.: Detecting overlapping communities with model-based overlapping seed expansion. In: 2010 International Conference on Advances in Social Networks Analysis and Mining (ASONAM), pp. 112–119. IEEE (2010)

8. Andersen, R., Lang, K.J.: Communities from seed sets. In: Proceedings of the 15th international Conference on World Wide Web, pp. 223–232. ACM (2006)
9. Celebi, M.E., Kingravi, H.A., Vela, P.A.: A comparative study of efficient initialization methods for the k-means clustering algorithm. Expert Systems with Applications. **40**, 200–210 (2013)
10. Arthur, D., Vassilvitskii, S.: k-means++: the advantage of careful seeding. In: Proceedings of the Eighteenth Annual ACM-SIAM Symposium on Discrete Algorithms, pp. 1027–1035. Society for Industrial and Applied Mathematics, Philadelphia (2007)
11. Page, L., Brin, S., Motwani, R., Winograd, T.: The PageRank citation ranking: Bringing ordering to the web. J. (1999)
12. Carbonell, J., Goldstein, J.: The use of MMR, diversity-based reranking for reordering documents and producing summaries. In: Proceedings of the 21st Annual International ACM SIGIR Conference on Research and Development in Information Retrieval, pp. 335–336. ACM (1998)
13. Mei, Q., Guo, J., Radev, D.: Divrank: the interplay of prestige and diversity in information networks. In: Proceedings of the 16th ACM SIGKDD International Conference on on Knowledge Discovery and Data Mining, pp. 1009–1018. ACM (2010)
14. Tong, H., He, J., Wen, Z., Konuru, R., Lin, C.Y.: Diversified ranking on large graphs: an optimization viewpoint. In: Proceedings of the 17th ACM SIGKDD International Conference on on Knowledge Discovery and Data Mining, pp. 1028–1036. ACM (2011)
15. Sun, Y., Han, J., Zhao, P.: RankClus: integrating clustering with ranking for heterogeneous information network analysis. In: Proceedings of the 12th International Conference on Extending Database Technology: Advances in Database Technology, pp. 565–576. ACM (2009)
16. Sun, Y., Han, J.: Ranking-based clustering of heterogeneous information networks with star network schema. In: Proceedings of the 15th ACM SIGKDD International Conference on Knowledge Discovery and Data Mining, pp. 797–806. ACM (2009)
17. Kücüktunc, O., Saule, E., Kaya, K.: Diversifing citation recommendations. J. ACM Transactions on Intelligent System and Technology (TIST) **5**(4), 55 (2014)
18. Li, R.H., Yu, J.X.: Scalable diversified ranking on large graphs. IEEE Transactions on J. Knowledge and Data Engineering **25**(9), 2133–2146 (2013)
19. Cheng, X.Q., Sun, B.J., Shen, H.W., Yu, Z.H.: Research Status and Trends of Diversified Graph Ranking. J. Proceedings of the Chinese Academy of Science **30**(2), 248–256 (2015)
20. Zhai, C.X., Cohen, W.W., Lafferty, J.: Beyond independent relevance: methods and evaluation metrics for subtopic retrieval. In: Proceedings of the 26th Annual International ACM SIGIR Conference on Research and Development in Information Retrieval, pp. 10–17 (2003)
21. Zhai, C.X., Lafferty, J.: A risk minimization framework for information retrieval. Information Processing & Management **42**(1), 31–55 (2006)
22. Lin, H., Bilmes, J., Xie, S.: Graph-based submodular selection for extractive summarization. In: Automatic Speech Recognition and Understanding Workshop (2009)
23. Zhu, X., Goldberg, A.B., Van Gael, J., Andrzejewski, D.: Improving diversity in ranking using absorbing random walks. In: HLT-NAACL, pp. 97–104 (2007)
24. Cheng, X.Q., Du, P., Guo, J.: Ranking on data manifold with sink points. IEEE Transactions on J. Knowledge and Data Engineering **25**(1), 177–191 (2013)
25. Agichtein, E., Brill, E., Dumais, S.T., et al.: Learning user interaction models for predicting web search result preferences. In: Proc. of SIGIR, pp. 3–10 (2006)

26. Lü, L., Zhang, Y.C., Yeung, C.H.: Leaders in social networks, the delicious case. PloS One **6**(6), e21202 (2011)
27. Dhillon, I.S., Guan, Y., Kulis, B.: Kernel k-means: spectral clustering and normalized cuts. In: Proceedings of the 10th ACM SIGKDD International Conference on Knowledge Discovery and Data Mining, pp. 551–556. ACM (2004)
28. Arfken, G.: Ill-Conditioned Systems. Mathematical Methods for Physicists, 3rd edn, pp. 233–234. Academic Press, Orlando (1985)
29. Liu, J., Liu, T.: Detecting community structure in complex networks using simulated annealing with k-means algorithms. J. Physica A: Statistical Mechanics and its Applications **389**(11), 2300–2309 (2010)
30. Lancichinetti, A., Radicchi, F., Ramasco, J.J., Fortunato, S.: Benchmarks for testing community detection algorithms on directed and weighted graphs with overlapping communities. J. Physical Review E **80**(1), 016118 (2009)
31. Lancichinetti, A., Fortunato, S., Kertész, J.: Detecting the overlapping and hierarchical community structure in complex networks. J. New Journal of Physics **11**(3), 033015 (2009)
32. Blondel, V.D., Guillaume, J.L., Lambiotte, R., et al.: Fast unfolding of communities in large networks. Journal of Statistical Mechanics: Theory and Experiment **2008**(10), P10008 (2008)
33. Rosvall, M., Bergstorm, C.T.: Maps of random walks on complex networks reveal community structure. J. Proceedings of the National Academy of Sciences **105**(4), 1118–1123 (2008)
34. Girvan, M., Newman, M.E.J.: Community structure in social and biological networks. J. Proc. Natl. Acad. Sci. **99**, 7821–7826 (2002)

Label Propagation Based Community Detection Algorithm with Dpark

Ting Wang$^{(\boxtimes)}$, Xu Qian, and Xiaomeng Wang

School of Mechanical Electronic and Information Engineering,
China University of Mining and Technology, Beijing, China
wangting33184@gmail.com, xuqian@cumtb.edu.cn, xiao_meng_wang@163.com

Abstract. Numerous methods for detecting communities on social networks have been proposed in recent years. However, the performance and scalability of the algorithms are not enough to work on the real-world large-scale social networks. In this paper, we propose Improved Speaker-listener Label Propagation Algorithm ($iSLPA$), an efficient and fully distributed method for community detection. It is implemented with Dpark, which is a Python version of Spark and a lightning-fast cluster computing framework. To the best of our knowledge, this is the first attempt at community detection on Dpark. It can automatically work on three kinds of networks: directed networks, undirected networks, and especially bipartite networks. In iSLPA, we propose a new initialization and updating strategy to improve the quality and scalability for detecting communities. And we conduct our experiments on real-world social networks datasets on both benchmark networks and *Douban* (http://www.douban.com) user datasets. Experimental results demonstrate that iSLPA has a comparable performance than SLPA, and have confirmed our algorithms is very efficient and effective on the overlapping community detection of large-scale networks.

Keywords: Community detection · Label Propagation Algorithm · Dpark · iSLPA

1 Introduction

Finding communities on real-world social networks has become one of the hottest research fields in social network analysis and complex networks in recent years, especially analyzing large-scale social network data is getting more complicated and challenging. However, most of the work has been done on non-overlapping or disjoint community detection in which one node can only belong to one community, such as Girvan-Newman algorithm [1]. In reality, network communities are not always disjoint and communities usually overlap with each other since one user can possibly join in or belong to multiple groups at the same time, in which one node can belong to two or more communities. For this reason, the research have been widely focused on overlapping community detection algorithms.

© Springer International Publishing Switzerland 2015
M.T. Thai et al. (Eds.): CSoNet 2015, LNCS 9197, pp. 116–127, 2015.
DOI: 10.1007/978-3-319-21786-4_10

Several methods for community finding have been proposed, such as the minimum-cut method, the hierarchical clustering method, and the modularity maximization. Also, there are a number of local based optimization methods utilizing seed expansion to grow natural communities [2], such as LFM [3], CPM [4], GCE [5], and our previous work PHSE [6]. Besides, statistical based method Order Statistics Local Optimization Method (OSLOM)[7] and Infomap [8] based on maps of random walks method show much promise as accurate. Another line of algorithms to solve this problem are modularity based community detection methods. However, modularity-based methods normally have poor performances, especially for larger systems and smaller communities. And modularity optimization has been proved to be resolution limit [9].

Besides these algorithms, Label Propagation Algorithm (LPA) [10] is by far one of the fastest community detection methods and is widely used in large scale networks. Also, LPA is scale independent for community detection because it doesn't involve modularity optimization [11]. Initially, every node is assigned with a unique label. At every step, each node updates its label to a new one which is most of its neighbors shares. The stop criteria is until every node has a label that is the maximum label of its neighbors. In this fashion, densely connected group of nodes can reach a consensus on a unique label and form a community quickly [12].

Many research of Label Propagation based method have been published in recent years, such as [11], [13], [14]. However, there are very few algorithms that can detect overlapping communities. COPRA [13] can detect overlapping communities by using label propagation technology. It sets a parameter v, which allows each vertex to belong to v communities at most. However, the parameter v is vertex-independent. It is hard for COPRA to adapt the situation of some vertices with a small number of community memberships and some others with a large number of community memberships.

Xie et al. propose Speaker-listener Label Propagation Algorithm (SLPA) [15][16], which uses memory list to store labels from each iteration, after post-processing the label list, one node can have multiple labels, which means it can detect the overlapping communities. However, SLPA is a sequential algorithm when updating the list of labels at each iteration, which makes it hard to achieve parallelism.

In this paper, we propose an improved Speaker-listener Label Propagation Algorithm (iSLPA), based on SLPA, a near linear algorithm to detecting overlapping communities on large-scale real-world social networks, which is using a new initialization and updating strategy and implemented with distribution computing framework Dpark, to achieve the fully distribution of detecting communities on large-scale networks. Also, it can automatically detect different kinds of social networks, such as directed, undirected and bipartite networks.

The rest of the paper is organized as follows. Section 2 describes the cluster computing framework Dpark. Section 3 describe the iSLPA algorithms. Section 4 provides a community detection results of iSLPA. We conclude with future work in Section 5.

2 Dpark

DPark[1] is a lightning-fast cluster computing framework based on Mesos, is a Python version implementation of Spark, similar to MapReduce but more flexible. It can relay on Python to do the distribution computing, and provides more function to give the iteration computing a better support.

Table 1. Common transformations supported by Dpark

Transformation	Meaning
map($func$)	Return a new distributed dataset formed by passing each element of the source through a function $func$.
filter($func$)	Return a new dataset formed by selecting those elements of the source on which $func$ returns true.
flatMap($func$)	Similar to map, but each input item can be mapped to 0 or more output items.
union ($otherDataset$)	Return a new dataset that contains the union of the elements in the source dataset and the argument.
groupByKey ([$numTasks$])	When called on a dataset of (K, V) pairs, returns a dataset of (K, Iterable$\langle V \rangle$) pairs.
reduceByKey ($func$, [$numTasks$])	When called on a dataset of (K, V) pairs, returns a dataset of (K, V) pairs where the values for each key are aggregated using the given reduce function $func$, which must be of type $(V, V) \Rightarrow V$. Like in groupByKey, the number of reduce tasks is configurable through an optional second argument.
join ($otherDataset$, [$numTasks$])	When called on datasets of type (K, V) and (K, W), returns a dataset of (K, (V, W)) pairs with all pairs of elements for each key.
flatMapValue ($func$)	Change to a new RDD, that is equivalent of $flatMap$ ($lambda$ (key, value): [(key, v) for v in $func$ (value)])

Dpark revolves around the concept of a resilient distributed dataset (RDD), which is a fault-tolerant collection of elements that can be operated on in parallel. Dpark has the feature of functional programming, including two types of operations on RDD: *Transformations* and *Actions*. Transformations create a new dataset from an existing one. The Table 1 lists some of the common transformations supported by Dpark. All transformations in Dpark are lazy, in that they do not compute their results right away. Instead, Dpark design *Class Dependency* to record the RDD *lineage*, which will let us to know the information of RDD's parent or ancestor. The transformations are only computed when an action, which return a value to the driver program after running a computation on the dataset, requires a result to be returned to the driver program, such as the actions of *reduce, collect, count, collectAsMap, saveAsTextFile*, and so on. By default, each transformed RDD may be recomputed each time you run an action on it. However, you may also persist an RDD in memory or on disk using

---

[1] https://github.com/douban/dpark

the *cache* method, in which case Dpark will keep the elements around on the cluster for much faster access the next time you query it. This *lazyness* design enables Dpark to run more efficiently.

3   Algorithm

The iSLPA is an overlapping community detection algorithm on large-scale real-world social networks based on SLPA, shown in Fig. 1. In bipartite network, edges only exist between the bi-side nodes, so, we only need to assign one side node with a unique label, based on the observation that number of communities is at most equal to the number of large side nodes. There are two kinds of label updating scheme based on its neighbors labels information of LPA based method: synchronous and asynchronous. Synchronous updating may lead to oscillation situation in bipartite or near bipartite structure networks. So, Raghavan *et al.* [10] suggest the asynchronous updating strategy to randomly update nodes in

---

Algorithm 1. iSLPA

Input: T: the user defined maximum iteration; r: post-processing threshold
Output: C: communities set

1 Initialization: Organize the graph as two side structure with the connection between, as *red* nodes and *blue* nodes.
2 IF the graph is bipartite: we only need to initialize the label list of one side as *label_red* or *label_blue*, suppose we initial *label_red* with a unique label, and *label_blue* as empty;
3 THEN Iteration: Iterate until the maximum iteration T is reached
4 1) Synchronously propagate label from *label_red* to *label_blue*:
5 a. Randomly select a label of each node in *label_red*.
6 b. Each node at *blue* get the updated labels list from its neighbors previously selected label at *label_red*.
7 c. Choose the most popular label as the updated label (if there are more than one popular labels, randomly choose one) and add the updated label to *label_blue*.
8 2) Synchronously propagate label from *label_blue* to *label_red*, same as step 1).
9 ELSE IF the graph is directed or undirected: initialize the label list of all nodes with a unique label as *label*.
10 THEN Iteration: Iterate until the maximum iteration T is reached
11 1) Synchronously propagate label from *red* to *blue*:
12 a. Randomly select a label in each *red* node's label list.
13 b. Each node at *blue* get the updated labels list from its neighbors previously selected label at *red*.
14 c. Choose the most popular label as the updated label (if there are more than one popular labels, randomly choose one) and add the updated label to *label*.
15 2) Synchronously propagate label from *blue* to *red*, same as step 1).
16 Post-processing: the post-processing based on the labels in the memories and the threshold r is applied to output the communities C.

Fig. 1. The Algorithm iSLPA

one iteration. However, the detecting results are unstable due to the randomness in asynchronous updating. As for SLPA, it is a sequential algorithm when updating the list of labels at each iteration, which makes it hard to achieve parallelism. So, we separate the graph as bi-side nodes with the connection between, this move will eliminate the oscillation during the update. And at each iteration, we take the synchronous updating way to updating each side nodes' label. In sum, in our method, we take a new initialization process and updating strategy to make the algorithm parallel when implemented the algorithm with Dpark, benefiting from the RDD structure, which can make the information of each node's neighbors locally.

Targeted on three kinds of networks, directed, undirected, and bipartite networks, Dpark implementation of the corresponding methods, Directed-iSLPA, Undirected- iSLPA, and Bipartite-iSLPA are elaborating in this section.

3.1   Directed-iSLPA

The iSLPA is suitable for the directed network input. We set the directed edge as a pair, and arrange the source node which links the others as *blue*, and the target node which is pointed to as *red*. Even it is not exactly a bipartite network, we also can get the information of the neighborhood who has influence on the others. We always believe the target nodes (*red* nodes) have the impact on the source nodes (*blue* nodes). So, we first initialize the all nodes' labels as *label*. In each iteration, we update *blue* nodes' labels based on its neighbors information, and *red* nodes' labels based on the neighbors information of *blue*. To get the overlapping detection results, *label* is set as a memory list for each node to store the labels in each updated iteration.

Fig. 2 illustrates our idea flow and functions used with Dpark. First, get label information from the original *rdd_follow*: (user, follower) data through functions *map*, *groupByKey*, *union* and *reduceByKey*. Second, use the method ⟨*propagateLabel*⟩ to run iteration to update the label memory list of each node, through functions *join*, *groupByKey*, *union* and *reduceByKey*. The process ⟨*propagateLabel*⟩ is following the Speaker-listener rule, about choosing the popular label in the neighbors. Random pick a label from each speaker's label list, group as listener's *pickedLabelList*, then using the method ⟨*getPopLabel*⟩ to find the most popular label, if there are several top labels that means their appearance frequencies are the same, then random pick one as the updated label. Third, using ⟨*getCommunityResult*⟩ to get the final communities detection results with the method ⟨*postProcessing*⟩ and ⟨*getCommunitites*⟩. ⟨*postProcessing*⟩ choses the labels which are frequently appearing in the node's label list through the threshold r, normally set $r = 0.1$ in our experiments. ⟨*getCommunitites*⟩ gets the communities results (community, user_list) from user's labels information (user, label_list) through the Dpark's function *flatMapValue* and *groupByKey*.

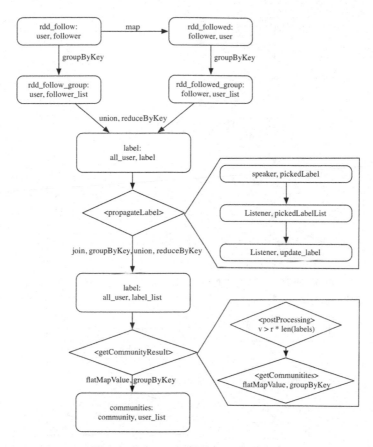

Fig. 2. Directed-iSLPA with Dpark

3.2   Undirected-iSLPA

For the consistency of the algorithm, in Fig. 3, we process the undirected network as follows. First, copy all reversed edges appended to the origin edges, and remove the duplicate edges using the method ⟨*removeDuplicates*⟩ through functions *union* and *groupByKey*. The rest progress is like **Directed-iSLPA**, running ⟨*propagateLabel*⟩ to iteratively update the nodes' label list, and to get final community detection results through ⟨*getCommunityResult*⟩. We deal with the iteration process only through propagating the *red* or *blue* nodes based on the last iteration labels information in the nodes' memory list.

3.3   Bipartite-iSLPA

First, we initialize the labels of one side (with larger size) of nodes marked as *red*, and the other side as *blue* through functions *map* and *groupByKey*, as shown in Fig. 4.

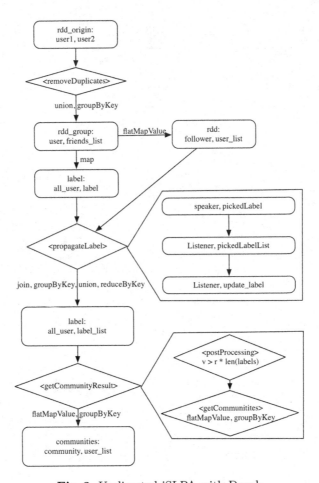

Fig. 3. Undirected-iSLPA with Dpark

Second, propagating red nodes' labels to the blue ones, that is to say, updating blue nodes' label memory list. The updating process is using flag $isBlueFlag$ to control the direction of propagation, if it is $True$ then propagating $label\_red$ to the $label\_blue$ through the method $\langle redToBlue\rangle$, otherwise, if flag $isBlueFlag$ $== False$, propagating $label\_blue$ to $label\_red$ through the method $\langle blueToRed\rangle$, both methods are using the functions $flatMapValue$, $join$, and $groupByKey$.

Finally, get community detection results through the method $\langle getCommunityResult\_bipartite\rangle$. Due to the structure of bipartite network, we need to filter the communities whose has only one side nodes using the method $\langle filCom\rangle$ with functions $filter$ and $union$.

All these improvements we designed using Dpark functions are making iSLPA more efficient and scalable. Next section will give the experiment results and evaluation of our method on real networks.

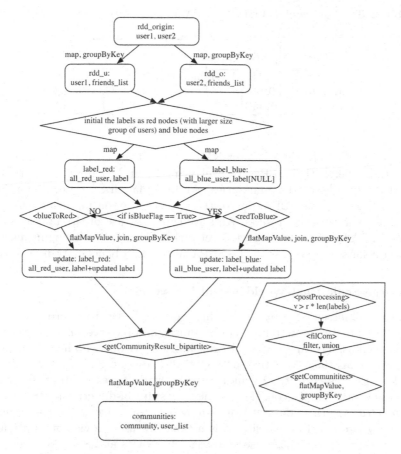

Fig. 4. Bipartite-iSLPA with Dpark

4   Experiment on Real-World Networks

The experiment results can be divided into two parts according to the datasets including benchmark networks and large-scale social networks.

4.1   Benchmark Networks

We conduct our experiments on four well known benchmark graphs, including Zacharys karate club network (*karate*)[17], Political blogs network (polblogs)[18] and Southern Women network (*women*)[19].

We take Normalized Mutual Information (NMI) metric to evaluate the communities results detected by different algorithms with the ground truth. We compare results of our method and SLPA. The statistics of networks and NMI evaluation of detection results are shown in Table 2.

In the case of the Zachary's karate (Fig. 5) and Southern Women (Fig. 6) networks, there are two communities with overlapping nodes are detected by both

Table 2. The NMI results of iSLPA, SLPA and the statistics of real networks

Datasets	NMI		Type	Nodes	Edges
	iSLPA	SLPA			
karate	**0.742**	0.511	undirected	34	78
political blogs	0.196	-	directed	1222	16782
southern women	**0.438**	0.342	bipartite	W:18, E:14	89

SLPA and iSLPA methods. The best NMI results of the iSLPA, 0.742 and 0.438 separately, are better than SLPA. The yellow nodes in the figures are overlapping nodes of the two detected communities. We find that SLPA detects more overlapping nodes than the iSLPA, due to the sequenced updating scheme of SLPA leads to a worse accuracy results. And even, in the case of the Political Blogs network, SLPA fails to detect directed networks. The iSLPA gets highly accuracy detection results compared with the SLPA method on benchmark networks.

4.2   Large-Scale Real-World Social Networks

Douban is a Chinese SNS website allowing registered users to record information and create content related to film, books, music, and recent events and activities in Chinese cities. The first dataset is a network of friends who has broadcast in public, which has 190,641 vertices and 8,901,291 edges, originally it is a directed network marked as *Directed*. An undirected network is created from the original directed network, simply considering the edges are bi-directional, and marked as *Undirected*. The third dataset is user-movie dataset, that is the users and the movies they cited on Douban site, it is a bipartite user-movie network marked as *Bipartite* which has 22,578 users and 12,128 movies and 892,638 edges.

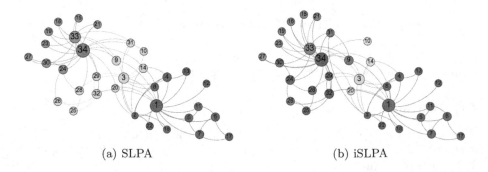

(a) SLPA   (b) iSLPA

Fig. 5. The detection result of Zacharys Karate Club by (a) SLPA and (b) iSLPA. The yellow nodes are the overlapping nodes of the two detected communitites. (For interpretation of the references to color in this figure legend, the reader is referred to the web version of this article.)

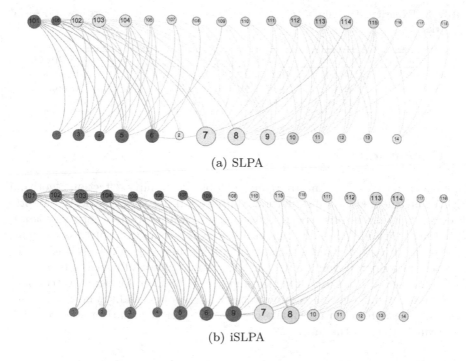

(a) SLPA

(b) iSLPA

Fig. 6. The detection result of Southern Women network by (a) SLPA and (b) iSLPA. The yellow nodes are the overlapping nodes of the two detected communitites. (For interpretation of the references to color in this figure legend, the reader is referred to the web version of this article.)

Table 3 gives description of the experiments including the vertex, edges and average degrees K of different social network datasets, and the detection results of iSLPA algorithm including number of communities and time consuming compared with SLPA. We set $r = 0.1$ and $iteration = 20$. The running time of iSLPA is faster than SLPA, later renamed to GANXiS[2], on three kinds of networks. Among the overlapping community detection algorithms [20], iSLPA has the fastest run-time with Dpark on Douban's networks. Other methods mentioned in the introduction, such as LFM, GCE, CFinder[3] - an implementation of CPM, was not able to run on the large-scale unbipartite networks. These algorithm are very computationally expensive and thus may not be suitable for detecting large-scale networks. The iSLPA has shown its potential for real time community analysis of large-scale networks.

---

[2] https://sites.google.com/site/communitydetectionslpa/
[3] http://www.cfinder.org/

Table 3. Experiment results of community detection on large-scale real-world social networks

Data	Vertexes	Edges	K	iSLPA	SLPA	Description
Directed	190,641	8,901,291	46.69	7.87min	-	friends
Undirected	190,641	8,901,291	46.69	9.03min	11.21min	friends
Bipartite	U:22,578 M:12,128	892,638	39.54	14.85min	19.15min	user-movie

5   Conclusion

This paper developed an improved and scalable community detection algorithm iSLPA implemented with distribution computing framework Dpark. It supports three kinds of unweighted social networks, which are directed networks, undirected networks and specially bipartite networks. Compared with the previous label propagation based algorithms, iSLPA performed competitively and is able to detect meaningful communities on large-scale real-world social networks. In the future, we will improve our algorithm to meet the need for detecting communities on many different kinds of networks with the universality and robustness, especially fewer research on bipartite networks, more comparison and research work will be done in the future.

References

1. Newman, M.E.J., Girvan, M.: Finding and evaluating community structure in networks. Phys. Rev. E **69**(2), 026113 (2004)
2. Lancichinetti, A., Fortunato, S.: Community detection algorithms: a comparative analysis (2009) cite arxiv:0908.1062
3. Lancichinetti, A., Fortunato, S., Kertész, J.: Detecting the overlapping and hierarchical community structure in complex networks. New J. Phys. **11**(3), March 2009
4. Palla, G., Derényi, I., Farkas, I., Vicsek, T.: Uncovering the overlapping community structure of complex networks in nature and society. Nature **435**(7043), 814–818 (2005)
5. Lee, C., Reid, F., McDaid, A., Hurley, N.: Detecting highly overlapping community structure by greedy clique expansion. In: Workshop on Social Network Mining and Analysis. (2010) cite arxiv:1002.1827
6. Wang, T., Qian, X., Xu, H.: An improved parallel hybrid seed expansion (PHSE) method for detecting highly overlapping communities in social networks. In: Motoda, H., Wu, Z., Cao, L., Zaiane, O., Yao, M., Wang, W. (eds.) ADMA 2013, Part I. LNCS, vol. 8346, pp. 385–396. Springer, Heidelberg (2013)
7. Lancichinetti, A., Radicchi, F., Ramasco, J.J., Fortunato, S.: Finding statistically significant communities in networks. CoRR abs/1012.2363 (2010)
8. Rosvall, M., Bergstrom, C.T.: Maps of random walks on complex networks reveal community structure. Proceedings of the National Academy of Sciences **105**(4), 1118–1123 (2008)
9. Ronhovde, P., Nussinov, Z.: Multiresolution community detection for megascale networks by information-based replica correlations. Phys. Rev. E **80**, 016109 (2009)

10. Raghavan, U.N., Albert, R., Kumara, S.: Near linear time algorithm to detect community structures in large-scale networks. Phys. Rev. E **76**, 036106 (2007)
11. Leung, I.X.Y., Hui, P., Liò, P., Crowcroft, J.: Towards real-time community detection in large networks. Physical Review E **79**(6), 066107+ (2009)
12. Liu, X., Murata, T.: Community detection in large-scale bipartite networks. In: IEEE Web Intelligence, pp. 50–57 (2009)
13. Gregory, S.: Finding overlapping communities in networks by label propagation. New Journal of Physics **12**(10), 103018+ (2010)
14. Liu, X., Murata, T.: How does label propagation algorithm work in bipartite networks? In: Proceedings of the 2009 IEEE/WIC/ACM International Conference on Web Intelligence and International Conference on Intelligent Agent Technology - Workshops, pp. 5–8. IEEE Computer Society, Milan (2009)
15. Xie, J., Szymanski, B.K., Liu, X.: SLPA: uncovering overlapping communities in social networks via a speaker-listener interaction dynamic process. In: 2011 IEEE 11th International Conference on Data Mining Workshops (ICDMW), pp. 344–349. Vancouver, BC (2011)
16. Xie, J., Szymanski, B.K.: Towards linear time overlapping community detection in social networks. In: Tan, P.-N., Chawla, S., Ho, C.K., Bailey, J. (eds.) PAKDD 2012, Part II. LNCS, vol. 7302, pp. 25–36. Springer, Heidelberg (2012)
17. Zachary, W.: An information flow model for conflict and fission in small groups. Journal of Anthropological Research **33**, 452–473 (1977)
18. Adamic, L.A., Glance, N.: The political blogosphere and the 2004 U.S. election: divided they blog. In: Proceedings of the 3rd International Workshop on Link Discovery, LinkKDD 2005, New York, NY, USA, pp. 36–43 (2005)
19. Davis, A., Gardner, B., Gardner, M.: Deep south: A social anthropological study of caste and class. University of Chicago Press, Chicago (1941)
20. Harenberg, S., Bello, G., Gjeltema, L., Ranshous, S., Harlalka, J., Seay, R., Padmanabhan, K., Samatova, N.: Community detection in large-scale networks: a survey and empirical evaluation. Wiley Interdisciplinary Reviews: Computational Statistics **6**(6), 426–439 (2014)

Clustering 1-Dimensional Periodic Network Using Betweenness Centrality

Norie Fu[1,2] and Vorapong Suppakitpaisarn[1,2,3](✉)

[1] Global Research Center for Big Data Mathematics,
National Institute of Informatics, Tokyo, Japan
[2] JST ERATO Kawarabayashi Large Graph Project, Tokyo, Japan
[3] Department of Computer Science, The University of Tokyo, Tokyo, Japan
vorapong@lager.is.s.u-tokyo.ac.jp

Abstract. In this paper, we propose a clustering method based on the infinite betweenness centrality for temporal networks specified by 1-dimensional periodic graphs. While the temporal networks have a wide range of applications such as opportunistic communication, there are not many clustering algorithms specifically proposed for them. We give a pseudo polynomial-time algorithm for temporal networks, of which the transit value is always positive and the least common divisor of all transit values is bounded. Our experimental results show that the centrality of networks with 125 nodes and 455 edges can be efficiently computed in 3.2 seconds. Not only the clustering results using the infinite betweenness centrality for this kind of networks are better, but also the nodes with biggest influence are more precisely detected when the betweenness centrality is computed over the periodic graph.

1 Introduction

In this paper, we propose a clustering method for temporal networks specified by 1-dimensional periodic graphs. Periodic graphs are infinite graphs that have a repetitive finite structure. That finite structure is called static graph.

The 1-dimensional periodic graph have a wide range of applications. Those include the model illustrated how people move in specific situations by Sekimoto et al. [16], and the model used for finding the optimal train schedule based on train demand by Orlin, Serafini, and Ukovich [1,13,17]. In this paper, we will focus on the application of the graphs to opportunistic communication where each object in sensor networks communicates with the others in every given period of time [10]. Each sensor is represented by a node in static graph, and each copy of the static graph represents the network in a specific period. A node i in a static graph representing a network in time t_1 is connected to a node j in a static graph for time t_2, if i can send information to j in $t_2 - t_1$ periods.

Because of the applications discussed in the previous paragraph, there are many works proposed algorithms for the graph. Those include a work by Orlin, who propose algorithms to determine weekly connected components [12],

© Springer International Publishing Switzerland 2015
M.T. Thai et al. (Eds.): CSoNet 2015, LNCS 9197, pp. 128–139, 2015.
DOI: 10.1007/978-3-319-21786-4_11

strongly connected components [12], Eulerian paths [12], minimum cost spanning trees [12], maximum flows [14], and minimum cost flows [15]. Beside that, an algorithm to test a planarity of a given periodic graph is proposed by Iwano and Steiglitz in [11]. In [6], we propose a shortest path algorithm for a class of periodic graphs. The algorithm improves the computation of a shortest path algorithm for an arbitrary periodic graph by Höfting and Wanke in [9].

Although there are many periodic extensions for many basic algorithmic problems in literature, there are not many data mining or machine learning techniques specifically proposed for them. In this paper, we will focus on clustering problem, one of the most common problems in data mining. We consider a clustering method based on betweenness centrality discussed in [3].

Betweenness centrality is known as one of the most common graph mining techniques used for extracting information from networks. Beside its application in clustering, we can also use the betweenness value to measure the influence of each node in the network [2,5].

It is also possible to find cluster a graph using the betweenness centrality of its static graph. However, we strongly believe that the static graph does not contain some important information we have in the periodic graph. A method which takes the graph with more information should help us find a better clustering results.

1.1   Our Contribution

The notation of betweenness centrality in the previous works is done only for finite graphs. As the number of nodes of an periodic graph is infinite, it is not clear if we can directly use the definition in our setting. In [7], we give a notation of betweenness centrality, we propose straightforward extension of the definition to the periodic case that preserve the meaning of the betweenness centrality. Beside that, we give a mathematical proof to show that the new definition is valid, using theoretical results on integer programming. Also, we give an algorithm to compute this betweenness centrality for a given network based on dynamic programming and Markov chains. The algorithm is proved to be polynomial time, when the input periodic graph is VAP-planar.

Although VAP-planar periodic graphs are used in many practical applications, graphs obtained from opportunistic communication do not usually have that property. That motivates us to consider a different class of periodic graphs in this paper. In that class, we assume that the graphs have the following properties.

1. Recall that a periodic graph contains a repetitive structure of static graphs. Let each static graph in the periodic graph be G_i and the transit value of an edge from G_i to G_j be $j - i$. We require that the transit values of all edges are positive.
2. We require that the weight of each edge is equal to its transit value.
3. We require that the least common multiple of all transit values is bounded by a constant K.

The transit value corresponds to the communication time between two nodes in our model. Clearly, the communication time must be positive. Also, we can consider the weight of each edge as the communication time between nodes, so it is natural to assume that the weight and the transit value are equal.

The least realistic requirement could be the third one. However, we observe that there are not usually many distinct values of communication time in most of the real-world datasets. Because of that, the least common divisor of the transit values is usually small.

In Section 3, we propose an algorithm that can find a betweenness centrality of a periodic graphs satisfying those three conditions. The ideas behind the algorithm are dynamic programming and Markov chains.

Since there is no algorithm for clustering the periodic graph proposed in literature, we compare our computation with the computation time of the fastest algorithm for finite graphs [18]. Although our asymptotic computation time looks much larger than the previous algorithm, the experimental results in Section 4 show that the computation time for both is not that different in practice.

Our algorithm takes 3.2 seconds for an opportunistic network with 125 nodes and 455 edges constructed from the data by Fournet and Barrat [4]. By using betweenness on periodic graph, we can find clusters with around 50% higher precision and recall, compared to clusters obtained from results from static graph. Beside that, we can spread information to 3% − 10% more nodes in periodic graph, if we use nodes with higher periodic betweenness instead of nodes with higher betweenness in static graph.

2   Problem Definition

2.1   Definition of Periodic Graphs

Definition 1 (Static Graph). *The tuple* $\mathcal{G} = (\mathcal{V}, \mathcal{E}, \mathrm{w})$ *of a vertex set* $\mathcal{V} = \{1, \ldots, n\}$*, a set of directed edges with vector labels* $\mathcal{E} = \{e^{(1)}, \ldots, e^{(m)}\} \subseteq (\mathcal{V} \times \mathcal{V}) \times \mathbb{Z}^d$*, and a weight function* $\mathrm{w} : \mathcal{E} \to \mathbb{R}_{>0}$ *is called a static graph.*

Definition 2 (Periodic Graph). *For a static graph* $\mathcal{G} = (\mathcal{V}, \mathcal{E}, \mathrm{w})$*, the periodic graph* $G = (V, E, \hat{\mathrm{w}})$ *generated by* $\mathcal{G}$ *is an infinite graph with weights of edges, such that* $V = \mathcal{V} \times \mathbb{Z}^d$*,* $E = \{((i, \mathbf{h}), (j, \mathbf{h} + \mathbf{g})) : \mathbf{h} \in \mathbb{Z}^d, ((i, j), \mathbf{g}) \in \mathcal{E}\} \subset V \times V$*, and* $\hat{\mathrm{w}} : E \to \mathbb{R}_{>0}$*,* $\hat{\mathrm{w}}((i, \mathbf{h}), (j, \mathbf{h} + \mathbf{g})) = \mathrm{w}(((i, j), \mathbf{g}))$*.*

If $\mathcal{G}$ has d-dimensional transit vectors, then we call G a *d-dimensional periodic graph*. Unless otherwise specified, we use $\mathcal{G} = (\mathcal{V}, \mathcal{E}, \mathrm{w})$ to denote a static graph, and $G = (V, E, \hat{\mathrm{w}})$ to denote the periodic graph generated by $\mathcal{G}$. By definition, a periodic graph can be any directed graph, but in this paper we consider only strongly connected 1-dimensional ones.

An example of a 1-dimensional periodic graph, together with its static graph, can be found in Fig. 1.

Definition 3 (Length of a Walk). *Given a walk* W *with edges* F *on* G*, we define the length of* W *as* $\sum_{e \in F} \hat{\mathrm{w}}(e)$*. Analogously on* $\mathcal{G}$*, we define the length of a walk* $\mathcal{W}$ *as* $\sum_{e \in \mathcal{F}} \mathrm{w}(e)$*.*

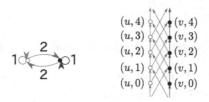

Fig. 1. A 1-dimensional periodic graph (right) and its associate static graph (left)

The *distance from s to t* in G, denoted by $d_G(s,t)$ (or simply $d(s,t)$ if the graph is omissible), is the length of a path from $s = (u, \mathbf{y})$ to $t = (v, \mathbf{z})$ in G such its length is minimized. That path is also known as a *shortest path*.

2.2 Periodic Betweenness Centrality

Let $H = (U, F)$ be a undirected graph. For any two vertices $s, t \in U$, we denote by $\sigma_{s,t}^H$ the number of shortest paths between s and t in H, and $\sigma_{s,t}^H(v)$ the ones that contains v.

The *betweenness centrality of a vertex v on a finite graph* $H = (U, F)$ is defined as $g^H(v) = \sum_{s \neq v \neq t} \frac{\sigma_{s,t}^H(v)}{\sigma_{s,t}^H}$. But we will abbreviate them as $g(v)$, $\sigma_{s,t}(v)$ and $\sigma_{s,t}$, when the graph that their are applied to results obvious.

Now, let $G = (V, E)$ be a 1-dimensional periodic graph, and fix the vertex ν whose betweenness centrality is to be computed. We can denote the distance-bounded vertices of μ by:

$$V_D(\mu) := \{\omega \in V : d_G(\mu, \omega) < D\} \cup \{\omega \in V : d_G(\omega, \mu) < D\}$$

where $D \in \mathbb{R}_{\geq 0}$. Unless otherwise specified, we abbreviate $V_D(\mu)$ by V_D when $\mu = \nu$. Let G_D the subgraph of G induced by a set of nodes V_D. Our betweenness centrality can be defined as follows.

Definition 4 (Periodic Betweenness Centrality). *For $\nu \in V$, the 1-dimensional periodic betweenness centrality of ν on G is*

$$\mathrm{pbc}^1(\nu) = \lim_{D \to +\infty} \frac{g^{G_D}(\nu)}{|V_D|^2}.$$

We note that the 1-dimensional periodic betweenness centrality is an extension of the betweenness centrality on a finite graph $H = (U, F)$, since the betweenness centrality of H can be obtained by multiplying it $|U|^2$. And since the main purpose of the betweenness is to compare the centrality of the vertices, scaling does not affect the result.

In [7], we show that, for all 1-dimensional periodic graph G and all nodes ν of G, the value of $\mathrm{pbc}^1(\nu)$ always converges to some positive real number. In the same paper, we give an algorithm that can output $\mathrm{pbc}^1(\nu)$ in finite time. The algorithm is when the periodic graph is VAP-planar.

3 Algorithm

Before introducing our algorithm, we give the following definition.

Definition 5 (Positive Periodic Graph). *Let $\mathcal{G} = (\mathcal{V}, \mathcal{E}, \mathbf{w})$ be a static graph of a 1-dimensional periodic graph G. If, for all $e^{(t)} = (i, j, \langle g \rangle) \in \mathcal{E}$, the value of g is positive, then G is a **positive periodic graph**.*

Since a cycle $\langle (i_1, i_2, \langle g_1 \rangle), (i_2, i_3, \langle g_2 \rangle), \ldots, (i_m, i_1, \langle g_m \rangle) \rangle$ must have $\sum_i g_i = 0$, and the summation of the value of g_i for all paths in a positive periodic graph is positive. We know that a positive periodic graph does not contain a cycle.

3.1 Dynamic Programming Idea

Recall the notation G_D defined in the previous section. For any $g' \in \mathbb{Z}$, we denote $V^{(g')} := \{(i, j, \langle g \rangle) \in V : g = g'\}$. Let assume without loss of generality that $\nu \in V^{(0)}$. We know that the number of shortest paths $\sigma_{s,t}^{G_D}(\nu) > 0$, only if $s \in V^{(\ell)} \cup \{\nu\}$ and $t \in V^{(u)} \cup \{\nu\}$ for some $\ell < 0$ and $u > 0$. Otherwise, $\sigma_{s,t}^{G_D}(\nu) = 0$.

In this subsection, we will give an idea how we calculate the value $\sigma_{s,t}^{G_D}(\nu) > 0$ for some specific $D \in \mathbb{Z}_+$, $s \in V^{(\ell)}$ and $t \in V^{(u)}$. To calculate that value, we will first compute the number of shortest paths from s to ν, denoted by $\sigma_{s,\nu}$, and the distance between s and ν, denoted by $d(s, \nu)$, in G_D. For $v \in V_D$, we denote $E_v := \{(u, u') \in E : u' = v\}$. Our idea behind the computation of those values are shown in Algorithm 1. We note that the function $\arg\min$ in Line 7 returns a set of all edges that minimize the value $d(s, u) + \mathbf{w}((u, u'))$.

Algorithm 1 is clearly slower than the fastest algorithm for the betweenness calculation proposed in [18]. However, the idea used in the algorithm can be extended to an infinite periodic graph in the following subsection. We will show the correctness and the computation time of the algorithm in the following theorem.

Theorem 1. *Algorithm 1 can calculate the value of $\sigma_{s,\nu}$ and $d(s, \nu)$ in $O(|E_D|)$ when E_D is the set of edges in V_D.*

Proof. The bottleneck of Algorithm 1 is Lines 6-8 of the algorithm. Because each edge will be considered only once in those lines, the computation time of the algorithm is $O(|E_D|)$.

We will prove the correctness of the algorithm by induction on the variable i. It is clear that there are no path from s to v when the node v in $\left(\bigcup_{\ell' \le \ell} V^{(\ell')} \cap V_D \right) \backslash \{s\}$, because our periodic graph is positive. By that, $\sigma_{s,v} = 0$ and $d(s, v) = \infty$, as assigned in Line 1. Because our positive periodic graph contains no cycle, it is clear that the only path from s to s is an empty set. Therefore, $\sigma_{s,s} = 1$ and $d(s, s) = 0$, as done in Line 2.

Input: A graph G_D, a node $s \in V_\ell$, and a node $\nu \in V^{(0)}$.
Output: $\sigma_{s,\nu}$ and $d(s, \nu)$

1 Set $\sigma_{s,v} \leftarrow 0$ and $d(s, v) \leftarrow \infty$ for all $v \in \left(\bigcup_{\ell' \leq \ell} V^{(\ell')} \cap V_D \right) \setminus \{s\}$.

2 Set $\sigma_{s,s} \leftarrow 1$ and $d(s, s) \leftarrow 0$.

3 $i \leftarrow \ell + 1$

4 **while** $V^{(i)} \cap V_D \neq \emptyset$ **do**

5   **forall the** $v \in V^{(i)}$ **do**

6         $d(s, v) \leftarrow \min_{(u,u') \in E_v} [d(s, u) + \mathrm{w}\,((u, u'))]$

7         $S_v \leftarrow \arg \min_{(u,u') \in E_v} [d(s, u) + \mathrm{w}\,((u, u'))]$

8         $\sigma_{s,v} \leftarrow \sum_{(u,u') \in S_v} \sigma_{s,u}$

9   **end**

10 **end**

Algorithm 1. An algorithm for calculating the number of shortest paths from s to ν and the distance between s and ν in a finite graph

Assume that Algorithm 1 can give a correct value of $\sigma_{s,v}$ and $d(s, v)$ for all $v \in \bigcup_{\ell' < i} V^{(\ell')} \cap V_D$. We know that a node $v \in V^{(i)} \cap V_D$ needs at least one edge to reach the node s. Because of that, $d(s, v)$ can be calculated as in Line 6 of the algorithm. All the shortest paths to the node v are the paths to some other nodes u added with an edge from u to v. The number of shortest paths is the summation of the number of shortest paths to each u such that $d(s, u) + \mathrm{w}((u, u'))$ is minimized, as calculated in Lines 7-8. □

Using the same method, we can calculate $\sigma_{s,t}$ and $d(s, t)$ for all $s \in \bigcup_{\ell' < 0} V^{(\ell')}$ and $t \in \bigcup_{\ell' > 0} V^{(\ell')}$. Also, by inversing the side of each edge, we can calculate the value of $\sigma_{\mu,t}$ and $d(\mu, t)$ for each $t \in \bigcup_{\ell > 0} V^{(\ell)}$. By those values, we know that there exists some shortest paths from s, t that pass ν, only if $d(s, t) = d(s, \nu) + d(\nu, t)$. If there is some shortest paths, it is clear that the number $\sigma_{s,t}^{G_D}(\nu)$ is equal to $\sigma_{s,\nu}^{G_D} \cdot \sigma_{\nu,t}^{G_D}$.

In short, we can calculate the betweenness centrality of ν by

$$g^{G_D}(\nu) = \sum_{s \neq \nu \neq t} p_{s,t,\nu} \frac{\sigma_{s,t}^{G_D}(\nu)}{\sigma_{s,t}^{G_D}} = \sum_{s \neq \nu \neq t} p_{s,t,\nu} \frac{\sigma_{s,\nu}^{G_D} \cdot \sigma_{\nu,t}^{G_D}}{\sigma_{s,t}^{G_D}},$$

when $p_{s,t,\nu} = 1$ if $d(s, t) = d(s, \nu) + d(\nu, t)$ and $p_{s,t,\nu} = 0$ otherwise.

3.2 Recurrence Relations

Recall that $V^{(\ell)}$ can be written in the form $\{(0, \ell), \ldots, (n, \ell)\}$. When $D \to \infty$ and Algorithm 1 do not terminate, we will find a betweenness centrality by solving a recurrence relation for $\sigma_{s,\nu}$, $\sigma_{\nu,t}$, $\sigma_{s,t}$ and $|V_D|^2$.

Let $g_{max} := \max\{g : (i, j, \langle g \rangle) \in E\}$. By Algorithm 1, we know that $\sigma_{s,(v,i)}$ can be written in the form

$$\sigma_{s,(v,i)} = \sum_{t=1}^{g_{max}} \sum_{(v',i-t) \in V^{(i-t)}} c_{i,t,v,v'} \sigma_{s,(v',i-t)}.$$

$c_{i,t,v,v'} = 1$ when $((v', i - t), (v, i)) \in S_{(v,i)}$, and $c_{i,t,v,v'} = 0$ otherwise.

Let $s \in V_\ell$ for some $\ell \in \mathbb{Z}_-$. In the following subsection, we will argue that, for all $i > \ell$, $c_{i,t,v,v'} = 1$ if $((v', i - t), (v, i)) \in E$ and $(v', i - t)$ is reachable from s. Otherwise, $c_{i,t,v,v'} = 0$. Since $\sigma_{s,(v',i-t)} = 0$ for any node v unreachable from s, we can still get the same solution even when we set $c_{i,t,v,v'}$ to 1. We can set $c_{i,t,v,v'} = 1$ if $((v', i - t), (v, i)) \in E$, and $c_{i,t,v,v'} = 0$ otherwise. Because the periodicity of our graph, we can calculate $\sigma_{s,(v,i)}$ by the following recurrence relation:

$$\sigma_{s,(v,i)} = \sum_{t=1}^{g_{max}} \sum_{(v',i-t) \in V^{(i-t)}} c_{t,v,v'} \sigma_{s,(v',i-t)},$$

for any $i \geq H + \ell$. Since we can calculate the value of $\sigma_{s,(v,i)}$ for all $v \in \bigcup_{i < H + \ell} V^{(i)}$ by Algorithm 1, we can solve the recurrence relation to find a closed form for $\sigma_{s,(v,i)}$.

The computation time for solving the recurrence relation is polynomial of $L := lcm(\{g : (i, j, \langle g \rangle)\})$ when $lcm(S)$ is the least common multiple of all members of S. When we require that $L \leq K$ for some constant K, we can find the solution of the recurrence relation in polynomial time.

When D is large enough, the number of shortest paths from (v', i') to (v, i) is equal to the number of shortest paths from $(v', i' - i)$ to $(v, 0)$, since the transition on a periodic graph does not change the number of paths. Hence, $\sigma_{(v',i'),(v,i)} = \sigma_{(v',i'-i),(v,0)}$. Let $s = (v', i')$. As we have the value of $\sigma_{(v',i'),(v,i)}$ for all v', v, i from the calculation on the previous paragraph, we can use those results to get $\sigma_{(v',j),(v,0)}$ all v', v, j. When $\nu = (v, 0)$, we can have the closed form for the number of shortest paths from all nodes to ν.

Using the similar idea, we can find closed forms for $\sigma_{(v,i),t}$, $\sigma_{s,t}$, and $|V_D|^2$. From those closed forms, we can calculate $pbc^1(\nu)$ defined in Section 2.

3.3   Properties of S_v

In this subsection, we will prove a property of the set S_v defined in Algorithm 1. Let $\mathbf{t}$ be a function from a path in E to $\mathbb{R}_+$ such that $\mathbf{t}(\langle e_1, \ldots, e_m \rangle) = \frac{\sum_{i=1}^{m} w(e_i)}{\sum_{i=1}^{m} g_i}$ when $e_i = (u_i, u_{i+1}, \langle g_i \rangle)$. Also, let $\mathcal{S}_{s,t} := \arg \min\{\mathbf{t}(P) : P \in \mathcal{P}_{s,t}\}$, when $\mathcal{P}_{s,t}$ is a set of all paths from s to t and $\arg \min$ returns a set of paths such that all members of the set minimize the value of $\mathbf{t}(P)$. By the notation, we have the following lemma.

Lemma 1. *Let $s \in V^{(\ell)}$ and $e := ((u', i'), (u, i), \langle g \rangle)$ be an edge in G. The edge e is in the set $S_{(u,i)}$, if and only if there is a path $P \in \mathcal{S}_{s,(u,i)}$ such that $e \in P$.*

Proof. We know that all paths $\langle e'_1, \ldots, e'_{m'} \rangle$ from $s \in V^{(\ell)}$ to (u, i) have $\sum_{i=1}^{m'} g'_i = i - \ell$ when $e'_i = (u'_i, u'_{i+1}, \langle g'_i \rangle)$. By that, the path P_1 has a smaller summation of weight than P_2, if and only if $\mathbf{t}(P_1) \leq \mathbf{t}(P_2)$. Because of that, P^* is the shortest path, if P^* minimizes $\mathbf{t}(P)$ and $P^* \in \mathcal{S}_{s,(u,i)}$. Since $e \in S_{(u,i)}$ if and only if e is in some shortest path from s to (u, i), we can prove this lemma. $\square$

With some further assumption, we can prove the following theorem.

Theorem 2. *If, for all $e = (u, u', \langle g_e \rangle) \in E$, $g_e = \mathrm{w}(e)$, then*

$$S_{(u,i)} = \{((u', i'), (u, i)) \in E : (u', i') \text{ is reachable from } s\}.$$

Proof. When $g_e = w(e)$ for all e, all paths P from s has $\mathbf{t}(P) = 1$. Because of that, $\mathcal{S}_{s,(u,i)}$ is a set of all paths from s to (u, i). The edge $((u', i'), (u, i))$ is included in one of the paths from s to (u, i), if and only if (u', i') reachable from s and there is an edge from (u', i') to (u, i). By Lemma 1, we know that $((u', i'), (u, i)) \in S_{(u,i)}$, if and only if $((u', i'), (u, i)) \in E$ and (u', i') is reachable from s. $\square$

Our result can be also applied to the case when $g_e \neq \mathrm{w}(e)$ for some $e \in E$. Recall that we denote the set of edges in the static graph by $\mathcal{E} = \{e^{(1)}, \ldots, e^{(m)}\}$. Let $e^{(i)} := (u_i, v_i, \langle g_i \rangle)$ and $\mathcal{E}_M := \{e^{(i)} \in \mathcal{E} : \mathrm{w}(e_i)/g_i = \min_{1 \leq j \leq m} \mathrm{w}(e_j)/g_j\}$. If the subgraph of the static graph $(\mathcal{V}, \mathcal{E}_M, \mathrm{w})$ is strongly connected, we can calculate the betweenness centrality by using only edges in $\mathcal{E}_M$. Due to the page limitation, we omit that proof in this paper.

We also omit an example of our computation due to the page limitation. Interested readers can find a brief example in [8], and we plan to publish an example that include all details of our computation in the full version of this paper.

4   Experimental Results

Our experimental settings and results are as follows.

4.1   Dataset

As temporal networks with periodic graphs are emerging research area, there are not so many published datasets with clustering information. We choose to construct a periodic graph based on a dataset collected for a research in [4][1]. In that paper, the authors install a devise on 125 high school students to detect all of their communication during 4 school days.

The data contains 28561 communication records. Each record consists of IDs of two students who make a communication, and a time stamp in which that communication occurs. We observe from the dataset that there is a clear periodic

---

[1] The dataset is published at http://www.sociopatterns.org.

(a)   (b)

Fig. 2. Edge-betweenness clusters using betweenness value on static graph (left) periodic graph (right)

pattern in those communication records. Every student communicate with their friends on daily basis (or even on hourly basis with closest ones).

We construct our static graph $\mathcal{G} = (\mathcal{V}, \mathcal{E}, w)$ from that observation. Each node in $\mathcal{V}$ represents a student. An edge $(i, j, \mathbf{g})$ is in the edge set $\mathcal{E}$, if student i communicates with student j once in every $\mathbf{g}$ hours, and the weight of an edge $(i, j, \mathbf{g})$ is equal to $\mathbf{g}$. As a result of this construction we get a static graph with 125 nodes and 455 edges.

By that static graph, we will get a periodic graph $G = (V, E, \hat{w})$, where $(i, \mathbf{h}) \in V$ represents a student i at time $\mathbf{h}$. An edge $((i, \mathbf{h}), (j, \mathbf{h} + \mathbf{g})) \in E$ represents the fact that the information known by i at time $\mathbf{h}$ will be known by j at time $\mathbf{h} + \mathbf{g}$, as i talks with j once in every $\mathbf{g}$ hours. This is due to the high school students have a fixed class schedule and they only share physical location with people for other classes (and can speak freely) in very specific situations such as lunch breaks or between-class breaks.

4.2 Computational Time

We implement our betweenness centrality algorithm and the fastest algorithm for finite graph in [18] using Python, and run the program on a personal computer with Intel(R) Core(TM) i7-3770 @ 3.40GHz CPU, Windows 8.1 64 bits, 16GB RAM. Our algorithm takes only 3.2 seconds for the periodic graph constructed in the previous subsection while the previous algorithm takes 0.4 seconds for computing betweenness for the static graph; resulting in only an 8-fold slower computation time when computing it on the infinite periodic graph.

4.3 Clustering Using $pbc^1(v)$

We can also find each edge-betweenness using the edge-partition technique, in that way the betweenness of that middle node will be the edge betweenness. One of the most common clustering method is to remove edges with highest betweenness, and group nodes that are in the same connected component into a cluster.

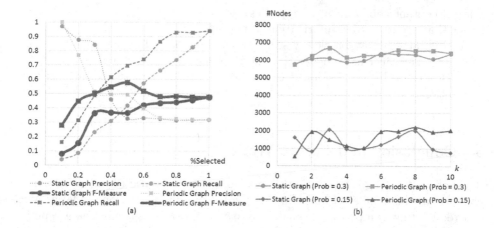

Fig. 3. Comparison between our results and previous works (a) clustering results (b) influence maximization

In this experiment, we set the number of removed edges to $p \times 455$ when p is a real number between 0.1 and 1.

In Fig. 3, we compare the clustering result obtained by removing edges with high infinite betweenness and the result obtained by removing edges with high betweenness in static graph. Clustering results is evaluated by precision, value, and F-measure calculated from the results and clusters given in the dataset. Although our precision is smaller than the value from the previous method in some p, our recall is significantly larger for all p. Because of that, our F-measure is also larger for all p. When $p = 0.5$, we improve precision by 51%, recall by 66%, and F-measure by 57%.

The clustering results is shown in Fig. 2. The color of each node represents a class of each student given in our dataset. Two nodes are considered to be in the same cluster, if they are connected in the result graphs. We can clearly seen from the figure that the pink nodes and the green nodes are put into the same cluster in the conventional clustering results, while all clusters are almost unicolor in our clustering results.

4.4 Maximizing Influence Using $pbc^1(v)$

In this subsection, want to model the way some information spread over the students (for example a rumour). For that, we select k students, with k being an integer between 1 to 10, and with probability $p \in \{0.15, 0.3\}$, the selected student will send information to node adjacent to them in $\mathcal{G}$. The node who receive the information will forward the information with the same probability after adding more content to it.

Because more content are added, students who did forward the information may forward the message again. To assure that large number of students can get a lot of content added during the process, we want to maximize the number of nodes that are forwarded information in periodic graph.

In Fig. 3(b), it can be clearly seen that nodes selected by periodic between centrality can affect more nodes than nodes selected by betweenness centrality in static graph. In our results, it can affects up to 20% more nodes than the conventional method when $k = 2$ and $p = 0.15$, and up to 9.9% when $k = 8$ and $p = 0.3$.

5   Conclusion and Future Work

It usually takes long computation time to extract information from a temporal network, as the number of nodes in the graph is usually exceptionally large. We can reduce that computation time if the network can be specified as a repetitive structure of a small graph, called static graph. In this paper, we propose an efficient algorithm that can compute betweenness centrality of that infinite network. The computation time of the algorithm is comparable to the time that the fastest method required for the static graph.

Currently, we are aiming to find more application of the betweenness centrality on the periodic graph, other than the clustering and the influence maximization. Also, we are planning to collect information to construct more periodic datasets, and use those datasets to show that our results are more preferable than the results obtained by using previous methods on static graph. Beside, we plan to find a mathematical model that can capture properties of opportunistic networks. We will use the model to generate a large periodic graph, before using that large graph to test if our algorithm is scalable enough in those practical settings.

Acknowledgement. The authors would like to thank Mr. Alonso J. Gragera Aguaza who kindly read our papers and rewrite some of the critical parts, Mr. Saran Tarnoi who kindly introduce us to the opportunistic network communication, and three anonymous reviewers who kindly give us several comments which significantly help improving the quality of this paper. The comments also give us an idea to extend this work in the future.

References

1. Anderegg, L., Eidenbenz, S., Gantenbein, M., Stamm, C., Taylor, D.S., Weber, B., Widmeyer, P.: Train routing algorithms: Concepts, design choices, and practical considerations. Proc. ALENEX 2003, pp. 106–118 (2003)
2. Cuzzocrea, A., Papadimitriou, A., Katsaros, D., Manolopoulos, Y.: Edge betweenness centrality: A novel algorithm for QoS-based topology control over wireless sensor networks. Journal of Network and Computer Applications **35**(4), 1210–1217 (2012)

3. Dunn, R., Dudbridge, F., Sanderson, C.: The use of edge-betweenness clustering to investigate biological function in protein interaction networks. BMC Bioinformatics **6**(1), 39 (2005)
4. Fournet, J., Barrat, A.: Contact patterns among high school students. PLoS ONE **9**(9), e107878 (2014)
5. Freeman, L.: A set of measures of centrality based on betweenness. Sociometry, 35–41 (1977)
6. Fu, N.: A strongly polynomial time algorithm for the shortest path problem on coherent planar periodic graphs. In: Chao, K.-M., Hsu, T., Lee, D.-T. (eds.) ISAAC 2012. LNCS, vol. 7676, pp. 392–401. Springer, Heidelberg (2012)
7. Fu, N., Aguaza, A.J.G., Suppakitpaisarn, V.: Betweenness centrality for 1-dimensional periodic graphs. (manuscript in preparation)
8. Fu, N., Suppakitpaisarn, V.: The betweenness centrality on 1-dimensional periodic graphs. In: CG Week 2014 - Workshop on Geometric Structures with Symmetry and Periodicity (2014)
9. Höfting, F., Wanke, E.: Minimum cost paths in periodic graphs. SIAM Journal on Computing **24**(5), 1051–1067 (1995)
10. Hossmann, T., Legendre, F., Carta, P., Gunningberg, P., Rohner, C.: Twitter in disaster mode: opportunistic communication and distribution of sensor data in emergencies. In: Proc. ExtremeCom 2011, pp. 1:1–1:6 (2011)
11. Iwano, K., Steiglitz, K.: Planarity testing of doubly periodic infinite graphs. Networks **18**, 205–222 (1988)
12. Orlin, J.: Some problems on dynamic/periodic graphs. Progress in Combinatorial Optimization, pp. 273–293 (1984)
13. Orlin, J.B.: Minimizing the number of vehicles to meet a fixed periodic schedule: An application of periodic posets. Operations Research **30**(4), 760–776 (1982)
14. Orlin, J.B.: Maximum-throughput dynamic network flows. Mathematical Programming **27**(2), 214–231 (1983)
15. Orlin, J.B.: Minimum convex cost dynamic network flows. Mathematics of Operations Research **9**(2), 190–207 (1984)
16. Sekimoto, Y., Watanabe, A., Nakamura, T., Kanasugi, H., Usui, T.: Combination of spatio-temporal correction methods using traffic survey data for reconstruction of people flow. Pervasive and Mobile Computing **9**(5), 629–642 (2013)
17. Serafini, P., Ukovich, W.: A mathematical model for periodic scheduling problems. SIAM Journal on Discrete Mathematics **2**(4), 550–581 (1989)
18. Yang, J., Chen, Y.: Fast computing betweenness centrality with virtual nodes on large sparse networks. PloS ONE **6**(7), e22557 (2011)

Existence of Optimal Network Clustering
in Social Contagion

Yu-Xiao Zhu[✉] and Jin-Hu Liu

Web Sciences Center, University of Electronic Science
and Technology of China, Chengdu, China
yuxiao-zhu@hotmail.com, ljinhu@126.com

Abstract. How does community structure affect the process of social contagion? Latest research shows the existence of optimal clustering, where global cascades require the minimal number of early adopters in classical threshold model. Is this interesting finding a general pattern in other model? Motivated by this, in this paper, we study the community structure in anther contagion model that involving social enhancement. Our results confirm the existence of optimal network clustering for cascade again. This work may shed some light on the in-depth understanding and application of the social contagion problem.

1 Introduction

Innovations and ideas spread through social networks [1–6]. In a social network, nodes are individuals while links represent the relations between different individuals [7]. The study of investigating the spreading process of information and behavior has attracted much attention of researchers from different fields in recent several years. This is so-called social contagion problem. Unlike epidemic contagion, social contagion should take social enhancement into account [8,9] which assuming the more times of exposures that one individual get, the higher probability will he/she adopt the information. For instance, an individual who has over one hundred friends adopting a particular behavior will be more likely to take this behavior than someone that only one friend in his/her circle has adopted the behavior. However, the difference of epidemic contagion and social contagion is still not fully clear yet. Moreover, it has been shown that when the effect of social reinforcement is strong, the contagion is strongly affected by network community structure [10]. According to recent studies, communities sometimes can, rather than inhibit, enhance spreading of information [11]. In the paper mentioned above, they performed experiments in classical linear threshold model [12]. Consider a network involving N nodes and E links, the states of all nodes can be classified as two types: active and inactive. Initially, certain fraction of nodes are selected as seeds and initialized in active state. At each time step, nodes' states are updated based on the following rule: if the fraction of its' neighbors that in active state has exceed the threshold (like 50%, 80% and so on), then the corresponding node will become active. Finally the states of all

© Springer International Publishing Switzerland 2015
M.T. Thai et al. (Eds.): CSoNet 2015, LNCS 9197, pp. 140–147, 2015.
DOI: 10.1007/978-3-319-21786-4_12

nodes will reach a steady state. One of the most interesting finding in that paper is that they demonstrate the existence of optimal clustering in classical linear threshold model. Here we aim to investigate whether this finding is a general pattern. Will it still valid in other social contagion model?

This article is organized as follows. In the next section, we will clearly introduce social enhancement and describe the details of social contagion model. Methods for how to generate networks with tunable community structure are presented in Section 3. In Section 4, we will describe our experiments and then show the corresponding results. Finally, we summarize our results in Section 5.

2   Contagion Model

In this paper, we adopt one modified threshold model [13]. In order to involving the social contagion into the process of information diffusion, each individual has a state of awareness value A_i (spans through 0, 1, 2, ..., M), which standing for the number of times that the corresponding individual exposed to the information. In this paper, we set the threshold function as a Kronecker delta function,

$$f(A_i) = \begin{cases} 0 & A_i < M \\ 1 & A_i = M \end{cases}$$

That's to say, individual will adopt the information when its awareness value reaches M. So at any time, the status of all individuals can classified into three types: ignorant, aware, adopted. In the ignorant state ($A_i = 0$), the node i know nothing about the information and also never heard about it from its neighbors. In the aware state ($0 < A_i < M$), the node has heard the information from its neighbors but still didn't adopt yet, so nodes in aware state will not spread information. In the adopted state ($A_i = M$), the node has adopt this information and will transmit to its neighbors that still in ignorant/aware state. To clearly illustrate this contagion model, the details of the spreading model are described as below:

Step1: Random select certain proportion of nodes as seeds (adopted state), set the awareness value of these seeds as M. While all the remaining nodes are considered as ignorant (the awareness value set as 0).

Step2: In each step, all adopted individuals will transmit the information to its neighbors that in ignorant/aware state with probability p. If infect successfully, the awareness value of its neighbor will plus one in this step. Note that, we assume an edge that has transmitted the information successfully will never transmit the same information again.

Step3: Repeat *Step2* until the status of all nodes will not change.

3   Networks with Community Structure

In order to systematically investigate the impact of community structure, we prepare an ensemble of networks with two communities with varying degree of strength. First, half of the nodes are randomly selected and assigned to community A, and the other half are assigned to community B. Then, $(1-\mu)E$ links are randomly distributed among node pairs in the same community and μE links are randomly distributed among node pairs that belong to different communities. The free parameter μ spans from 0 to 0.5, thus can tune the strength of community structure. Clearly, a large value of μ means more links between the two communities and thus weak community structure, while a small value of μ means less links that connect two communities and thus strong community structure. Figure 1 displayed visualization of three networks with different strength of community structure. As we can see, for the networks in same size, networks generated when $\mu = 0.1$ has strong community structure, while generated when $\mu = 0.2$ displayed weak community structure. Clearly, networks with $\mu = 0.5$ almost near total random network.

Let's assume one network with N nodes and E links involving with two same-sized communities A and B. That's to say, $N_A = N_B = \frac{N}{2}$. Then $(1-\mu)E$ links are randomly distributed among node pairs in the community A or B. While the remaining μE links are randomly distributed among node pairs that connect community A and B. So there are $\frac{1-\mu}{2}E$ links in each community. Clearly, for each community A and B, the intra-community link probability can calculated as below:

$$P_A = P_B = \frac{E_B}{\binom{N_B}{2}} = \frac{\frac{(1-\mu)E}{2}}{\frac{N_B(N_B-1)}{2}} = \frac{4(1-\mu)E}{N(N-2)}, \tag{1}$$

in which N_B and E_B denotes number of nodes and links in community B separately. So the average degree of intra-community

$$< k_{A_{intra}} >=< k_{B_{intra}} >= \frac{2E_B}{N_B} = \frac{2(1-\mu)E}{N}. \tag{2}$$

and the intra-degree distribution of community A and B can be calculated by:

$$\begin{aligned}
P(k_{A_{intra}}) &= \binom{N_A - 1}{k_{A_{intra}}}(P_A)^{k_{A_{intra}}}(1 - P_A)^{N_A - k_{A_{intra}} - 1} \\
&= \binom{\frac{N}{2} - 1}{k_{A_{intra}}}(\frac{4(1-\mu)E}{N(N-2)})^{k_{A_{intra}}}(1 - \frac{4(1-\mu)E}{N(N-2)})^{N_A - k_{A_{intra}} - 1},
\end{aligned} \tag{3}$$

$$\begin{aligned}
P(k_{B_{intra}}) &= \binom{N_B - 1}{k_{B_{intra}}}(P_B)^{k_{B_{intra}}}(1 - P_B)^{N_B - k_{B_{intra}} - 1} \\
&= \binom{\frac{N}{2} - 1}{k_{B_{intra}}}(\frac{4(1-\mu)E}{N(N-2)})^{k_{B_{intra}}}(1 - \frac{4(1-\mu)E}{N(N-2)})^{N_B - k_{B_{intra}} - 1}.
\end{aligned} \tag{4}$$

Based on analysis above, the total number of links between two communities $E_{AB} = \mu E$, while the total available number of links between two communities should be $N_A N_B$, so link probability for each node pairs that located in different communities

$$P_{AB} = \frac{E_{AB}}{N_A N_B} = \frac{4\mu E}{N^2}, \tag{5}$$

$\mu=0.1$ $\qquad\qquad\qquad$ $\mu=0.2$ $\qquad\qquad\qquad$ $\mu=0.5$

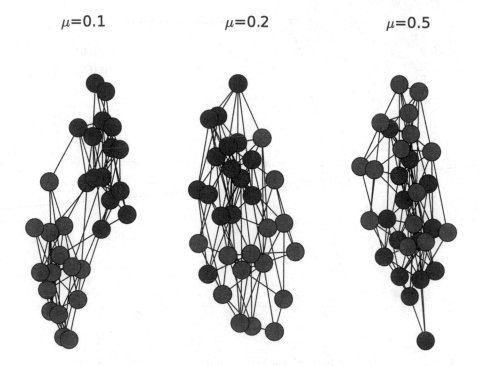

Fig. 1. Visualizations of same-sized ensemble networks with different μ. The number of nodes and links are 30 and 100 separately. All nodes are classified into two different communities with each community involving 15 nodes and nodes in different communities are labeled by different colors. Clearly, the ensemble network with smaller $\mu = 0.1$ has less links between two communities and show strong community structure. For network with bigger $\mu = 0.2$, it shows weak community structure compared with that of $\mu = 0.1$. While when we set $\mu = 0.5$, it almost approaches random network for all links are randomly chosen from all available node pairs.

while the degree distribution

$$P(k_{AB}) = \binom{N_A}{k_{AB}}(P_{AB})^{k_{AB}}(1 - P_{AB})^{N_A - k_{AB} - 1}$$
$$= \binom{\frac{N}{2}}{k_{AB}}(\frac{4\mu E}{N^2})^{k_{AB}}(1 - \frac{4\mu E}{N^2})^{\frac{N}{2} - k_{AB} - 1}. \tag{6}$$

4   Experiments and Results

To systematically study the impact of community structure, we employ the simple community model described in the previous section, which allow us to create a network with tunable community structures. The network generating process

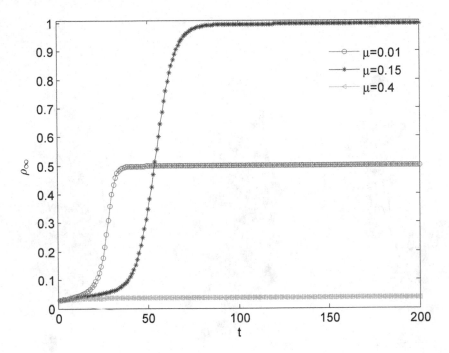

Fig. 2. Number of adopters in relationship with time step. The value shown in the figure are the average value of 10 networks with run 100 independent experiments on each network. The free parameter μ controls the strength of community structure. Smaller value of μ means less links to connect two communities and thus strong community structure. Initially, all the seeds were planted in one community and let contagion originates from one community. Strong community structures ($\mu = 0.01$) slow down the global cascade as the sparse connections between communities act as bottlenecks while speed up local spreading. More bridge links ($\mu = 0.15$) will make it's easier to spread information to another community, so that enhance global contagion. While when the ensemble network will almost be random ($\mu = 0.4$), it's like we just randomly select seeds in the whole network, which will disperse the ability to spread information out because the accumulation of successful transmission was suppressed here. The parameters for the simulation are: $N = 10000$, $E = 50000$, $N_A = 5000$, $\rho_0 = 0.05$, $M = 3$, $p = 0.2$.

produces graphs that have known community structure, but which are essentially random in other respects. Here we performed all the experiments on networks with 10000 nodes and 50000 links. The network was classified into two communities with each community involving 5000 nodes. For the information spreading process, we plant the seeds in, assuming that the contagion originates from one community. So in the beginning stage of spreading, the information is spreading in one community, it's called local contagion. While later information will gradually spread to another community through links that connect two communities and it's called global contagion.

Figure 2 displays number of final adopters with time in the networks with different community structures. Strong community structures ($\mu = 0.01$) slow

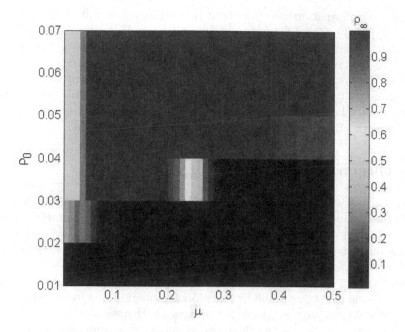

Fig. 3. Final adopters in relationship with number of initial seed and free parameter μ. Clearly, number of final adopters is in positive correlation with number of initial adopters if the effect of μ was teased out. While the relationship between ρ_∞ and μ is very complex. Both weak community structure ($\mu > 0.2$) and strong community structure ($\mu < 0.05$) will depress global cascade. While certain range of community structure ($0.05 < \mu < 0.2$) will enhance spreading cascade when enough seeds were supplied here ($0.02 < \rho_0 < 0.03$). The parameters for the simulation are: $N = 10000$, $E = 50000$, $N_A = 5000$, $M = 3$, $p = 0.2$. Moreover, our results show that this tendency is independent with M.

down the global cascade as the sparse connections between communities act as bottlenecks. At the same time, the local spreading is sped up. As μ increases (as modularity decreases, $\mu = 0.15$), there are more bridge links that connect two communities which can make it's easier to spread information to another community so that will enhance global spreading. While when μ was set too big ($\mu = 0.4$), the ensemble network will almost be random. In this case, it's like we just randomly select seeds in the whole network, which will disperse the ability to spread information out because the accumulation of successful transmission was suppressed here. From results shown here, strong community structure can enhance local contagion while suppress global contagion. The most interesting finding here is for certain strength range of community structure, it will enhance global contagion ($\mu = 0.15$). In depth, Figure 3 shows the detailed situation of final adopters in relationship with number of initial seed and free parameter μ. The phase diagram can be classified into three cases: no cascade, local cascade (information only adopted by individuals in originating community) and global cascade (information was spread out to another community from the originating community). Clearly, number of final adopters is in positive correlation with number of initial adopters. While the relationship between number of final adopters and μ is very complex. The most interesting here is for ρ_0 in certain range ($0.02 < \rho_0 < 0.03$), certain range of community structure will enhance global cascade ($0.05 < \mu < 0.2$). While strong community structure ($\mu < 0.05$) can only enhance local cascade. That's to say, if enough seeds were initialized, optimal strength of community structure exist for global cascade.

5   Conclusion and Discussion

Our analysis shows that in other social contagion model rather than classical linear threshold model there still exists optimal strength of community structure for facilitating global cascades. Unlike previous researches which point out that community structure can only enhance or hinder spreading contagion, our results show one picture that clearly demonstrate the complex influence of community structure in social contagion. Networks with strong community structure will enhance local contagion while hinder global contagion due to less links that connect different communities. While for networks with optimal strength of community structure, information can then spread outside the community effectively, thus enhance global contagion. Notice that the main contribution of this article does not lie in the finding. Instead, the significance of this work is to raise the serious question about the influence of community structure in social contagion. To us, this is a very important yet not completely understood problem in social contagion. As a starting point, we give some preliminary analysis, which is of course far from a satisfactory answer to the question. In fact, we think an in-depth understanding of this problem may shed light on this issue.

Acknowledgments. We thanks Ming Tang and Wei Wang for helpful discussion and useful suggestion. This work was funded by the Program of Outstanding PhD Candidate in Academic Research by UESTC: YBXSZC20131035.

References

1. Newman, M.E.J.: Networks: An Introduction. Oxford University Press (2010)
2. Easley, D., Kleinberg, J.: Networks, Crowds, and Markets: Reasoning About a Highly Connected World. Cambridge University Press (2010)
3. Anderson Roy, M., May Robert, M., Anderson, B.: Infectious Diseases of Humans: Dynamics and Control. Oxford University Press (1992)
4. William, G., Newill, V.A.: Generalization of Epidemic Theory. Nature **204**, 225–228 (1964)
5. Daley, D.J., Kendall, D.G.: Epidemics and rumours. Nature **204**, 1118–1118 (1964)
6. Zhu, Y., Zhang, X., Sun, G., Tang, M., Zhou, T., Zhang, Z.: PLoS ONE **9**(7), e103007 (2014)
7. Newman, M.E.J.: The Structure and Function of Complex Networks. SIAM review **45**(2), 167–256 (2003)

8. Damon, C.: The Spread of Behavior in an Online Social Network Experiment. Science **329**(5966), 1194–1197 (2010)
9. Backstrom, L., Huttenlocher, D., Kleinberg, J., Lan, X.: Group formation in large social networks: membership, growth, and evolution. In: Proceedings of the 12th ACM SIGKDD International Conference on Knowledge Discovery and Data Mining, New York, USA, pp. 44–54 (2006)
10. Weng, L., Menczer, F., Ahn, Y.-Y.: Virality Prediction and Community Structure in Social Networks. Scientific Report **3**, srep02522 (2013)
11. Nematzadeh, A., Ferrara, E., Flammini, A., Ahn, Y.-Y.: Optimal network clustering for information diffusion. Physical Review Letters **113**, 088701 (2014)
12. Watts Duncan, J.: A simple model of global cascades on random networks. Proc. Nat. Acad. Sci. **99**, 5766–5771 (2002)
13. Krapivsky Paul, L., Redner, S., Volovik, D.: Reinforcement-driven spread of innovations and fads. Journal of Statistical Mechanics: Theory and Experiment **2011**, P12003 (2011)

A Note on Clustering Difference
by Maximizing Variation of Information

Nam P. Nguyen[✉]

Department of Computer and Information Sciences,
Towson University, Towson, MD, USA
npnguyen@towson.edu

Abstract. In this paper, we investigate the problem of maximizing the difference between two partitions (or clusterings) of a complex network. Particularly, given the input network represented as an undirected graph and its initial partition X, we are interested in finding a partition Y such that the difference between X and Y, evaluated by the *Variation of Information* measure, is maximized. This problem is important in understanding fundamental properties of not only the network's structural organization (via its clusters) but also the internal and mutual interactions among those structures in response to adversarial perturbation. We propose an approximation algorithm to define the new partition Y with a guarantee ratio of $1 - \alpha - \beta$ (where α and β are constants derived from the network's initial partition), and present further optimization to improve the quality of the suggested approach.

1 Introduction

Partitioning, or clustering, a network is the process of grouping or rearranging the network' elements, commonly called its nodes and edges, into different subsets such that the elements in each subset are somehow similar to each other and elements in different subsets are disimilar in some sense. The fields of partitioning algorithms for classification in data mining and finding network' clusters in complex systems are mature and well-developed with many different methods and techniques proposed in the literature [1][2].

In this paper, we look at this classical clustering problem from a different angle: Instead of defining another network paritioning algorithm, we are interested in finding out the network's partition that maximally differs from its original clustering. Specifically, given a complex network and its initial partition X, we are interested in finding another partition Y of the network which is as much different from X as possible, i.e., the clustering difference between X and Y are maximized. We call this problem *Partition Difference Maximization* (PDM). PDM is fundamental in understanding not only the network's structural organization (via its clustering structures) but also the internal and mutual interactions among those structures in response to adversarial perturbation. A thorough analysis of PDM, as a result, will provide insights into how much the network entities can change as well as how different they can potentially be, compared to

© Springer International Publishing Switzerland 2015
M.T. Thai et al. (Eds.): CSoNet 2015, LNCS 9197, pp. 148–159, 2015.
DOI: 10.1007/978-3-319-21786-4_13

the original structures, in the worst-case scenario. Therefore, this problem lends itself effectively to the analysis of various important security aspects of complex networked systems such as network generation [3][4] community structure and network vulnerability [5][6], critical node selection [7], cascading failure [8], information diffusion in social networks [9][10].

In order to compare two partitions of a network, we need a measure that can evaluate either the similarity or dissimilarity between them. There are many different measures for this purpose suggested in the literature and they are generally classified into classes of pair-counting based and set-matching based measures [1]. Recently, information theoretic criteria have formed another fundamental class, and have been employed widely due to their strong mathematical foundation and the ability to distinguish non-linear similarities. Popular methods in this context include Entropy, Mutual Information (MI), Normalized Mutual Information (NMI), Adjusted-for-Chance and many other measures (see [11] for a summary). The seminal work of Meila [12] proposes Variation of Information (VI), a potential information theoretic measure to evaluate the *dissimilarity* between two partitions of a network. Essentially, VI aggregates the amount of lost and gained information about the first and second partitions when going from one to another. There are many advantages of using this metric (Section 3): (1) VI is a real metric on the space of all possible partitions of G; (2) VI is bounded by $\log N$ and can be normalized to be in the range of $[0, 1]$ using this quantity; (3) Small changes in a partition result in small VI-distances; and (4), VI can be efficiently computed in linear time. Therefore, this will be the measure we utilize for evaluating partitions in this paper.

In a nutshell, our contributions in this paper are:

- We investigate PDM, the problem of maximizing the Variation of Information between two partitions of a complex network: Given a network and its current partition X, PDM asks for another partition Y such that the dissimilarity $VI(X, Y)$ between X and Y is maximized. We believe that this is the first effort in optimizing VI as a measure for evaluating clustering difference.
- We propose an approximation algorithm to define the new partition Y with a performance guarantee ratio of $1 - \alpha - \beta$, where α and β are constants defined by the input partition X.
- We provide further optimization taking into account the marginal gains obtained in each round. This provides a better result in terms of solution quality. However, we are not yet able to obtain a better approximation guarantee for this fine-tuned algorithm rather than the ratio of $1 - \alpha - \beta$.

(Paper organization) In Section 2, we review the studies related to our work. Section 3 describes the notations, VI formulation and the main problem definition. Section 4 presents our proposed algorithms, their analysis and further optimization to fine-tune the objective function VI. Lastly, we conclude the paper in Section 5.

2   Related Work

Our work is mostly related to the category of Alternative Clustering Analysis in data mining with the common spirit of alternating given clusters in order to find a new clustering with better quality and of least similarity to the given ones [13]. A board classification of methods in this light of research include *Unguided Generation* with Meta Clustering [14], Eigenvectors of the Laplacian Matrix [15], the Decorrelated k-means and Convolutional EM [16], CAMI [17], etc. On the other hand, the *Guided Generation with Constraints* include MAXIMUS [18], Constrained Optimization Approach [19], ADFT [20], etc. Interested readers can see [13] for a complete survey.

Another approach for alternative clustering is based on information theoretic principles. This is mostly related to our work. The main difference of studies in this light of research is the objective function: *the measure to quantify the similarity/dissimilarity between partitions*. For instance, [21] proposed Conditional Information Bottleneck (CIB) and Ensembles based on the Mutual Information (MI) between clusters. In the same vein, NACI algorithm proposed by [22] also based on this concept of MI. Other popular methods are based on MI difference [23], NMI difference [5] and the Adjusted Mutual Information [11].

Among those methods, the most limitations that the utilized measures suffer are (1) None of them are real metrics on the space of possible clusters (2) They are non-monotone, i.e., they are not sensitive to any small change to the clusters (3) There do not exist any approximation guarantees for the optimization of these objectives. The seminal work of Meila [12] proposed Variation of Information, a potential information theoretic measure to evaluate the *dissimilarity* between two partitions of a network, that overcomes these limitations. This is the measure that we will use in this paper, and we believe that our work presents the first effort in optimizing VI as a measure for evaluating clustering difference.

3   Preliminary

In this section, we first describe the model and notations that will be used in the paper. We then describe the Variation of Information (VI) as a metric to evaluate the dissimilarity between two partitions. Finally, we define in detail the *Partition Difference Maximization* problem - our main focus in the paper.

3.1   Notations

Consider a complex network represented as an undirected graph $G = (V, E)$ where V is the set of nodes (or vertices) and E is the set of connections (or edges). Let X and Y be two partitions of V into disjoint clusters, i.e.,

$$X = \{X_1, X_2, ..., X_p\},$$

and respectively,

$$Y = \{Y_1, Y_2, ..., Y_q\},$$

where $X_i, Y_j \subseteq V$ and

$$\cup_{i=1}^{p} X_i = V \text{ and } X_i \cap X_s = \emptyset,$$

for all indices $0 \leq i, s \leq p$.
Similarly,

$$\cup_{j=1}^{q} Y_j = V \text{ and } Y_j \cap Y_t = \emptyset,$$

for all indicies $0 \leq j, t \leq q$.

For each $i \in [1, p]$ and $j \in [1, q]$, let x_i and y_j be the sizes of $X_i \in X$ and $Y_j \in Y$, i.e., $x_i = |X_i|$ and $y_j = |Y_j|$. Furthermore, without loss of generality, we assume that clusters in X are sorted by sizes:

$$x_1 \leq x_2 \leq ... \leq x_p.$$

For each pair of indicies $(i, j) \in [1, p] \times [1, q]$, let r_{ij} be the number of nodes in the intersection of X_i and Y_j, i.e., $r_{ij} = |X_i \cap Y_j|$. Finally, denote by N the number of nodes in V, i.e, $N = |V|$ and by $\mathcal{P}(V) \equiv \{X | X \text{ is a partition of } V\}$ the space of all possible partitions of V, respectively.

With these above notations, the following properties hold true for any network $G = (V, E)$:

1. $\displaystyle\sum_{i=1}^{p} x_i = \sum_{j=1}^{q} y_j = N.$

2. $\displaystyle\bigcup_{i=1}^{p}(X_i \cap Y_j) = Y_j \quad \forall j \in [1, q], \text{ and } \bigcup_{j=1}^{q}(X_i \cap Y_j) = X_i \quad \forall j \in [1, q].$

3. $\displaystyle\sum_{j=1}^{q} r_{ij} = x_i, \quad \sum_{i=1}^{p} r_{ij} = y_j, \text{ and } \sum_{ij} r_{ij} = N.$

3.2 Variation of Information

In order to evaluate the dissimilarity between partitions of V, we utilize Variation of Information (VI), a concept in Information Theory suggested in [12]. Conceptually, $VI(X, Y)$ is the total of two quantities: the conditional uncertainty about X in the presence of Y, and the same type of uncertainty about Y in the presence of X. As we show later on, this equals the aggregation of the amount of information about X (that we lose) and the amount of information about Y (that we still have to gain) when going from partition X to Y. $VI(X, Y)$ is 0 if X and Y are identical structures, and is $\ln N$ if X and Y are completely different partitions, e.g., $X = V$ and Y contains exactly N singleton clusters. Formally, given two partitions X and Y of V, the dissimilarity $VI(X, Y)$ between X and Y is defined as

$$VI(X, Y) = H(X) + H(Y) - 2I(X, Y), \tag{1}$$

where $H(X), H(Y)$ and $I(X, Y)$ are the entropy and Mutual Information of X and Y, respectively [24].

To formulate $H(X), H(Y)$ and $I(X, Y)$, we start out by considering cluster assignments X_i and Y_j, where X_i and Y_j are considered as cluster labels of a node in X and Y, respectively. Without loss of generality, we can further assume that X_i and Y_j are also values of two random "variables" X and Y, with individual distribution:

$$P(X_i) = P(X = X_i) = x_i/N,$$

$$P(Y_j) = P(Y = Y_j) = y_j/N,$$

and the joint distribution:

$$P(X_i, Y_j) = P(X = X_i; Y = Y_j) = r_{ij}/N.$$

The entropy (or uncertainty) of X and Y is defined below. Note that we will use the natural base e for all logarithmic functions hereafter:

$$H(X) = -\sum_{i=1}^{p} P(X_i) \log P(X_i) = -\sum_{i=1}^{p} \frac{x_i}{N} \ln \frac{x_i}{N},$$

$$H(Y) = -\sum_{j=1}^{q} P(Y_j) \log P(Y_j) = -\sum_{j=1}^{q} \frac{y_j}{N} \ln \frac{y_j}{N}.$$

The Mutual Information $I(X, Y)$ is defined as

$$I(X, Y) = \sum_{i=1}^{p} \sum_{j=1}^{q} P(X_i, Y_j) \log \frac{P(X_i, Y_j)}{P(X_i) P(Y_j)} = \sum_{i=1}^{p} \sum_{j=1}^{q} \frac{r_{ij}}{N} \ln \frac{N r_{ij}}{x_i y_j}.$$

Finally, $VI(X, Y)$ defined in Eq. (1) is written as:

$$VI(X, Y) = \frac{1}{N} \Big(\sum_{i=1}^{p} x_i \ln x_i + \sum_{j=1}^{q} y_j \ln y_j - 2 \sum_{i=1}^{p} \sum_{j=1}^{q} r_{ij} \ln r_{ij} \Big). \qquad (2)$$

Eq. (2) reveals that $VI(X, Y)$ essentially aggregates the amount of information about X (that we loose via the negative term of r_{ij}'s) and the amount of information about Y (that we still have to gain via the second term of y_j's) when going from partition X to Y.

There are many advantages of using VI over other measures for evaluating partition difference [12]: (1) First and foremost, VI satisfies non-negativity, symmetry and subadditivity, indicating that *it is a real metric on* $\mathcal{P}(V)$ *- the space of all possible partitions of* G. Hence, VI is ideal to evaluate partition difference; (2) VI is not bounded by a constant value. Rather, it is bounded by $\ln N$ and can be normalized by this quantity; (3) The product of two partition X and Y is "colinear" with the two partitions. This implies small changes in a partition result in small VI-distances; and lastly, (4) VI can be efficiently computed in an order of $O(N + pq)$, which can be regarded as linear time in term of N.

3.3   Problem Definition

With all the necessary notations taken into account, our main problem is defined as follow:

Definition 1 (Partition Difference Maximization - PDM). *Given a network represented as an undirected graph $G = (V, E)$ and its initial disjoint partition $X = \{X_1, X_2, ..., X_p\}$, find another disjoint partition Y of V such that the dissimilarity measure $VI(X, Y)$ is maximized, i.e., find the partition Y such that*

$$Y = \underset{Y' \in \mathcal{P}(V)}{\operatorname{argmax}} \{VI(X, Y')\}.$$

In an essence, PDM problem asks for a partition $Y \in \mathcal{P}(v)$ which incurs a maximum dissimilarity to X measured by $VI(X, Y)$. The solution to PDM will provide valuable insights to a larger set of problems such as network structural vulnerability [25][5] or critical node detection [6]. Unlike the case of other measure, such as Normalized Mutual Information, the trivial all-singleton scenario for Y (i.e., every node is a cluster of itself in Y) does not always yield the maximal dissimilarity in VI. Rather, we are interested in the case where the information contained in clusters of Y spreads out more while barely overlaps with clusters of X (which, in turn, makes y_j's big and r_{ij}'s small in Eq. (2)).

Many variations of problems in this direction can also be defined based on PDM. For instance, when X and Y are *overlapping partitions* (i.e., clusters in X (and in Y) can overlap), or in a different angle, *finding partition Y on the graph G after some of its critical nodes are deleted under adversarial attacks*. Vice versa, another variation can include the *finding of k critical nodes of G* such that the clustering Y, defined by some given partitioning methods on G excluding these k nodes, are of maximal difference from X.

4   Method

In order to find the partition Y that incurs the maximum dissimilarity $VI(X, Y)$, it is important to investigate the behavior of VI measure and also its properties in general. Recall that, given the network $G = (V, E)$ and two partitions X and Y, the dissimilarity $VI(X, Y)$ (Eq. (2)) is written as

$$VI(X, Y) = \frac{1}{N} \left(\sum_{i=1}^{p} x_i \ln x_i + \sum_{j=1}^{q} y_j \ln y_j - 2 \sum_{i=1}^{p} \sum_{j=1}^{q} r_{ij} \ln r_{ij} \right).$$

Since x_i's and N are given as parts of input parameters, we treat them as constants hereafter. As a result, maximizing $VI(X, Y)$ is equivalent to maximizing the two later terms. This is also our main objective to optimize

$$\max \left\{ \sum_{j=1}^{q} y_j \ln y_j - 2 \sum_{i=1}^{p} \sum_{j=1}^{q} r_{ij} \ln r_{ij} \right\} \tag{3}$$

Note that although $\sum_{j=1}^{q} y_j \ln y_j$ is a convex function, $-2 \sum_{i=1}^{p} \sum_{j=1}^{q} r_{ij} \ln r_{ij}$, on the other hand, is a concave function. This fact makes the main objective neither convex nor concave. At a glance, the optimization presented in Eq. (3) requires y_j's to be big meanwhile keeping the overlaps r_{ij} as small as possible. However, the bigger the values y_j's are the larger the clusters Y_j's are. Thus, the higher chance Y_j's will overlap with clusters X_i's in X, implying potentially big values of r_{ij}'s. Moreover, since $r_{ij} = |X_i \cap Y_j|$ involves both clusters X_i and Y_j, the last term in Eq. (3) penalties the main objective significantly if Y_j and X_i heavily overlap. This is the main source difficulty in globally optimizing the objective function.

4.1   Algorithm

We next describe our first algorithm trying to maximize $VI(X,Y)$. The idea behind our approach is simple, intuitive yet effective: since we are penalized by the overlaps of X_i's and Y_j's, let us try not to get penalized, i.e., we will choose Y_j in such a way that $y_j = |Y_j|$ is as large as possible meanwhile keeping $r_{ij} \ln r_{ij} = 0$. This requires r_{ij} has to be exactly equal to 1 for all pairs (i,j).

To satisfy this requirement, we will first mark all nodes in V unassigned. We then select exactly one unassigned node from every cluster $X_i \in X$, and let Y_1 be the set of selected nodes. By this selection, the newly created cluster Y_1 contains exactly p nodes from $X_1, X_2, ..., X_p$ and satisfies $r_{i1} = |X_i \cap Y_j| = 1$ for all $i \in [1,p]$. We continue defining Y_2, Y_3, etc, in this very same manner until all nodes in X_{p-1} are assigned to new clusters in Y. Lastly, we make each left-over unassigned node in X_p a singleton cluster in Y. By doing in this way, every cluster Y_j of Y spreads out evenly over all clusters of X, and has exactly one common element with any cluster X_i. This ensures $r_{ij} = 1$ for all pairs of indices (i,j). This idea leads us to the approach presented in Alg. 1.

Algorithm 1. A bounded solution to PDM

Input: Network $G = (V, E)$ and a disjoint partition $X = \{X_1, X_2, ..., X_p\}$ of V;
Output: A disjoint partition Y of V;

   1. Mark all nodes in V as "unassigned", set $Y = \emptyset$ and $q = 1$.
   2. Select an unassigned node (if there is any) in each set $X_1, X_2, ..., X_p$. Mark all the selected nodes "assigned" and include them in Y_q.
   3. If there are still unassigned nodes, increase q and go back to Step 2.
   4. If all nodes are assigned, return partition $Y = \{Y_1, Y_2, ..., Y_q\}$.

4.2   Performance Analysis

We next analyze the performance of the suggested approach. We will show that Alg. 1, despite its simple nature, achieves a constant approximation factor that depends only on the input parameters, and is a very efficient approach with linear time and space complexity. An an algorithm said to obtain an *approximation guarantee of* $\rho(n)$, (or equivalently is bounded by $\rho(n)$ if) for all input of size n, the cost c of the solution obtained by the algorithm is within a factor $\rho(n)$ of the cost of an optimal solution [26].

Lemma 1. *Alg. 1 produces a partition* Y *containing* $q = x_p$ *clusters, among which there are* x_{p-1} *real clusters of sizes greater than 1. Moreover, all clusters* $Y_j \in Y$ *satisfy* $r_{ij} = 1$ *for all pairs* $(i, j) \in [1, p] \times [1, q]$.

Proof. The proof of this proposition follows naturally from the algorithm and is omitted here.

Lemma 2. *Let* $a \geq 3$ *be a constant. Then* $f(x) = \left(\frac{a}{x}\right)^x$ *is an increasing function for* $x \in (0, 1]$.

Proof. Rewrite $f(x) = a^x x^{-x}$. Taking the derivative of $f(x)$ yields

$$\frac{df}{dx} = a^x x^{-x} (\ln a - \ln x - 1) > 0$$

for all $x \in (0, 1]$ and $a \geq 3$. Thus, the conclusion follows.

Theorem 1. *Given the network* $G = (V, E)$ *and its initial partition* X, *Alg. 1 provides a partition* Y *of* V *satisfying*

$$\max\left\{\frac{2}{N}, (1 - \alpha - \beta) \ln N\right\} \leq VI(X, Y) \leq \ln N$$

where $N = |V|$, $\alpha = \frac{x_p - x_{p-1}}{N} \in [0, 1)$ *and* $\beta = \log_N \frac{p}{2}$ *are constants derived from the input parameters. In order words, Alg. 1 achieves a constant approximation guarantee for PDM problem.*

Proof. We first give an estimate for $\sum_{i=1}^{p} x_i \ln x_i$. Using Log Sum inequality [24] yields

$$\sum_{i=1}^{p} x_i \ln x_i \geq \left(\sum_{i=1}^{p} x_i\right) \ln \frac{\left(\sum_{i=1}^{p} x_i\right)}{p} = N \ln N - N \ln p.$$

Due to Lemma 1, the last term involving r_{ij}'s in Eq. (3) is canceled (because $\ln r_{ij} = 0$). Thus, a lower bound on $\sum_{j=1}^{q} y_j \ln y_j$ is of desire. We again utilize the convexity of $x \ln x$ function and Log Sum inequality to yield

$$\sum_{j=1}^{q} y_j \ln y_j = \sum_{y_j > 1} y_j \ln y_j \geq \left(\sum_{y_j > 1} y_j\right) \ln \frac{\left(\sum_{y_j > 1} y_j\right)}{x_{p-1}}$$

Note that we divide by x_{p-1} instead of x_p for the number of clusters in Y since there are x_{p-1} non-singleton clusters in Y (Lemma 1). To find the summation of non-singletons y_j's, we compute

$$\sum_{y_j > 1} y_j \ln y_j = p x_1 + (p - 1)(x_2 - x_1) + (p - 2)(x_3 - x_2) + \dots + 2(x_{p-1} - x_{p-2})$$

$$= 2 x_{p-1} + (x_1 + x_2 + \dots + x_{p-2}) = N - (x_p - x_{p-1})$$

$$= (1 - \alpha) N.$$

Thus,

$$\sum_{y_j>1} y_j \ln y_j \ge (1-\alpha)N \ln \frac{(1-\alpha)N}{x_{p-1}}$$

$$= (1-\alpha)N \ln N - N \ln \left(\frac{x_{p-1}}{1-\alpha}\right)^{1-\alpha}.$$

Hence, the main objective in Eq. (2) is simplified to

$$VI(X,Y) = \frac{1}{N}\left(\sum_{i=1}^{p} x_i \ln x_i + \sum_{y_j>1} y_j \ln y_j\right)$$

$$\ge (2-\alpha)\ln N - \ln p\left(\frac{x_{p-1}}{1-\alpha}\right)^{1-\alpha}$$

Due to Lemma 2, $\left(\frac{x_{p-1}}{1-\alpha}\right)^{1-\alpha}$ is a decreasing function for $\alpha \in [0,1)$, and will attain its maximum value x_{p-1} at $\alpha = 0$. Moreover, since $x_{p-1} \le \frac{x_p + x_{p-1}}{2} \le \frac{N}{2}$, we can estimate $VI(X,Y)$ by

$$VI(X,Y) \ge (2-\alpha)\ln N - \ln(p \times x_{p-1})$$

$$\ge (2-\alpha)\ln N - \ln\left(p\frac{N}{2}\right)$$

$$= (1-\alpha)\ln N - \ln\frac{p}{2}$$

$$= (1-\alpha-\beta)\ln N$$

On the other side, it has been show by Melia [12] that $VI(X,Y)$ is upper and lower bounded by $\ln N$ and $\frac{2}{N}$ in general, respectively. Hence, the conclusion follows.

4.3   Complexity Analysis

It is straightforward that Alg. 1 visits every node in the network exactly once in order to produce the resulting partition Y. Thus, this algorithm is $O(N)$, i.e., is of linear time and space complexity.

Remark

The analysis of the suggested approach implies that if the sizes of the largest and second largest clusters in X are not very different from each other, i.e., $(x_p - x_{p-1})$ approaches 0, then the proposed approach is suboptimal (only different by a constant factor $\log_N \frac{p}{2}$) in VI optimization. The derived guarantee factor also mathematically emphasizes the following intuitions that have been observed in prior studies [25][5]: the sizes of the tops clusters as well as the number of clusters in the initial partition do contribute significantly into the structural difference of network' clusterings.

Algorithm 2. A bounded solution to PDM with marginal gain

Input: Network $G = (V, E)$ and a disjoint partition $X = \{X_1, X_2, ..., X_p\}$ of V;
Output: A disjoint partition Y of V;

1. Mark all nodes in X_i sets as "unassigned", $Y = \emptyset$ and $q = 1$.
2. Select an unassigned node (if there is any) in every set $X_1, X_2, ..., X_p$. Mark all the selected nodes "assigned" and put them in Y_q.
3. For the current cluster Y_q containing y_q nodes, select z^* (Eq. (4)) more "unassigned" nodes from the largest cluster X_p and include them to Y_q.
4. If there are still unassigned nodes, increase q and go back to Step 2.
5. If all nodes are assigned, return partition $Y = \{Y_1, Y_2, ..., Y_q\}$.

4.4 Further Optimization

As the original approach presented in Alg. 1 tries to zeroes out the penalty terms $\ln r_{ij}$'s, it assigns left-over "unassigned" nodes in X_p (the largest cluster of X) to singleton clusters. This approach, however, disregards the potential marginal gains that incur when the size of the largest dominates the size of the second largest cluster in the network, i.e., when $x_p \gg x_{p-1}$. These marginal gains can be achieved by further increasing the current y_j and pay the penalty r_{pj} specifically for only X_p, as long as the marginal gains keep increasing positively.

Mathematically, let z be the number of more "unassigned" nodes that we can select from X_p to include in Y_j. The marginal gain for each cluster Y_j can sequentially be formulated as a function of z:

$$g(z) = (y_j + z)\ln(y_j + z) - 2(z + 1)\ln(z + 1).$$

Note that because we only select more nodes from X_p, we have to pay the penalty only for r_{pj} where j is the currently selected cluster. In this context, Alg. 1 corresponds exactly to the case of $g(0)$. Our improvement objective, therefore, is to further maximize $g(z)$ for every cluster Y_j at Step 2 of Alg. 1.

To find the maximum value of $g(z)$, we compute

$$\frac{dg}{dz} = \ln \frac{y_j + z}{e(z + 1)^2},$$

where e is the base of the natural Logarithm function. We then solve $\frac{dg}{dz} > 0$ for z. This yields the range $z \in [0, z^*]$ where

$$z^* = \left\lfloor \frac{1 - 2e + \sqrt{4e(y_j - 1) + 1}}{2e} \right\rfloor, \tag{4}$$

In this range of z, $g(z)$ is an strictly increasing function, and attains the maximum value at $g(z^*)$, which is the optimal marginal gain for each cluster Y_j. This further optimization leads us to the second approach with the benefit of marginal gains. Step 3 in Alg. 2 represents this change. Unfortunately, we do not yet have a way to further analyze and provide an estimate for this marginal gain over the prior approach.

5   Conclusion

We study PDM, which is the problem of maximizing the difference between two partitions, or clusterings, of a network. This problem is crucial in understanding fundamental properties of network' structures and their internal and mutual interactions in response to adversarial perturbation. We propose an approximation algorithm to define the new partition Y with a constant guarantee ratio of $1 - \alpha - \beta$, and present further optimization to improve the quality of the suggested methods.

References

1. Berkhin, P.: Survey Of Clustering Data Mining Techniques. Technical report, Accrue Software, San Jose, CA (2002)
2. Fortunato, S.: Community detection in graphs. Physics Reports **486**(3–5), 75–174 (2010)
3. Atwood, J., Ribeiro, B., Towsley, D.: Efficient network generation under general preferential attachment. Computational Social Networks **2**(1), 7 (2015)
4. Atabati, O., Farzad, B.: A strategic model for network formation. Computational Social Networks **2**(1), 1 (2015)
5. Alim, M.A., Nguyen, N.P., Dinh, T.N., Thai, M.T.: Structural vulnerability analysis of overlapping communities in complex networks. In: 2014 IEEE/WIC/ACM International Joint Conferences on Web Intelligence (WI) and Intelligent Agent Technologies (IAT), vol. 1, pp. 5–12, August 2014
6. Dinh, T.N., Xuan, Y., Thai, M.T., Pardalos, P.M., Znati, T.: On new approaches of assessing network vulnerability: hardness and approximation. IEEE/ACM Trans. Netw. **20**(2), 609–619 (2012)
7. Ventresca, M., Aleman, D.: Efficiently identifying critical nodes in large complex networks. Computational Social Networks **2**(1), 6 (2015)
8. Peters, K., Buzna, L., Helbing, D.: Modelling of cascading effects and efficient response to disaster spreading in complex networks. IJCIS **4**(1/2), 46–62 (2008)
9. Kim, H., Beznosov, K., Yoneki, E.: A study on the influential neighbors to maximize information diffusion in online social networks. Computational Social Networks **2**(1), 3 (2015)
10. Fidler, D.: Power and loyalty defined by proximity to influential relations. Computational Social Networks **2**(1), 2 (2015)
11. Vinh, N.X., Epps, J., Bailey, J.: Information theoretic measures for clusterings comparison: Variants, properties, normalization and correction for chance. J. Mach. Learn. Res. **11**, 2837–2854 (2010)
12. Meil, M.: Comparing clusteringsan information based distance. Journal of Multivariate Analysis **98**(5), 873–895 (2007)
13. Aggarwal, C.C., Reddy, K.C.: Data Clustering: Algorithms and Applications. CRC Press (2013)
14. Caruana, R., Elhawary, M., Nguyen, N., Smith, C.: Meta clustering. In: 2013 IEEE 13th International Conference on Data Mining, pp. 107–118 (2006)
15. Dasgupta, S., Ng, V.: Mining clustering dimensions. In: Frnkranz, J., Joachims, T. (eds.), ICML, pp. 263–270. Omnipress (2010)
16. Jain, P., Meka, R., Dhillon, I.S.: Simultaneous unsupervised learning of disparate clusterings. Stat. Anal. Data Min. **1**(3), 195–210 (2008)

17. Hong, X., Bailey, D.J.: Generation of alternative clusterings using the cami approach. In: SIAM SDM (2010)
18. Bae, E., Bailey, J., Dong, G.: A clustering comparison measure using density profiles and its application to the discovery of alternate clusterings. Data Min. Knowl. Discov. **21**(3), 427–471 (2010)
19. Qi, Z., Davidson, I.: A principled and flexible framework for finding alternative clusterings. In: Proceedings of the 15th ACM SIGKDD International Conference on Knowledge Discovery and Data Mining, KDD 2009, pp. 717–726. ACM, New York (2009)
20. Cui, Y., Fern, X.Z., Dy, J.G.: Non-redundant multi-view clustering via orthogonalization. In: Proceedings of the 2007 Seventh IEEE International Conference on Data Mining, ICDM 2007, pp. 133–142. IEEE Computer Society, Washington, DC, (2007)
21. Gondek, D., Hofmann, T.: Non-redundant clustering with conditional ensembles. In: Proceedings of the Eleventh ACM SIGKDD International Conference on Knowledge Discovery in Data Mining, KDD 2005, pp. 70–77. ACM, New York (2005)
22. Dang, X.-H., Bailey, J.: A hierarchical information theoretic technique for the discovery of non linear alternative clusterings. In: Proceedings of the 16th ACM SIGKDD International Conference on Knowledge Discovery and Data Mining, KDD 2010, pp. 573–582. ACM, New York (2010)
23. Vinh, N.X., Epps, J.: Mincentropy: a novel information theoretic approach for the generation of alternative clusterings. In: Proceedings of the 2010 IEEE International Conference on Data Mining, ICDM 2010, pp. 521–530. IEEE Computer Society, Washington, DC (2010)
24. Cover, T.M., Thomas, J.A.: Elements of Information Theory. Wiley-Interscience (1991)
25. Nguyen, N.P., Alim, M.A., Shen, Y., Thai, M.T.: Assessing network vulnerability in a community structure point of view. In: Proceedings of the 2013 IEEE/ACM International Conference on Advances in Social Networks Analysis and Mining, ASONAM 2013, pp. 231–235. ACM, New York (2013)
26. Vazirani, V.V.: Approximation Algorithms. Springer (2003)

A Point-of-Interest Recommendation Method Based on User Check-in Behaviors in Online Social Networks

Erzhong Zhou[✉], Jiajin Huang, and Xinxin Xu

Beijing University of Technology, Beijing 100124, People's Republic of China
{zez2008,hjj}@emails.bjut.edu.cn, xuxx@faropt.com

Abstract. Point-of-interest (POI) recommendations aim at identifying candidate POIs and ranking them in a descent order according to the probabilities of a user visiting them. The paper takes the scalability of information and user personalization into consideration to improve POI recommendation service, and proposes a personalized POI recommendation method based on user check-in behaviors in online social networks. First, the user's travel experience in the target region is used to reduce the range of candidate POIs. At last, the proposed method ranks the candidate POIs to meet the user's personalized need by combining the user preference, attraction of a POI on the target user, and social recommendations from friends. Experimental results show that the proposed method is feasible and effective.

Keywords: POI recommendation · User preference · Attraction of a POI · Social recommendation

1 Introduction

With the combination of mobile phone and Internet, the location-based services become more and more attractive. If users login the location-based social websites and send some messages, their check-in data are generated and their friends can know users' current positions and behaviors in an electronic map in addition to feelings. The user's check-in behavior is proven more susceptible to the social link and geographic position [2,8]. For example, based on the check-in data from Gowalla website, the research shows that more than 30% of new places visited by a user have been visited by a friend or friends of friends [1,2]. Hence, studying users' check-in behaviors can contribute to Point-of-interest (POI) recommendation. A POI is a terminology referred in a geographic information system for denoting a concrete entity such as a restaurant or gym. The POI recommendation has been one of important issues in location-based services.

The main tasks of the POI recommendation include how to identify candidate POIs for a target user and rank the POIs according to the probabilities of a user visiting these POIs. However, the scalability of information and user personalized needs have been the main obstacles for POI recommendation methods.

M.T. Thai et al. (Eds.): CSoNet 2015, LNCS 9197, pp. 160–171, 2015.
DOI: 10.1007/978-3-319-21786-4_14

Hence, the paper takes the scalability of information and user personalization into consideration to raise the performance of POI recommendations.

The POI recommendation first needs to acquire the user's current location and personalized characteristics. As for the representation of the geographic location, the location can be described in different granularities. For example, a POI is often specified by the geographic longitude and latitude in a geographic information system. In addition, a street or block can also be used to determine the user location. The paper analyzes the influence of a geographic location from different geographic levels. For example, the regions frequently visited by users are also identified for evaluating the geographic influence of an unvisited POI. In terms of the personalization, the travel characteristics of users include their preferences and travel experience. The user check-in behaviors in online social websites not only reflect the travel characteristics of users, but also reflect the influences of other factors such as the popularity of a POI and social link. Following the analysis, we propose a personalized POI recommendation method based on user check-in behaviors in online social networks. The rest of the paper is organized as follows. Section 2 introduces the related work. Section 3 provides a POI recommendation strategy by analyzing the travel characteristics of different kinds of tourists. Section 4 elaborates the proposed POI recommendation method. Section 5 presents the experimental results for validating the performance of the proposed method. Section 6 gives conclusions and future work.

2 Related Work

The check-in data generated by users in the location-based social network record their trajectories in geographic coordinates, which is the most obvious difference from a traditional online one. Recently, there is a great progress for personalized POI recommendations. The links between entities in the location-based social network, such as social ties between users, interactions among users and POIs, and proximity between POIs, are extracted for POI recommendations. For example, the user-based collaborative filtering method speculates the preference of the target user from similar users. Those POIs that are popular with similar users are more likely to be recommended to the target user [11]. Symeonidis and colleagues found that users prefer to accept the opinions from their friends [7]. Based on the social influence, Ye and colleagues proposed a friend-based collaborative filtering approach for POI recommendations [9]. However, the data scarcity is obvious for collaborative filtering methods because users can seldom visit a great number of POIs. Especially, most users often live at the same area with their friends. Hence, the data scarcity has a great impact on collaborative filtering methods with respect to those unvisited POIs. Ye and colleagues found that users prefer to visit POIs near those that they have visited [10]. The distance between POIs is consequently taken into consideration [14]. In order to overcome the data scarcity and meet personalized needs of users, more researchers evaluate the probability of a user visiting a POI from the user preference, popularity of a POI and social influence [12,13]. Based on the geographic and social influences,

Zhang and colleagues evaluated the sequential influence by the location-location transition graph and additive Markov chain to recommend POIs [15]. Following such a multi-factor recommendation strategy, this paper not only considers the popularity of a POI but also evaluates the geographic influence of the POI on the target user. In addition, the user's forgetting factor is also taken into account when the social influence is evaluated.

3   Problem Analysis and Strategy

This section lays emphasis on how to implement the personalized information organization and management that aim to facilitate the identification and ranking of POIs. Geographic locations and user characteristics are essential requirements for POI recommendations. Hence, POI recommendations must first focus on the identification of a user's location and identity. The following conditions are taken into consideration: 1) how does a tourist choose a POI in an unfamiliar region when the tourist takes a short trip? 2) how does a tourist choose a POI in a familiar region when the tourist takes a short trip? 3) how does a tourist choose a POI when the tourist takes a long trip?

As for above conditions, the moving trajectories of a tourist need to be analyzed from the macroscopic and microscopic views. First, it is undoubtedly that the local people are more proficient in finding those local interesting POIs than newcomers. Meantime, newcomers often pay more attention to the popularity of a POI or opinions from their friends. Hence, the trajectories of local people are often concentrated in their living and working areas, and the trajectories of the newcomers are often concentrated around the famous scenic spots. At last, each tourist has different preferences on different POIs, and the geographic location of the same POI has different impacts on different tourists. Based on the above analysis, it is obvious that a tourist's experience and preferences play an important role in tour planning. Hence, we have following hypotheses: 1) when a user visits a familiar region, the user is more likely to visit those POIs near familiar ones; 2) when a user arrives in an unfamiliar region, the user is more likely to visit the POIs that are either famous or highly recommended by friends.

Based on the above hypotheses, it is important to identify the familiar regions of the target user. The paper first partitions the target region according to the distribution of POIs, and then those sub-regions are ranked by evaluating user's travel experience. At last, the user's familiar regions are identified. In other words, the candidate POIs are identified according to the user's travel experience. As shown in Figure 1, the candidate POIs are limited in the familiar region if a local person visits a familiar region. Oppositely, the candidate POIs are limited within a certain radius when a newcomer or a local person visits an unfamiliar region. Wang and colleagues found that the new POI visited by a user is often located within 10 kilometers around those POIs once visited by the user [8]. Meantime, the distribution of POIs can reflect the range of human activities. Hence, the paper implements the personalized information organization and management through the granularity of the geographic partition.

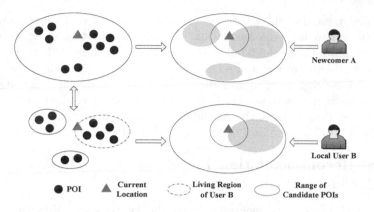

Fig. 1. Identification of candidate POIs

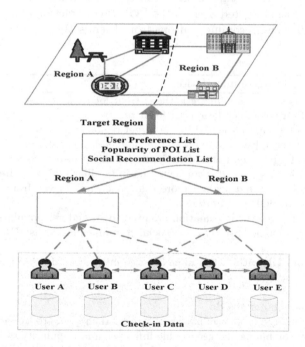

Fig. 2. Personalized information organization and management

The POI recommendation methods consider more factors so that those methods are more and more sophisticated. In order to improve the performance of POI recommendations, the work flow is separated into the offline and online phases. The online work includes the identification and ranking of candidate POIs for the user's need. The proposed method ranks the candidate POIs to meet the users' personalized needs by combining the user's preference on the POI, attraction of the POI on the target user, and social recommendations from friends. Hence, as shown in Figure 2, the offline work is as follows: 1) extract the information oriented to the whole region such as the user preference, popularity of the POI and social recommendation; 2) partition the target region into the sub-regions and identify the familiar sub-regions for each user; 3) rearrange the information oriented to the user's familiar sub-region from original information set. After identifying users' personalized needs, we can do the following online work: 1) judge if the user can stay in the visited region according to the current trajectory and then identify the candidate POIs; 2) compute the recommendation degree of each candidate POI and rank those POIs.

4   POI Recommendation Based on User Check-in Behaviors

The proposed method first identifies the candidate POIs according to the user's current location and travel experience. At last, the POIs are ranked by combining the preference of the targeted user on the POI, the attraction of the POI for the target user, and social recommendations from friends. The process is presented in Algorithm 1.

Algorithm 1.. POI recommendation

Input: u_id is the target user, loc_id is the current location of u_id.
Output: L_i is POI recommendation list for u_id.
1. Extract the information oriented to the target region such as popularity of POI list PL, user preference list HL and social recommendation list FL;
2. Partition the target region and identify the familiar sub-regions for each user;
3. **if** current trajectory of u_id is concentrated in familiar region A **then**
4.   Retrieve all POIs that are not more than given miles away from loc_id in A as the candidate POI set $Candiates$;
5.   Rearrange the related information oriented to A, and get popularity of POI list PL_i, user preference list HL_i and social recommendation list FL_i;
6. **else**
7.   Retrieve all POIs within the radius of given miles around loc_id as the candidate POI set $Candiates$;
8. **end if**
9. **for** POI p_j in $Candidates$ **do**
10.   Compute the geographic influence of p_j and then evaluate the attraction of p_j on u_id by combining the geographic influence and popularity of p_j;
11.   Compute the recommendation of p_j for u_id by combing the preference of u_id on p_j, attraction of p_j on u_id and social recommendation of p_j for u_id;
12. **end for**
13. Rank the candidate POIs in a descent order, and choose the top N POIs to constitute L_i;
14. Return L_i.

4.1   Identification of Candidate POIs

Based on the strategy mentioned in Section 3, it is important to judge if the moving trajectory of the target user is concentrated in corresponding familiar regions.

Hence, the paper needs to evaluate user's travel experience in different regions in advance. If a user stays longer in a region, the user is likely to have high experience. Zheng and colleagues adopt the idea that a tourist can have high travel experience if the tourist visits more famous places in the given region [17]. Meantime, the region is more likely to be familiar one for the target user if the region includes most of POIs visited by the user. The paper adopts Zheng and his colleagues' strategy to consider the relation between the travel experience and the popularity of a POI. In addition, the recency and frequency of user check-in behaviors are also counted to evaluate user's travel experience.

The popularity of a POI is often measured by using information entropy presented as follows [3]:

$$E(l_i) = - \sum_{u \in U_i} (\frac{C_{u,i}}{\sum_{u \in U_i} C_{u,i}} \times \log \frac{C_{u,i}}{\sum_{u \in U_i} C_{u,i}}) \tag{1}$$

where l_i denotes the ith POI in region A, U_i denotes all users who have visited l_i, and $C_{u,i}$ denotes the number of visiting l_i of user u. The information entropy is low if the POI belongs to a private place. In other words, the popularity of the POI is low. Based on the information entropy, the travel experience of user u in region A is evaluated by:

$$AC_{u,A} = \sum_{l_i \in A} \sum_{j=1}^{C_{u,i}} e^{-\Delta t_j} \times E(l_i) \tag{2}$$

where Δt_j denotes the time interval between the date when user u visited l_i for the jth time and current time. Owing to different travel patterns, the paper categorizes users into two categories, namely local people and newcomers. As for local people, their check-in behaviors are more likely to happen in their living and working areas [10]. As for newcomers, their check-in behaviors are often concentrated in scenic spots. Hence, the paper chooses two most familiar regions for a local person and chooses one most familiar region for a newcomer according to their travel experience.

The distribution of POIs is irregular because of the dispersed population and other factors. Meantime, the density-based spatial clustering of applications with noise (DBSCAN) method is often used to analyze human activities [16]. Hence, the proposed method adopts the DBSCAN method to partition the target region. The DBSCAN method first identifies the core point of each cluster according to the minimum of sample points within a given distance. Then related core points whose distance from the target point is not more than the given distance are clustered. Meantime, the corresponding clusters of related core points are merged into a new cluster. Based on the DBSCAN method, POIs in the target region are clustered into different sub-regions. However, some famous POIs such as some historical places often lie in the suburbs so that those POIs are possible to be classified into the isolated points. As for such a problem, the proposed method clusters those isolated POIs again, which is stated as follows: first, the isolated POIs are ranked in a descent order; then the top one POI is selected as the center of a new cluster and other POIs within the given radius are classified into the new cluster. The above process is repeated until all isolated POIs are classified into a corresponding cluster.

4.2 User Preference

It is very important to evaluate the user's preference on POI recommendations because more researchers find that the influence of the user's preference is greater than the ones of other factors [10,12,13]. The number of a user visiting the same POI can reflect the user's preference on such a POI. Meantime, the spatio-temporal distribution of visiting the given POI and the total number of users who visit the same POI can also reflect the user's preference. Namely, the target user is more likely to have a strong preference on a POI if the user frequently visits the POI and other users seldom visit that POI. According to the strategy mentioned above, Wang and his colleagues evaluate the user's preference by the following equation [5]:

$$Pre(u_k, l_i) = \frac{C_{k,i}}{\sum_{l_j \in L_k} C_{k,j}} \times \frac{d_{k,i}}{d_k} \times \log \frac{\sum_{u \in U_i} C_{u,i}}{|U_i|} \tag{3}$$

where $C_{k,i}$ denotes the number of user u_k visiting POI l_i, L_k denotes the total POIs visited by user u_k, l_j denotes the jth POI in L_k, $d_{k,i}$ denotes the number of days when u_k visited l_i, d_k denotes the number of days when u_k stayed in the target region, U_i denotes all users who have visited l_i, u denotes a given user in U_i, and $|U_i|$ denotes the total number of users who have visited l_i. Equation 3 is also adopted by our method.

4.3 Attraction of a POI

A popular POI is more likely to be chosen for the users who do not visit it. Meantime, users prefer to visit those POIs in their familiar regions. Hence, we need to take the popularity and geographic influence of a POI into consideration. In order to evaluate the geographic influence of a POI on the target user, the distance between the candidate POI and visited one is counted. The geographic influence of the POI is evaluated by the following equation:

$$D(u_k, l_i) = \prod_{l_k \in L_k} e^{-dis(l_i, l_k)} \tag{4}$$

where $dis(l_i, l_k)$ denotes the distance between POIs l_i and l_k, L_k denotes all POIs visited by user u_k, and l_k denotes the kth POI in L_k. If the user has a long trip, the geographic influence of a newly visited POI is obviously small according to Equation 4. Based on the popularity and geographic influence of a POI, the attraction of POI l_i on user u_k is computed by the following equation:

$$Pop(u_k, l_i) = E(l_i) \times \frac{Pre(u_k, l_j)}{e^{dis(l_i, l_j)}} \times D(u_k, l_i) \tag{5}$$

where $l_i = argmin_{l_k \in L_k} dis(l_i, l_k)$, which is the nearest POI visited by u_k in terms of POI l_i.

4.4   Social Recommendation

Friends can also influence the user's choice. A user is more likely to visit POIs visited by his/her friends [1,2]. The social influence used by the friend-based collaborative filtering approach is defined by [9]:

$$R_{u_k,i} = \frac{\sum_{u_j \in U'_k} w_{j,k} \times C_{j,i}}{\sum_{u_j \in U'_k} w_{j,k}} \tag{6}$$

where U'_k denotes the friends of user u_k, u_j denotes the jth friend in U'_k, $C_{j,i}$ denotes the count of user u_j visiting POI l_i, and $w_{j,k}$ denotes the similarity between users u_j and u_k. If two users share more same POIs, they are more similar in terms of their interests. Based on this hypothesis, the interest similarity between users is defined by Adam as follows [6]:

$$w_{j,k} = \frac{|L_j \cap L_k|}{\min(|L_j|, |L_k|)} \tag{7}$$

where $|L_j \cap L_k|$ is the number of POIs visited by both u_k and u_j, $|L_j|$ is the total number of POIs visited by user u_j, and $\min(x,y)$ is the minimum between parameters x and y.

However, it is undoubtedly that the social influence will become weak with the lapse of time. Hence, the forgetting factor needs to be considered. Following the friend-based collaborative filtering approach [9], the social influence is evaluated by:

$$Sr(u_k, l_i) = \frac{\sum_{u_j \in U'_k} w_{j,k} \times \sum_{m=1}^{C_{j,i}} e^{-\triangle t_m}}{\sum_{u_j \in U'_k} w_{j,k}} \tag{8}$$

where $\triangle t_m$ is the number of time intervals between the date when user u_j visited POI l_i for the mth time and current time. Mao and colleagues find that two users are easy to be friends if users live very closely [9]. Hence, friend set U'_k includes the friends of friends. Meantime, user moving trajectories are recorded in the website so that the user can draw on those online records for a tour plan. Hence, friend set U'_k also includes the target user in our method.

4.5   POI Recommendation Model

The POI recommendation model is the linear combination of factors mentioned above. The probability score of user u_k visiting POI l_i is calculated by:

$$R(u_k, l_i) = (1 - \alpha - \beta) \times Pre(u_k, l_i) + \alpha \times Pop(u_k, l_i) + \beta \times Sr(u_k, l_i) \tag{9}$$

where parameters α and β range from 0 to 1, and the sum of α and β is not more than 1. In order to properly tune above parameters, the evaluation results of the user preference (Pre), entropy of a POI (E), attraction of a POI (Pop), and social recommendations from friends (Sr) need to be normalized. Let Pre, E, Pop, Sr be x, the normalization equation is defined as follows [4]:

$$norm(x) = \frac{1}{1 + e^{-x}} \tag{10}$$

5   Experiments

Experimental data are from Gowalla website [1]. There are a total of 6442890 check-in records and 950327 social links generated by 196591 users from February 2009 to October 2010 in Gowalla. The experiment extracts 154803 check-in records generated in Los Angeles by 6550 users. A check-in record is composed of user ID, check-in time, longitude, latitude and location ID. A social link is composed of user ID and friend's user ID. In order to validate the performance of the proposed method, the experiment observes the top N recommended POIs for test users. Test data are composed of each test user's last check-in location in Los Angeles. 429 test users are satisfied with following conditions: 1) they have more than 5 check-in records; 2) they have at least two check-in records for the last trip; 3) the distance between last two check-in locations is not more than 5 kilometers. The precision and recall of the proposed method are observed, and related equations are defined as follows [13]:

$$precision@N = \frac{\sum_{u \in U_t} |Q_u \cap M_u|}{\sum_{u \in U_t} |M_u|} \qquad (11)$$

$$recall@N = \frac{\sum_{u \in U_t} |Q_u \cap M_u|}{\sum_{u \in U_t} |Q_u|} \qquad (12)$$

where Q_u denotes the true check-in location for user u, M_u denotes the top N recommended POIs for user u, and U_t denotes the test users. The parameter N is set as 3, 5, 10, 20, 30 and 50 respectively. When the DBSCAN algorithm is used, the minimum of sample points is set as 15 and the distance is set as 0.5 kilometer. The Los Angeles region is partitioned into 2256 sub-regions according to the distribution of check-in locations.

The experimental data do not record the user's identity so that the users need to be categorized. The number of check-in records and number of days of visiting POIs are counted to categorize users into the local people and newcomers. According to users' check-in patterns, the user category is judged by:

$$LR(u, A) = \frac{ca(u, A)}{ta(u)} \times \log(\frac{cd(u, A)}{10} + 0.1) \qquad (13)$$

where $ca(u, A)$ denotes the number of check-in records of user u in region A, $ta(u)$ denotes the total number of check-in records of user u in Gowalla dataset, and $cd(u, A)$ denotes the number of days when user u stayed in region A. The time span of check-in data is 21 months. Given that a local user is so busy that most of check-in behaviors happened on the weekend, the duration of check-in behaviors should be more than 30 days. Hence, the user is classified into a local person if the target user meets following conditions: 1) stay more than 30 days in the target region; 2) more than half of user's check-in records in Gowalla dataset are generated in the target region.

In order to select optimized parameters, parameters α and β in Equation 9 are assigned different values that are shown in Table 1. According to the performance

Table 1. Parameters tuning for the proposed method

Group ID	α	β
1	0.1	0.3
2	0.3	0.6
3	0.4	0.4
4	0.6	0.3
5	0.5	0.1
6	0.3	0.1

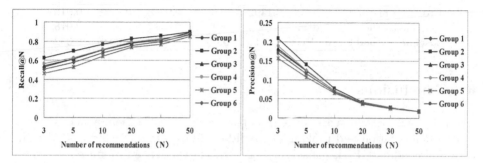

Fig. 3. Performance comparisons when tuning parameters

comparison shown in Figure 3, the parameter α is set as 0.3 and parameter β is set as 0.6.

In order to validate the performance of the proposed method, two baseline methods are chosen. Baseline method 1 only applies Equation 9 to POI recommendations, which does not consider user's travel experience to reduce the range of candidate POIs. Baseline method 2 applies the friend-based collaborative filtering approach, which computes the similarity between the user and his/her friend by the following equation [10]:

$$w_{j,k} = \eta \times \frac{|F_j \cap F_k|}{|F_j \cup F_k|} + (1 - \eta) \times \frac{|L_j \cap L_k|}{|L_j \cup L_k|} \qquad (14)$$

where F_j denotes the friend set of user u_j, L_j denotes the POI set which user u_j has visited and parameter η ranges from 0 to 1. F_j also includes the friends of user u_j's friends and the parameter η is set as 0.5 in our experiment.

As shown in Tables 2 and 3, the proposed method and baseline method 1 almost overlap with respect to the precision and recall. Hence, it is obvious that the proposed method can effectively reduce the range of candidate POIs according to user's travel experience, which is useful for the large scale data. Baseline method 2 computes the probability of a user visiting a POI by identifying the similar users who not only have similar interests but also have similar travel experiences. However, users may not check in each time, although users often habitually visit some familiar POIs. Such a phenomenon strictly affects the performance of baseline method 2, and the proposed method reduces that effect by combining the user preference, the attraction of a POI, and social recommendations from friends.

Table 2. Comparison of recall among three methods

N	3	5	10	20	30	50
Proposed method	0.6224	0.6946	0.7692	0.8298	0.8555	0.8974
Baseline method 1	0.6224	0.6923	0.7692	0.8322	0.8578	0.8998
Baseline method 2	0.0466	0.0559	0.0699	0.0979	0.1142	0.1375

Table 3. Comparison of precision among three methods

N	3	5	10	20	30	50
Proposed method	0.2078	0.1392	0.0772	0.0418	0.0288	0.0182
Baseline method 1	0.2075	0.1385	0.077	0.0417	0.0287	0.0181
Baseline method 2	0.0155	0.0112	0.007	0.0049	0.0038	0.0028

6   Conclusions

The proposed method lays emphasis on the user personalization, which is shown in the identification and ranking of candidate POIs. On one hand, the identification of candidate POIs considers user's travel experience to reduce the search range. On the other hand, the candidate POIs are ranked for meeting user's personalized needs by combining the user preference, attraction of a POI, and social recommendations from friends. Experimental results show that the proposed method is feasible and effective. However, the proposed method does not analyze the user's travel motivation. If the target region is labeled with some tags such as shopping, recreation or social contact, the range of candidate POIs can be further reduced. Identifying the user's travel motivation is our future work.

Acknowledgments. The study was supported by National Natural Science Foundation of China (61420106005) and Beijing Postdoctoral Research Foundation (2014ZZ-31).

References

1. Cho, E., Myers, S.A., Leskovec, J.: Friendship and mobility: user movement in location-based social networks. In: Proceedings of the ACM SIGKDD International Conference on Knowledge Discovery and Data Mining, pp. 1082–1090 (2011)
2. Jin, L., Long, X.L., Zhang, K., Lin, Y.R., Joshi, J.: Characterizing users' check-in activities using their scores in a location-based social network. Multimedia Systems (in press)
3. Li, X.H., Ceikute, V., Jensen, C.S., Tan, K.L.: Effective online group discovery in trajectory databases. IEEE Transcations on Knowledge and Data Engineering **12**(25), 2752–2766 (2013)
4. Li, X.Y.: Research on personal identity recognition method based on multi-biometric. Tianjing university, Tianjing (2010)
5. Ren, K.J.: Information rectrieval and user data mining based on geographic information. Dalian University of Technology, Dalian (2013)

6. Sadilek, A., Kautz, H., Bigham, J.P.: Finding your friends and following them to where you are. In: Proceedings of the 5th ACM International Conference on Web Search and Data Mining, pp. 723–732 (2012)
7. Symeonidis, P., Krinis, A., Manolopoulos, Y.: Geosocialrec: explaining recommendations in location-based social networks. In: Catania, B., Guerrini, G., Pokorný, J. (eds.) ADBIS 2013. LNCS, vol. 8133, pp. 84–97. Springer, Heidelberg (2013)
8. Wang, H., Terrovitis, M., Mamoulis, N.: Location recommendation in location-based social networks using user check-in data. In: Proceedings of the ACM International Symposium on Advances in Geographic Information Systems, pp. 364–373 (2013)
9. Ye, M., Yin, P.F., Lee, W.C.: Location recommendation for location-based social networks. In: Proceedings of the 18th SIGSPATIAL International Conference on Advances in Geographic Information Systems, pp. 458–461 (2010)
10. Ye, M., Yin, P.F., Lee, W.C., Lee, D.L.: Exploiting geographical influence for collaborative point-of-interest recommendation. In: Proceedings of the 34th International ACM SIGIR Conference on Research and Development in Information Retrieval, pp. 325–334 (2011)
11. Ying, J.C., Chen, H.S., Lin, K.W., Lu, E.H.C., Tseng, V.S., Tsai, H.W., Cheng, K.H., Lin, S.C.: Semantic trajectory-based high utility item recommendation system. Expert Systems with Applications 41(10), 4762–4776 (2014)
12. Ying, J.J.C., Lu, E.H.C., Kuo, W.N., Tseng, V.S.: Urban point-of-interest recommendation by mining user check-in behaviors. In: Proceedings of the ACM SIGKDD International Conference on Knowledge Discovery and Data Mining, pp. 63–70 (2012)
13. Ying, J.J.C., Kuo, W.N., Tseng, V.S., Lu, E.H.C.: Mining user check-in behavior with a random walk for urban point of interest recommendations. ACM Transactions on Intelligent Systems and Technology 5(3), 1–26 (2014)
14. Yuan, Q., Cong, G., Ma, Z.Y., Sun, A., Magnenat-Thalamann, N.: Time-aware point-of-interest recommendation. In: Proceedings of the 36th International ACM SIGIR Conference on Research and Development in Information Retrieva, pp. 363–372 (2013)
15. Zhang, J.D., Chow, C.Y., Li, Y.H.: LORE: exploiting sequential influence for location recommendations. In: Proceedings of the 22nd ACM SIGSPATIAL International Conference on Advances in Geographic Information Systems, pp. 103–112 (2014)
16. Zhang, K., Pelechrinis, K.: Understanding spatial homophily the case of peer influence and social selection. In: Proceedings of the 23rd International Conference on World Wide Web, pp. 271–281 (2014)
17. Zheng, Y., Xie, X.: Learning travel recommendations from user-generated GPS traces. ACM Transactions on Intelligent Systems and Technology 2(1), 1–29 (2011)

Investigating Correlation Between Strength of Social Relationship and Interest Similarity

Yan Yu[1](✉) and Lingfei Mo[2]

[1] Computer Science Department, Southeast University Chengxian College,
Nanjing 210088, China
yuyanyuyan2004@126.com
[2] School of Instrument Science and Engineering, Southeast University,
Nanjing 210009, China

Abstract. Recent years have seen exponential growth of microblog which provides users with a new communication and information sharing platform. Some recommendation approaches have been proposed by leveraging the social relationships in microblog based on the principle of homophily to improve the accuracy of recommendation. To prove the feasibility of users social relationships as the bases of recommendation in microblog, we investigate the correlation of strength of social relationship and user interest similarity in microblog by using real-world data set. We observe that strength of social relationship shows strong positive correlation with user interest similarity in microblog. We believe our investigation presents substantial impact for social recommendation research in microblog and will benefit future research in both recommender systems and other related social implications.

Keywords: Mociroblog · Interest similarity · Strength of social relationship

1 Introduction

Microblog, such as Twitter and Google+, has become a popular Internet service. Essentially, microblog enables an easy and lightweight way of communication, which allows people to write short messages and then broadcast and share them through the participating online social networks. The short message can be anything like news, daily activities, and opinions. Microblog has noticeably changed the way of information consumption, which has surely emerged as a mainstream social network medium globally. Users form an explicit social network by following other user in microblog. A user as a follower automatically receives the messages posted by the users he/she follows, known as followee or social friend.

Microblog presents new opportunities to improve the accuracy of recommender systems [1]. In real life, people tend to resort to friends in their social networks for advice before consuming a service. Findings in the fields of sociology and psychology indicate that humans tend to associate and bond with

© Springer International Publishing Switzerland 2015
M.T. Thai et al. (Eds.): CSoNet 2015, LNCS 9197, pp. 172–181, 2015.
DOI: 10.1007/978-3-319-21786-4_15

similar others [2]. Microblog provides novel ways for people to communicate and build virtual communities, which not only makes it easier for users to share their opinions with each other, but also serve as a platform for developing new recommender system algorithms. Many recommendation approaches have been proposed by leveraging the topological structure of formed social networks to improve the accuracy of recommendation [3, 4, 5, 6, 7] in microblog. The basic assumption behind these approaches is based on the principle of homophily [2]. Homophily shows the tendency of individuals to associate and bond with similar others. Individuals in homophilic relationships share common characteristics.

Some researchers have investigated the connections of social relationships and users' interest similarities in online social networks to provide fundamental support to the research of friend-based social recommendation problem in online social networks [8, 9, 10]. Ziegler et al. in [8] investigated the correlation between user's trust and similarity on an online community. Their experiments showed that user's trusted peers are more similar to their sources of trust than arbitrary peers. Lee and Brusilovsky in [9] investigated several features of trust netowrk by using real life data collected from a social web system. They found that users formed by trust network exhibit significantly higher similarity than non-connected users. Ma in [9] conducted several experiments on two friend communities obtained from real world recommender systems to investigate the correlations between social friend relationships and user interest similarities. They observed that social friend relationships cannot reflect user's interest similarity in these communities. They also found that users' similarities with their friends are diverse.

Although previous researches have investigated the correlation between social relationship and user's interest similarity in online social networks, there is still left open question needed to be further explored. In the previous work, all social relationships between users are the same, regardless whether the corresponding users have a stronger connection or a weaker connection. In fact, literature states that there may be stronger and weaker connections between users in online social networks [11, 12, 13]. Strength of social relationship is an important concept in social network analysis. Understanding strength of social relationship might apply it to make better recommendation [11].

Hence, the research question we explore in this paper is: Is there a positive correlation between strength of social relationship and user interest similarity in microblog? We conduct several experiments on a real-world data set to address this question. Our main contribution in this paper is that we observe that strength of social relationship shows strong positive correlation with user interest similarity in microblog.

The remainder of this paper is organized as follows. Section 2 introduces several related work in the literature. Section 3 conducts experiments on a real-world data set to investigate the connection between strength of social relationship and user interest similarity in microblog, followed by the conclusion and future work in Section 4.

2   Related Work

In this section, we review two research directions which are relevant to our work: user interest analysis and strength of social relationship.

2.1   User Interest Similarity Analysis

Ziegler et al. in [8] believed that recommendations based on trust relationship only make sense when trust relation can reflect user's interest similarity to some extent. In order to investigate the correlation between user trust and similarity, they performed two experiments based on data obtained from an online community focusing on books, which allows users to express which other users they trust as well as which books they appreciate. Their experiments showed that user's trusted peers are more similar to their sources of trust than arbitrary peers when the community's trust network is tightly bound to some particular application.

Lee and Brusilovsky in [9] investigated several features of self-defined trust networks by using real life data collected from a social web system to examine how similar users interests are in these networks. They measured users interest similarity by calculating similarity of items and meta-data users share. Their study showed that users formed by a self-defined trust network exhibit significantly higher similarity on items and meta-data than non-connected users. The similarity is highest for directly connected users and decreases with the increase of distance between users. They believed that self-defined trust relationship in social web system can be valuable for improving the accuracy of recommendation algorithm.

Ma in [9] argued that trust relationship is just one kind of social relationships. Many recommenders systems are designed for users to interact with their friends in the real life. Social friendships are quite different from trust relationships in many aspects. Thus, the previous hypotheses on trust relationships may not be held in friend-based recommender systems. They conducted several experiments on two friend communities obtained from real world recommender systems to investigate the correlations between social friend relationships and user interest similarities. They observed that social friend relationships cannot reflect user interest similarities in these communities. Average similarity between a user and his/her friends is even correlated with the average similarity between this user and some other randomly selected users. They also found that a users similarities with his/her friend are diverse in the social friend communities. That meant that some friends of a user are quite similar with this user while the other friends are not similar with this user. Network topology, connected components, and number of co-friends all affect the users interest similarities.

However, as mentioned in Section 1, all social relationships between users are the same in online social networks, regardless whether the corresponding users have a stronger connection or a weaker connection. Understanding strength of social relationship might apply it to make better recommendation [11]. In this paper we aim to analyze the connection between strength of social relationships and user interest similarity in microblog.

2.2 Strength of Social Relationship

In the online social networks, all social relationships between users are the same, regardless whether the corresponding users have a stronger connection or a weaker connection. However, literature states that there may be stronger and weaker relationships between users in online social networks [11, 12, 13]. Strength of social relationship is an important concept in social network analysis. Understanding strength of social relationship might apply it to make better recommendation [11].

Strength of a social relationship is a quantifiable property that characterizes the link between two users. By definition, strength of a social relationship is a combination of amount of time, the emotional intensity, the intimacy and reciprocal services which characterize the tie [12]. Granovetter proposed four dimensions of strength of social relationship: amount of time, intimacy, intensity and reciprocal services [14]. In theory, strength of social relationship has at least seven dimensions and many manifestations. In practice, relatively simple proxies have substituted for them. In different contexts the strength of a social relationship may have different definitions and measures.

Users in microblog differ substantially from other online social networks, such as ones in Facebook or LinkedIn, where social relationships can only be established with the consent of both to-be connected users. In contrast, a social relationship in microblog is asymmetric. In other words, a user can follow a followee without the followees consent. The asymmetry of social ties in microblog has made microblog social networks called hybrid networks [15, 16, 17]. Kwak et al. [18] found that 77.9% of users' social relationships are not reciprocal in Twitter. Reciprocity is a source of social cohesion [19, 20, 21]. When two individuals attend to one another, the bond is reinforced in each direction and both people will find the tie rewardings [22].

In microblog, besides following other users, a user can also interact with other users in microblog. If a user wants to notify another user about his/her message, he/she would use an '@' to notify the other user. A user can repost a message and append some comments to share it with his/her followers. A user can add some comments to a message. The number of interaction between two users is often considered the strength of their relationship. Researcher has found that pairs of individuals with strong tie exhibit greater similarity than those with weak ties in online social networks [19]. In microblog, fantastic fans may repost stars many more times than other persons they follow.

3 Experiment Analysis

In this section, we conduct experiments to investigate the correlation between strength of social relationship and interest similarity in microbblog. More specifically, the first experiment we conduct is to explore whether the reciprocal social relationship indicates positive connection with user interest similarity in microblog. The second experiment we conduct is to evaluate how the number of interaction between a pair of users in social relationship can affect the interest similarity between these two users in microblog.

We first describe the data set used in this section. We then define interest similarity metric for evaluation. Last, we give detailed experimental analysis.

3.1   Data Set Description

The data set we use in this paper is from Tencent Weibo, which is a Chinese microblog website launched by Tencent in April, 2010. Tencent Weibo has become one of leading microblog platforms in China. Similar to Twitter, a user in Tencent Weibo can broadcast a short message and follows other users on the website. Besides following other users, a user can also interact with other users in Tencent Weibo. A user can @, repost and comment on other users. The data set we use in this paper is from KDD Cup 2012 Track 1, which is a prediction task that involves predicting whether or not a user will follow a recommended user. The data set is a sampled snapshot of Tencent Weibo, including user profiles, social graph, interaction, and so on.

In this paper, we only use social graph and interaction for our analysis. The data set includes 2,320,895 users, 50,655,143 social relationships, 899,899 @ interactions, 8,790,544 repost interactions and 2,179,510 comment interactions between users in certain number of recent days. Without loss of generality while keeping our test meaningful and manageable, we randomly sample 15,000 users from the KDD Cup 2012 Track 1 data set as our target users to investigate their interest similarities with their social friends. In order to reduce noises, we require that each target user needs to have at least five claimed social relationships.

3.2   Definition of Interest Similarity

In microblog, users build following relationships not only for communicating with their friends or acquaintances in real life but also for seeking information they interest [16, 17]. So the following relationship can reflect user's interest in the microblog. Interest similarity we explore in this paper is based on the homophily principle of shared interest [15]. In microblog, shared interests can be represented by u→k← v, where user u and user v both follow user k. User u and v sharing interests is surely one kind of similarity in microblog. In this paper, we use Jaccard Similarity Coefficient to define the interest similarity between two users u and v based on the followees they follow in common [7].

Mathematically, we can construct a directed graph G (V, E), where V represents a set of users in microblog and E represents a set of social relationships among these users. A directed edge $<u, v> \in E$ exists between user u and v if u follows v. The set of out-neighbors of user u is $\Gamma_+(u) = \{v \in V | <u, v> \in E\}$, and the out-degree of u is $| \Gamma_+(u) |$, where $| \cdot |$ denotes the size of the set. Thus, the interest similarity $s_{u,v}$ between user u and user v is defined as:

$$s_{u,v} = \frac{\mid \Gamma_+(u) \cap \Gamma_+(v) \mid}{\mid \Gamma_+(u) \cup \Gamma_+(v) \mid} \tag{1}$$

3.3   Reciprocal Social Relationship

Unlike some online social networks, a followed user has the option but not the requirement to similarly follow back. Thus, social relationships in microblog may be asymmetric. We distinguish two kinds of following relationships between two users in microblog: unidirectional and reciprocal. If user u follows user v, but v does not follow u, we call the social relationship between u and v as unidirectional and we call v is u's unidirectional social friends. If user u and user v both follow each other, we consider them reciprocal friends and we call v is u's reciprocal friend. Reciprocity is a source of social cohesion. When two individuals attend to one another, the bond is reinforced in each direction and both people will find the tie rewardings [22]. Hence, the first experiment we conduct is to compare the differences of interest similarities between unidirectional and reciprocal social relationship. We define uni_s_u as the average interest similarity between user u and his/her unidirection social friends, $reci\_s_u$ the average interest similarity between u and his/her reciprocal social friends:

$$uni\_s_u = \frac{\sum_{v \in Uni(u)} s_{u,v}}{\mid Uni(u) \mid} \tag{2}$$

where Uni(u) represents the list of unidirectional social friends of user u follows.

$$reci\_s_u = \frac{\sum_{v \in Reci(u)} s_{u,v}}{\mid Reci(u) \mid} \tag{3}$$

where Reci(u) represents the list of reciprocal social friends of user u.

For any target user u who has reciprocal social relations, we calculate the value of uni_s_u and $reci\_s_u$ and compared both values. Figure 1 plots the correlations between average unidirectional social friends interest similarity and reciprocal social friends interest similarity on data set. Every data point in Figure 1 represents a target user, where the x-axis indicates average unidirectional social friends interest similarities and y-axis represents the related average reciprocal social friends interest similarities. We observe that the plot in Figure 1 exhibits strong biases towards the upper-left region, which indicates that reciprocal social relationship and user interest similarity has high correlation in microblog. Users involved in a reciprocal social relationship shows significantly larger similarity than users involved in a unidirectional social relationship.

We further compare the correlation between interest similarity in social and in non-social friend relationship. We define s_u as average interest similarity between user u and his/her social friends, n_s_u as average interest similarity of user u with his/her non-social friends:

$$s_u = \frac{\sum_{v \in \Gamma_+(u)} s_{u,v}}{\mid \Gamma_+(u) \mid} \tag{4}$$

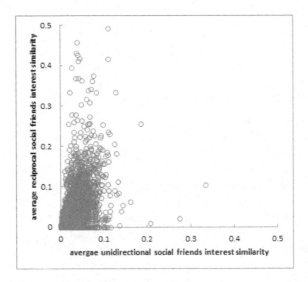

Fig. 1. The correlation between unidirectional and reciprocal social friend interest similarity

$$n\_s_u = \frac{\sum_{v \in R(u)} s_{u,v}}{|R(u)|} \tag{5}$$

where R(u) indicates the list of randomly selected users whom user u does not follow. R(u) has the same size with $\Gamma_+(u)$, and R(u)∩Γ_+(u)=ϕ.

For each target user u, we calculate s_u and n_s_u, and compare these two average interest similarities for each target user in great detail. Figure 2 plots the correlations between average social friends interest similarities and non-social friends interest similarities of target users on date set. In Figure 2, each point represents a target user, where the x-axis indicates average social friends interest similarities and the y-axis specifies the related average non-social friends interest similarities. We find that the plot in Figure 2 shows biases towards lower-right region, which indicates that social relations has correlation with user interest similarity in microblog However, we also observe that the biases towards the lower-right region are not strong, which shows that the connection between social relationship and user interest similarity is not strong in microblog. Users tend to look for interesting information by following other uses in microblog [9]. So social relationship has correlation with user interest similarity. However, the correlation is not strong. We can understand this from three aspects. First of all, users trend to follow celebrities in microblog, because these celebrities are well known to people and they are more likely to be reliable and stable information sources [20]. Consequently, even if two users don't have social relationship, they may both follow the same celebrity. Secondly, users play a dual role in microblog as they are both information sources and seekers [10]. What a user follows may not wholly be similar with what he/she is followed. Lastly, there exist some spammers in microblog. One of the most common ways for spammers to gain

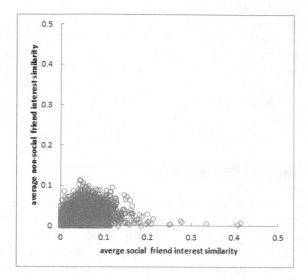

Fig. 2. The connections between social and non-social friends interest similarity

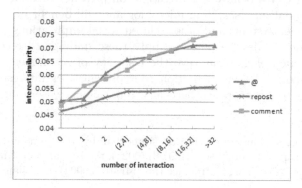

Fig. 3. Interest similarity conditioned on the number of interaction

popularity and get more spam targets is to follow a huge number of users and wait for them to follow back [21]. Hence, we cannot observe strong correlations between social following relationships and user interest similarity.

3.4   Number of Interaction

In this subsection, we perform the second experiment to analyze the correlation between the number of interaction of a pair of users in social relationship and their interest similarity in microblog. More specifically, the peer similarity $s_{u,v}$ is measured according to Equation 1 for social relationship $<u,v>$ between user u and user v in our data set. Then, the number of @, repost and comment from user u to v is counted respectively. Figure 3 shows the result.

The number of interaction is divided into eight groups in the x-axis of Figure 3, in which "(4,8]" indicates the number of interaction is greater than 4 but less or equal to 8. We observed from Figure 3 that the user interest similarity increases as their number of interaction increases. One interpretation for this observation is that many interactions from a user may indicate the strong interest in the message posted by his/her social friend. The number of interaction between two users is often considered the strength of their relationship. Researcher has found that pairs of individuals with strong tie exhibit greater similarity than those with weak ties in online social networks [19]. In microblog, fantastic fans may repost stars many more times than other persons they follow.

4   Conclusion and Future Work

To prove the feasibility of users social relations as the bases of recommendation in microblog, we examined the correlations of strength of social relation and user interest similarity in microblog. Using Tencent Weibo data set, we find strong positive connection between reciprocal social relationship and user interest similarity in microblog. As to users interaction, we find that the number of interaction between two users in social relation indicates the degree of user interest similarity. We believe our investigation presents substantial impact for social recommendation research in microblog and will benefit future research in both recommender systems and other related social implications

In our future studies we plan to apply these findings to recommendation application in microblog to improve the accuracy of recommendation. At the same time, a more concrete definition to represent user interest similarity is needed. We plan to incorporate other features to calculate user interest similarity more accurately.

Acknowledgments. This research was supported by the Fundamental Research Funds for the Central Universities (2242014R30021), the Research on University's Natural Science Projects of Jiangsu Province (no. 14KJB520004), and Jiangsu's Qing Lan Project (11038).

References

1. Yang, X., Guo, Y., Liu, Y., Steck, H.: A Survey of Collaborative Filtering Based Social Recommender Systems. Computer Communications **41**, 1–10 (2014)
2. McPherson, M., Smith-Lovin, L., Cook, J.M.: Birds of a Feather: Homophily in Social Networks. Annual Review of Sociology **27**, 415–444 (2001)
3. Lin, J., Sugiyama, K., Kan, M.Y., Chua, T.S.: Addressing cold-start in app recommen-dation: latent user models constructed from twitter followers. In: Procedings of the 36th International ACM SIGIR Conference on Research and Development in Information Retrieval, pp. 283–292. ACM Press, New York (2013)
4. Ting, I.H., Chang, P.S., Wang, S.L.: Understanding Microblog Users for Social Recom-mendation Based on Social Networks Analysis. Journal of Universal Computer Science **18**, 554–576 (2012)
5. Yamaguchi, Y., Amagasa, T., Kitagawa, H.: Recommending fresh URLs using twitter list. In: Proceedings of the Seventh International AAAI Conference on Weblogs and Social Media, pp. 733–736. AAAI Publication, New York (2013)

6. Yu, Y., Wang, X.: Link Prediction in Directed Network and Its Application in Microblog. Mathematical Problems in Engineering **2014**, 1–9 (2014)
7. Yu, Y., Qiu, R.G.: Followee Recommendation in Microblog Using Matrix Factorization Model with Structural Regularization. The Scientific World Journal **2014**, 1–11 (2014)
8. Ziegler, C.N., Golbeck, J.: Investigating Interactions of Trust and Interest Similarity. Decision Support System **43**, 460–475 (2007)
9. Lee, D., Brusilovsky, P.: Does trust influence information similarity? In: Proceedings of the ACM RecSys09 Workshop on Recommender System and the Social Web, pp. 71–74. ACM Press, New York (2009)
10. Ma, H.: On measuring social friend interest similarities in recommender system. In: Proceedings of the 37th International ACM SIGIR Conference on Research and Develop-ment in information Retrieval, pp. 465–474. ACM Press, New York (2014)
11. Gilbert, E., Karahalios, K.: Predicting tie strength with social media. In: Proceedings of the 27th International Conference on Human Factors in Computing Systems, pp. 211–220. ACM Press, New York (2009)
12. Petróczi, A., Nepusz, T., Bazsó, F.: Measuring Tie-strength in Virtual Social Networks. Connections **27**, 39–52 (2006)
13. Xiang, R., Neville, J., Rogati, M.: Modeling relationship strength in online social net-works. In: Proceedings of the 19th International Conference on World Wide Web, pp. 981–990. ACM Press, New York (2010)
14. Granovetter, M.S.: The Strength of Weak Ties. The American Journal of Sociology **78**, 1360–1380 (1973)
15. Yin, D., Hong, L., Davison, B.D.: Structural link analysis and prediction in microblogs. In: Proceedings of the 20th ACM International Conference on Information and Knowledge Management, pp. 1163–1168. ACM Press, New York (2011)
16. Java, A., Song, X., Finin, T., Tseng, B.: Why we twitter: understanding microblogging usage and communities. In: Proceedings of the 9th WebKDD and 1st SNA-KDD 2007 Workshop on Web Mining and Social Network Analysis, pp. 56–65. ACM Press, New York (2007)
17. Krishnamurthy, B., Gill, P., Arlitt, M.: A few chirps about twitter. In: Proceedings of the 1st Workshop on Online Socail Networks, pp. 19–24. ACM Press, New York (2008)
18. Kwak, H., Lee, C., Park, H., Moon, S.: What is twitter, a social network or a news media? In: Proceedings of the 19th International Conference on World Wide Web, pp. 591–600. ACM Press, New York (2010)
19. Wilson, C., Boe, B., Sala, A., Puttaswamy, K.P., Zhao, B.Y.: User interactions in social networks and their implications. In: Proceedings of the 4th ACM European Conference on Computer Systems, pp. 205–218. ACM Press, New York (2009)
20. Wilson, M., Nicholas, C.: Topological analysis of an online social network for older adults. In: Proceeding of the 2008 ACM Workshop on Search in Social Media, pp. 51–58. ACM Press, New York (2008)
21. Zhou, Y., Chen, K., Song, L., Yang, X., He, J.: Feature analysis of spammers in social networks with active honeypots: a case study of chinese microblogging networks. In: 2012 IEEE/ACM International Conference on Advances in Social Networks Analysis and Mining, pp. 728–729. IEEE Computer Society, Washington D.C. (2012)
22. Golder, S.A., Yardi, S.: Structural predictors of tie formation in twitter transitivity and mutuality. In: 2010 IEEE Second International Conference on Social Computing, pp. 88–95. IEEE Press, New York (2010)

Biased Respondent Group Selection Under Limited Budget for Minority Opinion Survey

Donghyun Kim[1], Wei Wang[2], Matthew Tetteh[1], Jun Liang[4], Soyoon Park[3], and Wonjun Lee[3(✉)]

[1] Department of Mathematics and Physics, North Carolina Central University, Durham, NC, USA
donghyun.kim@nccu.edu, mtetteh@eagles.nccu.edu
[2] School of Mathematics and Statistics, Xi'an Jiaotong University, Xi'an, China
wang_weiw@163.com
[3] Department of Computer Science, University of Texas at Dallas, Richardson, TX, USA
{sogehao,wlee}@korea.ac.kr
[4] Department of Computer Science and Engineering, Korea University, Seoul, South Korea
jxl130131@utdallas.edu

Abstract. This paper discusses a new approach to use the information from a special social network with high homophily to select a survey respondent group under a limited budget such that the result of the survey is biased to the minority opinions. This approach has a wide range of potential applications, e.g. collecting complaints from the customers of a new product while most of them are satisfied. We formally define the problem of computing such group with better utilization as the p-biased-representative selection problem (p-BRSP). This problem has two separate objectives and is difficult to deal with. Thus, we also propose a new unified-objective which is a function of the two optimization objectives. Most importantly, we introduce two polynomial time heuristic algorithms for the problem, where each of which has an approximation ratio with respect to each of the objectives.

1 Introduction

Recently, the value of the information from online resources such as online social networks are getting more recognized and thus lots of research efforts are made to maximize its utilization [1–3]. Following the trend, online survey is also being recognized as a critical tool to make a wide range of significant marketing and political decisions. Due to the reason, a huge amount of investment is being made for various researches on online survey [4]. There are several motives that promote online survey [5]. Most of all, it costs much less and produces results

This work was supported in part by US National Science Foundation (NSF) CREST No. HRD-1345219. This research was jointly supported by National Natural Science Foundation of China under grants 11471005.

much faster than its counterpart. How to select a meaningful survey respondent group has been a tough but critical question to deal with to make a traditional off-line survey method more effective and reliable, and this is still true for online survey. Here, the definition of "meaningful" can differ based on the purpose of the survey. In most cases, a survey aims to learn the general opinions from the public of interest by sampling, and thus it is significant to elect a group of unbiased respondents among the public using a proper method, e.g. random sampling.

Previously, Kim et. al. [6] introduced a new strategy to elect a survey respondent group to perform efficient biased survey and collect more minority opinions with the assistance from an artificial social network graph constructed by them. They argued that in opposition to the widely accepted belief, biased survey could be useful, and discussed an example to support their claim. In the example, they pointed out that when the majority of the people, who purchased a new smartphone, are very satisfied, the opinions of unsatisfied users of the new smartphone, who can be classified as minority opinion holders, could provide useful information to the new smartphone's product quality manager, who is more interested in complaints. Based on this observation, Kim et. al. introduced a new strategy to select a respondent group more suitable for such survey in the sense that from which the diversified voices from unsatisfied users (minority opinions) can be heard more loudly. Most importantly, their strategy only requires the expected similarity of the opinions between each pair of users on the issue to construct the biased respondent group. Kim et. al.'s approach is rather localized and consumes less resources, and thus is more practical in big data environment compared to its alternative straightforward approach which analyzes the sentiment of each individual on the issue first, and then identifies the minority opinions as this certainly requires global analysis.

To achieve the goal, they first compute a new social network graph G in which each node represents a member of the society, and there is an edge between two nodes only if the opinions of the two people, who are represented by the nodes, are similar enough on the subject of interest. In the literature, such social network graph, in which two nodes are neighboring only if they are sharing close opinion, is told to have high level of homophily [7]. Once such a graph is constructed, the algorithm attempts to compute a smaller size inverse k-core dominating set D of G, which is a subset of the nodes in $G = (V, E)$ such that all nodes in V is either in D or neighboring to a node in D (domination property), and the degree of each node in D in the induced graph by D in G is at most k (inverse k-core property). Note that the dominating property on D is necessary to ensure that D has a representation over all of the members in the society, and the inverse k-core property is enforced to make sure to obtain more minority opinions with greater diversity. Most of all, in the simulation, the authors have shown that their approach is in fact effective.

Meanwhile, online survey may not be completely free-of-cost, even though it is usually much cheaper than the traditional off-line survey approaches. For instance, a recent study conducted by Singer and Ye shows that a reasonable compensation can certainly improve the response rate of the survey [8]. However, we notice the approach proposed by Kim et. al. does not provide any explicit way to control the cost of their approach (the size of the group returned by their algorithm), and this can be a very critical issue in order to make the approach more practical. Motivated by this observation, in this paper, we introduce two new approaches to perform effective biased survey using homophily rich graph under limited budget. The main contributions of this paper can be summarized as follows.

(a) For the first time in the literature, we discuss the motivation for selecting a survey respondent group to better capture more diversified minority opinions under limited budget. We formulate the problem of our interest as a new optimization problem two independent objectives. We also discuss how the two objectives can be combined into one objective function.

(b) We introduce two polynomial time algorithms to solve the proposed problem. The first algorithm is based on Kim et. al. [6]'s approximation algorithm for the minimum inverse k-core dominating set problem, and can be considered as its generalization. This algorithm has a proven approximation ratio with respect to Objective 1. The second one is a simpler greedy algorithm and has the best possible approximation ratio with respect to Objective 2.

The rest of this paper is organized as follows. Section 2 discusses some preliminaries. We introduce the two new approaches for the problem of our interested in Section 3. Section 4 concludes this paper and presents future works.

2   Preliminaries

2.1   Notations and Definitions

In this paper, $G = (V, E)$ represents a social network graph with a node set $V = V(G)$ and an edge set $E = E(G)$. We assume the relationship between each pair of members is symmetric, which is true in homophily high social network graph, and thus the edges in E are bidirectional. Also, we use n to denote the number of nodes in V, i.e. $n = |V|$. For any subset $D \subseteq V$, $G[D]$ is a subgraph of G induced by D. For a pair of nodes $u, v \in V(G)$, $Hopdist(u, v)$ is the hop distance between u and v over the shortest path between them in G. Given a node v in G, $deg(v, G)$ is the degree of v in G. For any $V' \subseteq V$ in $G = (V, E)$, $deg(V', G)$ is $\max_{v \in V'} deg(v, G)$. Also, $deg(G)$ is $\max_{v \in V} deg(v, G)$. $dia(G)$ is the diameter of G, which is the length of the longest shortest path between any pair of nodes in the graph G. For each node $v \in V$, $N_{v,V}(G)$ is the set of nodes in V neighboring to v in G. In other words, the nodes in $N_{v,V}(G)$ are the 1-hop neighbors of v in G. Similarly, $N_{v,V}^d(G)$ is the set of nodes in V, which are at most d-hops far from v in G. Note that we will use $N_{v,V}(G)$ and $N_{v,V}^1(G)$ interchangeably. Given a graph G, a subset $D \subseteq V$ is a *dominating set (DS)* of G if for each node $u \in V \setminus D$, $\exists v \in D$ such that $(v, u) \in E$. In general, a subset $D \subseteq V$ is a *d-hop dominating set (d-DS)* of G if for each node $u \in V \setminus D$,

$\exists v \in D$ such that $Hopdist(v, u) \leq d$. In graph theory, the *minimum dominating set problem (MDSP)* is to find a minimum size DS in a given G. Also, the goal of the *minimum d-hop dominating set problem (MdDSP)* is to find a minimum size d-DS in G. Given a graph G, a subset $D \subseteq V$, and a positive integer k such that $0 \leq k \leq \Delta$, where Δ is the degree of G, D is an *inverse k-core* in G if for each $v \in D$, $|N_{v,D}(G)| \leq k$. Generally speaking, D is an *inverse (k,d)-core* in G if for each $v \in D$, $|N_{v,D}^d(G)| \leq k$. Given $\langle G, k \rangle$, the minimum inverse k-core dominating set problem is to find a minimum size inverse k-core dominating set of G. Similarly, given $\langle G, k, d \rangle$, the minimum inverse (k, d)-core dominating set problem is to find a minimum size inverse k-core d-hop dominating set of G.

2.2   Formal Definition of Problem

In this paper, we are interested in selecting a survey respondent group whose size is p, which is a positive constant determined by the available budget, such that (a) more members with minority opinions are selected (biased to minority opinions), and (b) the group can well-represent the overall minority opinions (well-representation of diversified minority opinion). In the following, we explain the desirable properties of the group to be elected for our purpose and their implications in terms of graph theory.

Property 1: Higher Bias to Minority Opinion Holders. Previously, Kim et. al. [6] introduced a way to construct a homophily high social network graph, in which there exists an edge between two nodes only if the opinions of the members represented by the two nodes are similar enough. They also found that in a homophily high social network graph, a node with lower node degree tends to be a minority opinion holder. This implies that a group with size p possibly includes more minority opinion holders (and thus the group is more biased) when the average degree of the selected nodes (or their total node degree) in the given social network graph is lower. In this paper, we will assume a homophily high social network graph G as an input of our algorithms and thus, prefer to have a node subset V' with size p such that $\sum_{v \in V'} deg(v, G)$ becomes as small as possible as an output of our algorithm.

Property 2: Better Representation of Minority Opinion Holders. In a homophily high social network graph G, a pair of nodes are connected in G only if their expected opinions on the subject of interest are similar enough. In the literature, the minimum size dominating set problem is widely used to select an efficient representative group. For instance, in [6], Kim et. al. were looking for a minimum size dominating set of G with certain properties to elect a group of survey respondents which can represent the rest.

Unfortunately, there are two issues to extend this approach to the problem of our interest directly. First, depending on the input graph G, a dominating set (or 1-hop dominating set) with the enforced size constraint p may not exist. Second, we may ignore the majority opinion holders in the process of selecting the representatives for minority opinion holders in contrast to the fact that a dominating set implicitly does not ignore them.

One way to address the first concern is to relax the 1-hop domination constraint and allow a representative of a node to be multiple hops far from the node. In this way, the size of the dominating set can be reduced. However, as the hop distance between two nodes in G generally implies the degree of difference on the opinions between them, and thus as the hop distance grows, the effect of the representation becomes smaller. Consequently, it would be more desirable to find a subset of nodes, V' from V with size p such that the maximum hop distance from a minority opinion holder to its nearest node in V' becomes minimized. To address the second concern, the concept of nodes with "minority opinion holders" should be more clearly defined. Based on [6], a node with lower degree has a better chance to be a minority opinion holder. Therefore, we may attempt to identify those minority opinion holders by computing the degree of each node in G (by following Property 1) and consider those nodes with smaller node degree as minority opinion holders, where the concept of "smaller" is dependent on the context and can be specified by the survey organizer.

Property 3: Greater Diversification of Minority Opinion Holders. In practice, there can be a number of different minority opinions, and thus minority opinions are quite diversified. Therefore, it is important to construct a size p representative group in a way that more diversified minority opinions can be collected. Now, we have the following remark.

Remark 1. We argue that Property 3 is already included in Property 2. For instance, given a connected graph G with one huge complete subgraph (majority opinion holders) and two non-adjacent smaller size complete subgraphs (minorities), if we select two representatives from the same smaller size complete subgraph, the total hop distance discussed in Property 2 will be greater compared to the case in which one representative is selected from each smaller size complete subgraph.

Based on Properties 1, 2, 3, and Remark 1, we formally define our problem of interest.

Definition 1 (p-BRSP). *Given a homophily high social network graph $G = (V, E)$, a subset $S \subset V$, which is the group of nodes in V whose node degree is no greater than a threshold level (and therefore are suspected as nodes representing minority opinion holders), a positive integer $p \leq |V| = n$, the p-biased-representative selection problem (p-BRSP) is to find a subset $V' \subseteq V$ whose size is p from G such that*

(a) **Objective 1:** *the total node degree of V' is minimum, or equivalently*
$$\sum_{v \in V'} deg(v, G) \text{ is minimum, and}$$

(b) **Objective 2:** *the maximum hop distance between a node in V to its nearest node in V', or equivalently* $\max_{u \in S \setminus V'} \arg \min_{v \in V'} Hopdist(u, v)$ *is minimum.*

Algorithm 1. Greedy-MI(k,d)CDSA (G,S,p)

1: **for** $d = 1$ to $dia(G)$ **do**
2:   **for** $k = 0$ to $deg(G)$ **do**
3:     Prepare an empty set $D_{(d,k)}$, i.e. $D_{(d,k)} \leftarrow \emptyset$.
4:     For each $v_i \in V$, prepare a counter n_i which is initialized to 0, i.e. $n_i \leftarrow 0$.
5:     Suppose $X_j = \{v_i | v_i \in V \text{ and } n_i = j\}$.
6:     **while** $X_0 \cap S \neq \emptyset$ **do**
7:       Find $v_i \in V \setminus \left((\bigcup_{j \geq k} X_j) \cup D_{(d,k)} \right)$ so that $|N^w_{v_i, X_0 \cap S}(G)|$ is maximized,
      where $1 \leq w \leq d$. A tie can be broken by selecting a node with smaller node degree.
8:       Set $D_{(d,k)} \leftarrow D_{(d,k)} \cup \{v_i\}$.
9:       **for** each node $v_j \in N^w_{v_i, V}(G)$ with any $1 \leq w \leq d$ **do**
10:         $n_j \leftarrow n_j + 1$.
11:       **end for**
12:     **end while**
13:   **end for**
14: **end for**
15: Output the $D_{(d,k)}$ whose size is p and which minimizes the objective function in Eq. (1).

Meanwhile, it is uncertain that which of the requirements is more significant. As a result, we redefine the problem such that its objective is to minimize

$$\alpha \times \frac{\sum_{v \in V'} deg(v, G)}{\sum_{v \in W} deg(v, G)} + (1 - \alpha) \times \frac{\max_{S \in V \setminus V'} \arg\min_{v \in V'} Hopdist(u, v)}{dia(G)}, \quad (1)$$

from some $0 \leq \alpha \leq 1$, which is determined by the operator of the survey, where W is the set of the first p nodes in G with largest node degree. During the rest of paper, we discuss how to quality solutions of p-BRSP with the objective function in Eq. (1).

3   Two Polynomial Time Algorithms for p-BRSP

In this paper, we introduce two new heuristic algorithms for p-BRSP along with some interesting theoretical analysis.

3.1   First Approach: Greedy-MI(k,d)CDSA

Previously, Kim et. al. introduced Greedy-MIkCDSA, a simple greedy strategy for the minimum inverse k core dominating set problem (MIkCDSP). Given an MIkCDSP instance $\langle G, k \rangle$, Greedy-MIkCDSA first prepares an empty set D, which will eventually include the output, an inverse k-core dominating set (IkCDS) of G. For each node $v_i \in V$, the algorithm creates a counter n_i which is initialized to 0. The counter will be used to track the number of neighbors of v_i in D. Depending on the counter, the algorithm creates a partition of the nodes in V, $X_0, X_1, \cdots$, where X_j is the subset of nodes in V whose counter is j. This means that initially X_0 is equal to V and each of the rest is empty. Clearly, the number of the subsets is bounded by n. Then, the algorithm iteratively picks a node v_i from $V \setminus \left((\bigcup_{j \geq k} X_j) \cup D \cup Q \right)$, i.e. v_i is a node which is

(a) with a counter n_i whose value is less than k (i.e. has less than k neighbors in DS),
(b) not selected as a DS node yet, and
(c) without any neighboring node w_l which is in D (otherwise v_i is already dominated) and, at the same time, in X_j for some $j \geq k$ (otherwise adding v_i to D will violate the k-inverse-core property),

such that the number of neighbors of v_i in X_0 is the maximum, where $Q = \{w_1, \cdots, w_q\}$ such that $w_l \in Q$ has at least one neighbor in $(\bigcup_{j \geq k} X_j)$. Any tie can be broken arbitrarily. The algorithm eventually terminates when all nodes in V is either in D or dominated by some node in D while maintaining $G[D]$ as an inverse k-core.

The main idea of Greedy-MIkCDSA is still applicable to p-BRSP. However, due to the size constraint p, we need to employ d-hop dominating set instead of 1-hop dominating set. To extend the main idea of Greedy-MIkCDSA which utilizes 1-hop dominating set to d-hop dominating set, there are several challenges to deal with at the same time. That is, we must find a subset V' of nodes with size p such that V' can d-hop dominates all nodes in S. At the same time, we need to adjust k properly, otherwise, there might be no feasible solution. Note that there can be more than one $\langle V', d, k \rangle$ computed in this way, and we need to find the one which can minimize the objective of p-BRSP in Eq. (1).

Algorithm 1 is the formal definition of this modified algorithm, namely Greedy-MI(k, d)CDSA. The core idea of our modification is that we vary d and k (Line 1 and Line 2 of Algorithm 1) and compute an inverse k-core d dominating set whose size is p. From Line 3 to Line 12, the strategy of Kim et. al's algorithm for the minimum inverse k-core dominating set problem is used to compute a smaller size inverse (k, d) core dominating set $D_{(d,k)}$ of G in a greedy manner. Note that $D_{(d,k)}$ intends to d-hop dominate the nodes in S which is a subset of V rather than the whole nodes in V. In Line 15, the algorithm returns the best $D_{(d,k)}$, which minimizes the objective function in Eq. (1) among all feasible ones.

Theorem 1. *Algorithm 1 is correct.*

Proof. This algorithm always returns a feasible solution and therefore correct as with $d = dia(G)$ and $k = deg(G)$, we can always find a single node in G which is an inverse k-core d dominating set of G.

Note that the potential function of Algorithm 1 is submodular and thus the algorithm has an approximation factor of $O(\log \delta)$ with respect to Objective 1, where *delta* is the maximum degree of the input graph. The proof of this claim is very similar to the proof of Theorem 2 in [10]. The only difference is that now we are considering d-hop domination instead of 1-hop the ratio becomes $O(\log \delta^d) = O(d \log \delta) = O(\log \delta)$.

3.2   Second Approach: Simple-p-RSPA

Now, we propose a simpler greedy algorithm for p-BRSP. Before discussing our new strategy, we first introduce a related problem, namely the p-center problem with degree constraint (p-CDC), and propose a 2-approximation algorithm for it, where 2 is the best possible. Then, this algorithm is used to design our second strategy, the simple-p-RSP algorithm (Simple-p-RSPA).

Definition 2 (p-CDC). *Given a graph $G = (V, E)$, a positive integer p, a subset $S \subset V$ representing the group of people with minority opinion, and a degree constraint W, the p-center problem with degree constraint (p-CDC) is to find a subset of nodes D satisfying (a) $|D| \leq p$, and (b) $\sum_{v \in D} deg(v, G) \leq W$ such that the furthest distance from a node in S to its nearest node in D becomes minimum.*

Now, we present a 2-approximation algorithm for the p-CDC problem, which is best possible unless $P = NP$. The main idea of the algorithm and corresponding analysis are motivated by Hochbaum's algorithm for the p-center problem [9]. However, p-CDC has a additional degree-sum constraints, and the objective function is also different from that of p-center (note we try to minimize the maximum hop-distance of a node in S (instead of V) to its nearest center in D). So the method in [9] cannot be applied directly here.

Let $\Gamma = (V, E')$ be a complete weighted graph constructed from G, in which the edge weight, $cost(e)$ of $e = (u, v)$, is the length of shortest-path between nodes u and v, and the node weight of v is the degree of v in G for every $v \in V$. Now, we order the edges of Γ in the following way: $cost(e_1) \leq cost(e_2) \leq \cdots \leq cost(e_m)$; where $m = \binom{n}{2}$ is the number of edges in the complete graph Γ.

Let $G_i = (V, E_i)$ with $E_i = \{e_1, e_2, \cdots e_i\} = \{e \mid cost(e) \leq cost(e_i)\}$. Let H_i be the subgraph of G_i induced by the 1-hop neighbors of S, together with S, i.e., $H_i = G_i[\cup_{v \in S} N_{v,V}(G_i) \cup S]$. Let H_i^2 be the square of H_i, i.e., H_i^2 is a graph obtained from H_i such that two nodes are adjacent in H_i^2 if and only if the hop-distance between them is no more than two in H_i. The algorithm is as follows:

(a) Step 1. Compute $H_1^2, H_2^2, \cdots, H_m^2$ and $(G_1[S])^2, (G_2[S])^2, \cdots, (G_m[S])^2$.

(b) Step 2. For each $i = 1, 2, \cdots, m$, compute a Maximal Independent Set (MIS) M_i of $(G_i[S])^2$ (which is also an independent set of H_i^2) as follows: At each time, choose a node $v \in S$ with the lightest node weight, then remove all the neighbors of v together with v in H_i^2, i.e., $N_{v,V}(H_i^2)$, from $(G_i[S])^2$; in the remaining graph, repeat the same process until there is no node left in $G_i^2[S]$.

(c) Step 3. Choose the smallest index i (say j) such that $|M_i| \leq p$ and $w(M_i) = \sum_{v \in M_i} deg(v) \leq W$.

(d) Step 4. Output $S = M_j$ as the centers.

Now, we show the algorithm describe above is a 2-approximation for p-CDC.

Lemma 1. M_i *dominates* S *in graph* $(G_i[S])^2$ *for every* i.

Proof. Note M_i is a maximal independent set of $(G_i[S])^2$, it is also a dominating set of $(G_i[S])^2$. Since if there is one node, say v in S, which is not dominated by M_i, then $M_i \cup \{v\}$ is also an independent set; contradicts the fact that M_i is a maximal indecent set.

Lemma 2. *Let* D_i^* *be a subset of* V *with minimum size which dominates* S *in graph* G_i, *then* $|D_i^*| \geq |M_i|$.

Proof. Since M_i is an independent set in H_i^2, the hop distance of any two nodes $u, v \in M_i$ is at least three in H_i. Thus all the stars $S(u) = \{v \in V \mid (u,v) \in E(H_i)\}$ centered at $u \in M_i$ are pairwise disjoint each other. For each star $S(u)$, at least one vertex has to be selected into D_i^* in order to dominate S. Therefore, we have $|D_i^*| \geq |M_i|$.

Lemma 3. *Let* WD_i^* *be a subset of* V *with minimum total weight which dominates* S, *then* $w(WD_i^*) \geq w(M_i)$.

Proof. The proof is similar to that of Lemma 3. Now the key point is that by the construction of M_i, for each star $S(u)$ ($u \in M_i$), we have $w(u) \leq w(v)$ for any $v \in N_{G_i}(v) = \{v | (u,v) \in E(H_i)\}$. Note $S(u) \cap S(v) = \emptyset$ for $u, v \in M_i$ and $u \neq v$. Thus, at least one node in each $S(u)$ ($u \in M_i$) has to be selected into WD_i^*. Note u is the lightest node in $S(u)$. It follows that $w(WD_i^*) \geq w(M_i)$.

Theorem 2. *Above algorithm is a 2-approximation for p-CDC.*

Proof. Let i^* be the smallest index such that there exists a subset D_{i^*} of G_{i^*} that dominates S such that $|D_{i^*}| \leq p$ and $w(D_{i^*}) \leq W$. Then we have $OPT = cost(e_{i^*})$, where OPT is the optimal value of the p-CDC problem. By our algorithm, for each i ($i = 1, 2, \cdots, j-1$), we have either $|M_i| > p$ or $w(M_i) > W$. It follows from Lemma 2 and Lemma 3 that either $|D_i^*| \geq |M_i| > p$ or $w(WD_i^*) \geq w(M_i) > W$. Thus we have $i^* > i$ for $i = 1, 2, \cdots, j-1$, i.e., $j \leq i^*$ and $cost(e_j) \leq OPT$. Since M_i is a maximal independent set of $(G_i[S])^2$, it also a dominating set of $(G_i[S])^2$. So in $(G_i[S])^2$, the stars centered at each $u \in M_i$ span all the nodes in $S = V((G_i[S])^2)$. Let v be any node in a star centered at some $u \in M_i$. Then v is at most two hops away from u in $G_i[S]$. By triangle inequality, $cost(e) \leq 2cost(e_j)$ for any edge $e = (u, v)$ in the star. Note $cost(e_j) \leq OPT$. We have $cost(e) \leq 2OPT$. This completes the proof.

Next, we discuss how to use the 2-approximation algorithm for the p-CDC problem to construct Simple-p-RSPA. There is a major challenge to apply the 2-approximation algorithm for the p-CDC problem to our problem of interest, since we are looking for a subset of nodes with size exactly p. If we enforce this, then the algorithm may not produce a feasible solution with insufficient W. To address this concern, it is necessary for us to find a valid W. To this purpose, we may set W to be $W_i = \sum_{v \in V_i} deg(v_i, G)$, where V_i is the subset of the first

i nodes with largest node degree, for each $i = n, n - 1, \cdots, 1$ and apply the modified 2-approximation algorithm for the p-CDC problem. Finally, we choose the one out of all feasible outputs such that the objective of p-BRSP in Eq. (1) is minimized. Note that this final result still has the approximation factor of 2 with respect to Objective 2.

4 Concluding Remarks

This paper introduces a new application of the information which can be extracted from social network information. The main focus of this paper is to use the information for biased survey so that more amount of minority opinions can be heard. We formalize the problem of our interest as a new optimization problem with two separate objectives. Then, we propose two heuristic algorithms for the problem, each of which has an approximation factor with respect to each of the objectives. As a future work, we will conduct simulations to evaluate the performance of the proposed algorithms. We are also interested in using real data to see if our approach is in fact effective. We also plan to use apply approach to identify the users with less satisfaction and compensate them so that the negative reputation of a new product can be suppressed. We believe this can compensate the existing approaches which focus on how to compensate users to spread positive reputation [11,12].

References

1. Anifantis, E., Stai, E., Karyotis, V., Papavassiliou, S.: Exploiting Social Features for Improving Cognitive Radio Infrastructures and Social Services via Combined MRF and Back Pressure Cross-layer Resource Allocation. Computational Social Networks **1**(4) (2014)
2. Ai, C., Zhong, W., Yan, M., Gu, F.: A Partner-matching Framework for Social Activity Communities. Computational Social Networks **1**(5) (2014)
3. Ventresca, M., Aleman, D.: Efficiently Identifying Critical Nodes in Large Complex Networks. Computational Social Networks **2**(6) (2015)
4. Vehovar, V., Manfreda, K.L.: Overview: online surveys. In: Fielding, N.G., Lee, R.M., Blank, G. (eds.) The SAGE Handbook of Online Research Methods, pp. 177–194. SAGE, London (2008)
5. Duffy, B., Smith, K., Terhanian, G., Bremer, J.: Comparing Data from Online and Face-to-face Surveys. International Journal of Market Research **47**(6), 615–639 (2005)
6. Kim, D., Zhong, J., Lee, M., Li, D., Li, Y., Tokuta, A.O.: Efficient Respondents Selection for Biased Survey using Homophily-high Social Network Graph. Optimization Letters (OPTL) (under 3rd review)
7. Bisgin, H., Agarwal, N., Xu, X.: A Study of Homophily on Social Media. World Wide Web **15**(2), 213–232 (2012)
8. Singer, E., Ye, C.: The Use and Effects of Incentives in Surveys. The Annals of the American Academy of Political and Social Science **645**(1), 112–141 (2013)
9. Hochbaum, D.S., Shmoys, D.B.: A Best Possible Heuristic for the k-Center Problem. Mathematics of Operations Research **10**(2), 180–184 (1985)

10. Kim, D., Zhong, J., Lee, M., Li, D., Tokuta, A.O.: Efficient respondents selection for biased survey using online social networks. In: Cai, Z., Zelikovsky, A., Bourgeois, A. (eds.) COCOON 2014. LNCS, vol. 8591, pp. 608–615. Springer, Heidelberg (2014)
11. Lu, Z., Fan, L., Wu, W., Thuraisingham, B., Yang, K.: Efficient Influence Spread Estimation for Influence Maximization under the Linear Threshold Model. Computational Social Networks **1**(2) (2014)
12. Kim, H., Beznosov, K., Yoneki, E.: A Study on the Influential Neighbors to Maximize Information Diffusion in Online Social Networks. Computational Social Network **2**(3) (2015)

An Algorithm for Friendship Prediction on Location-Based Social Networks

Gao Xu-Rui[1], Wang Li[1]([✉]), and Wu Wei-Li[1,2]

[1] College of Computer Science and Technology, Taiyuan University of Technology,
Taiyuan 030024, China
{l_lwang,gxr_lm}@126.com
[2] Department of Computer Science, University of Texas,
Dallas Mail Station EC 32601, North Floyd Road Richardson,
Austin, TX 75083, USA
weiliwu@utdallas.edu

Abstract. With the development and widely utilization of mobile device, location-based social network (LBSN) has become one important platform for many novel applications. Location information will help discover latent friend relationship and then guide trip, recommend goods and so on. In this paper we propose an algorithm for friendship prediction. Firstly, we adopt the *information gain* to measure the contribution of different features to human friendship, and extract user social relationship, check-in distance and check-in type as three key features. Secondly, we take the prediction problem as a classification problem and choose SVM (support vector machine) to classify it. At last some experiment results show our algorithm valid to some extent.

Keywords: Location-based social network · Friendship prediction · SVM

1 Introduction

The location-based online social networks have seen soaring popularity, attracting millions of users [1]. Many location-based online social networking applications, such as Foursquare, Mingle, Gowalla, etc [2], have offered amazing, novel and valuable services for users based on the sharing of location information by checking-in on those websites.

LBSN offers a new social networking platform for making friends, sharing information, searching contents with the location labels, and making communication with nearby friends [3]. Different from usual online social networking, LBSN not only maintains contacts in the virtual network, but also supports face to face communication in real world, recommend valued information to users. It has attracted a great number of users and aroused great concern of researchers. Community mining [4], privacy protection [5], friend prediction [6] and location recommendation are main related research topics.

© Springer International Publishing Switzerland 2015
M.T. Thai et al. (Eds.): CSoNet 2015, LNCS 9197, pp. 193–204, 2015.
DOI: 10.1007/978-3-319-21786-4_17

In this paper, firstly we use *information gain* to evaluate the contribution of different features to the friendship. Then, we extract user social relationship, check-in distance and check-in type as key factors for friendship prediction. Base on the selected features we present a friendship prediction method. Finally, we take real check-in data from two location-based social networks (Gowalla and Brightkite) to validate effectiveness of feature selection and the accuracy of friendship prediction.

The paper is organized as follows: section 1 introduces the research background and significance; section 2 describes the related work of friendship prediction on LBSN; section 3 focuses on the datasets and feature selection; section 4 puts forward the method for prediction, parameter optimization and evaluation; section 5 summarizes our studies.

2   Related Work

Friendship prediction has become one of the major studies on LBSN. Check-in data carries a wealth of user information, and it can be used to improve the accuracy of friendship prediction, which also benefits friend or trajectory recommendation in the future. Generally, the network connections between users on LBSN and their online behaviors, to some extent, reflect their behaviors in real life. In other words, people who have the same interests and geographically close, or have the same social circle, often easily become friends [7].

Lee et al.[8] presented multi-Layered friendship prediction model and quantified the correlation between users' friendship with their mobility characteristics, social graph properties, and user profiles. Ma et al. [9] proposed one method to judge the existence of social relationship and the type of relationship. Quercia D et al. [10] put forward a directed social network link prediction approach based on topic model, which analyzes node semantic information, synthesizes network node attributes and structural characteristics for link prediction. Wang et al. [11] took the personal factors, global factors and time factor as main features for friendship prediction. They pointed out that if only considering personal factors or global factors, some opposite conclusions may appear. For example, the probability of tourists visiting New York's Times Square is far less than that of the locals. Crandall et al. [12] demonstrated how temporal and spatial co-occurrences between people help to infer social ties among them, and the main goal was to put forward a generative model which explains empirical data. Liben-Nowell et al.[13] described how the probability of friendship between two individuals can be related to the geographic distance between them.

Table 1 describes the relevant characteristics used in the friendship prediction. From which, we can easily notice that the social relationship and check-in distance are two key factors. But, in the factor of social relation, people mainly consider individual features of common neighbor nodes; ignore the relation between common neighbor nodes and other neighbor nodes. Moreover, in some cases, users check in different places, but they have the similar check-in type, which meaning they may be friends with similar interests.

Table 1. Features of Friendship Prediction

Literatures	Social Relation	Check-in Distance	User Profile	Semantic Information
Schwartz(93,ACM)	$\checkmark$		$\checkmark$	
Liben-Nowell(05,$PNAS$)		$\checkmark$		
Crandall(10,$PNAS$)	$\checkmark$	$\checkmark$		
Scellato(11,ACM)	$\checkmark$	$\checkmark$		
Quercia(12,ACM)		$\checkmark$		$\checkmark$
Ma(13,$JNUDT$)			$\checkmark$	$\checkmark$
Wang(14,$ICDM$)	$\checkmark$	$\checkmark$		

Base on these analyses, in this paper, we extract user social relationship, check-in distance and check-in type as main characteristics to predict friendship.

3   Feature Selection on LBSN

3.1   Dataset

Gowalla and Brightkite are two classical online location-based social networking services. In Stanford Network Analysis Project(SNAP) project, two data sets collected from Gowalla and Brightkite respectively are offered freely and they have been used by many literatures researching on LBSN. Then, in this paper we choose the two data sets as experiment data for friendship prediction research. The records of Gowalla dataset are collected from Feb. 2009 to Oct. 2010, whereas the Brightkite dataset is collected from Apr. 2008 to Oct. 2010. The total number of check-ins for Gowalla is 6.4 million and 4.5 million for Brightkite. Each dataset also contains a social network of friendships, which serves as the ground truth evaluation [2,14]. The relevant dataset statistics is given in Table 2.

Table 2. Dataset Statistics

	Gowalla	Brightkite
Nodes	196,591	58,228
Edges	950,327	214,078
Check-ins	6,442,890	4,491,143

In the remainder of the paper we will use word "check-in" to refer to an event when the time and the location of a particular user is recorded. Additionally, we calculate the distance between friends for two datasets (see Fig.1), and observe that the probability of friendship decreases with the increasing of their distances. Meanwhile, check-in records obey long-tailed distribution.

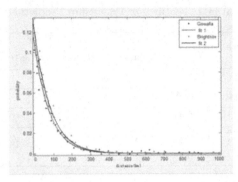

Fig. 1. Distribution of distance between friends

3.2   Feature Selection

Recently, the main way for friendship prediction is to select features that affect friendship and establish a prediction model. Identifying key features is one of the most important issues that decides the prediction precision and recall. *Information gain(IG)* can evaluate the contribution feature attributes to the target attribute on the whole, so we use it to extract appropriate feature attributes.

In general terms, the expected *information gain* is the change in information entropy X from a prior state to Y state that takes some information [15], as given in equation (1):

$$IG(X,Y) = H(X) - H(X|Y)$$

$$H(X) = \sum_{i=1}^{m} -p(x_i)log^{p(x_i)}$$

$$H(X|Y) = \sum_{i,j} p(x_i, y_j)log\frac{p(y_j)}{p(x_i, y_j)} \qquad (1)$$

We take Gowalla dataset as an example, calculate *information gain* values of different attributes and the result is shown in Table 3.

Table 3. *IG* for Different Attributes

Attribute	IG Value
Social Relationship	0.837
Check-in Distance	0.484
Check-in Type	0.294
Common check-in location	0.014

Feature of Social Relationship. In published studies, social relationship is one of important factors for new friendship formation and many different measurement methods were put forward, such as Jaccard coefficient [16], Adamic-Adar Index(AA) [17] or Resource Allocation Index(RA) [18] and so on. However, most of these approaches only consider individual features of common neighbor nodes, ignore the relation between common neighbor nodes and other neighbor nodes.

According to empirical observation, we divide the edges into four categories: 1. the edges between the common neighbor nodes , assigned the weight of a; 2. the edges between the common neighbors and the user nodes , assigned weight of b; 3. the edges between the common neighbor nodes and other neighbors of the user nodes predicted, assigned the weight of c; 4. the edges between the common neighbors and its own neighbors, allocated the weight of d. To assess the relationship between i and j, the four types play different roles. Ranking them by importance, the sequence is $a > b > c > d$. In this paper, we simplify it and set $a=2$, $b=1.5$, $c=1$, and $d=0.5$. The similarity of social relationship between i and j is equation(2).

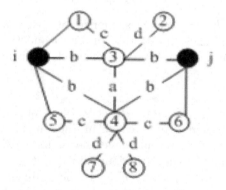

Fig. 2. Network with different neighbor relationships

$$sim(i,j) = \frac{a \cdot c_a + b \cdot c_b + c \cdot c_c + d \cdot c_d}{c_a + c_b + c_c + c_d} \tag{2}$$

Where, c_a, c_b, c_c, c_d represent the corresponding number of four categories of edges respectively. According to calculation method above, the similarity of user i and user j in Fig.2 is computed as follow:

$$sim(i,j) = \frac{2 + 1.5 \times 2 + 1 \times 1 + 0.5 \times 1}{5}$$
$$+ \frac{2 + 1.5 \times 2 + 1 \times 2 + 0.5 \times 2}{7}$$
$$= 2.45$$

We compute the similarities of (i,j) and $(1,2)$ in Fig.2 by different methods and the result is shown in Table 4. Obviously, our proposed approach is superior to traditional methods in distinguishing different relationship between node pairs.

Table 4. Similaritiy Analysis

	Jaccard	AA	RA	Our Approach
(i,j)	0.29	2.61	0.34	2.45
$(1,2)$	0.33	1.43	0.20	1.00

Feature of Check-in Distance. From Table 3, we can see that excepting for social relationship, check-in distance also affects the friendship. In this paper, we build the following equation (3) to characterize the feature.

$$a_{i,j} = \frac{\sum_{m=1}^{M} \frac{\sum_{n=1}^{N} d(l_{im}, l_{jn})}{N}}{\sum_{m=1}^{M} \sum_{n=1}^{N} d(l_{im}, l_{jn})} \tag{3}$$

Where, $l_{i1}, l_{i2}, \ldots, l_{iM}$ is check-in sequence for user i, $d(l_{im}, l_{jn})$ is the distance between the mth check-in location of user i and the nth for user j. We define $a_{i,j}$ as the characteristic property of check-in distance among different users.

Feature of Check-in Type. The check-in types of users reflect their own interests. Sometimes, different users check in different places, but the types of places are similar. It may show they have the similar preferences, and then it brings more possibility to establish friendship. $(t_{i1}, t_{i2}, \ldots, t_{iM})$ is the set of check-in types of user i; $(c_{i1}, c_{i2}, \ldots, c_{iM})$ is the number of check-in for each location; C_i is the total number of check-in. Let L denote the total number of users. Then $(T_{k1}, T_{k2}, \ldots, T_{kL})$ is the number of check-in for each user in the location k and $p_i(k)$ is the check-in probability for user i in the location k, shown as equation (4).

$$p_i(k) = \frac{T_{ki}}{C_i} \tag{4}$$

We introduced the concept of location information entropy [19] to describe the location type, shown as equation (5). The more check-ins in public places is, the smaller probability of each user check-in is. Thus, the larger information entropy shows fairly open for this place. In the experiment, we will remove such locations, and count the same check-in types in the remaining location types.

$$E(t_k) = \sum_{i=1}^{L} -p_i(k) log^{p_i(k)} \tag{5}$$

The final feature is described by the ratio between the same types and the total types. We define t_i as the check-in type set of user i, t_j as the check-in type set of user j. Defining a_t as the feature of check-in type among different users, shown as equation (6).

$$a_t(i,j) = \frac{t_i \cap t_j}{t_i \cup t_j} \qquad (6)$$

4   Friendship Prediction Based on LBSN

4.1   SVM Based Prediction Method

Since the result of friendship prediction is "Yes" or "No", we can make it as a classification problem. SVM can improve generalization performance; solve high dimensional and nonlinear problems; avoid local minimum problems. Therefore, we use support vector machine to construct model. In machine learning, SVM is supervised learning model with associated learning algorithms that analyze data and recognize patterns, used for classification and regression analysis. The SVM training algorithm builds a model that assigns new examples into one category or the other, making it a non-probabilistic binary linear classifier [20].

4.2   Parameter Optimization

Grid search is one traditional way of performing parameter optimization, which is simply an exhaustive searching through a manually specified subset of the parameter space of a learning algorithm [21]. A grid search algorithm must be guided by some performance metrics, typically measured by cross-validation on the training set.

In our experiment, SVM classifier equipped with a RBF kernel has at least two parameters that need to be tuned for good performance on unknown data: a regularization constant C and a kernel parameter g. Grid search then trains a SVM with each pair (C, g) in the Cartesian product of these two datasets and evaluates their performance on a K-fold cross validation set. Finally, the grid search algorithm outputs the settings that achieved the highest score in the validation procedure.

C is a trade-off between training error and the flatness of the solution. The larger C is the less the final training error will be. But if you increase C too much you risk losing the generalization properties of the classifier, because it will try to fit as best as possible all the training points (including the possible errors of your dataset) [22]. In addition, a large C usually increases the time needed for training. Generally, we select (C, g) which achieves the highest classification accuracy and gets the smallest parameter C as the best parameters.

In Fig.3 and Fig.4, we plot the optimal (C, g) for two datasets. Also, we can find the optimal (C, g) from Table 5 . For Gowalla, we select $C = 0.35355$ and $g = 0.17678$. And in Brigthtskite dataset, we choose $C = 0.25$ and $g = 0.25$.

4.3   Evaluation Indicators

We use precision, recall, $F1$ measure and AUC value for model evaluation. And regarding the friendship as positive; non-friendship as negative, given in Table 6.

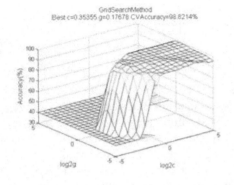

Fig. 3. Grid Search, Gowalla   **Fig. 4.** Grid Search, Brightkite

Table 5. Parameter Optimization

(a) Gowalla				(b) Brightkite		
K	(C,g)	CV Accurancy		K	(C,g)	CV Accurancy
K=2	C=0.70711 g=0.125	97.7528%		K=2	C=0.17678 g=0.125	96.9621%
K=3	C=0.35355 g=0.17678	98.8214%		K=3	C=0.25 g=0.25	97.2368%
K=5	C=1.4142 g=0.70711	98.8214%		K=5	C=0.25 g=0.70711	96.9621%
K=8	C=1.4142 g=0.25	98.8214%		K=8	C=0.35355 g=0.5	97.2368%
K=10	C=0.35355 g=0.088388	97.7528%		K=10	C=0.125 g=0.70711	96.9621%

The precision and recall are defined as follows:

$$precision = \frac{TP}{TP+FP}$$
$$recall = \frac{TP}{TP+FN} \tag{7}$$

Table 6. Test Results

	Condition Positive	Condition Negative
Result Positive	True Positive (TP)	False Positive (FP)
Result Negative	False Negative (FN)	True Negative (TN)

However, precision and recall indicators may meet the contradictory situation, we also employ $F1$ measure, which is the harmonic mean of precision and recall [23], shown in equation (8).

$$F1 = \frac{2 \times precision \times recall}{precision + recall} \tag{8}$$

In addition, we use the area under the curve (AUC) value to better evaluate the classification performance. Usually, the value of AUC is between 0.5 and 1.0 and a larger AUC represents better performance [24].

4.4 Experimental Results

We mix social relation, check-in distance and check-in type for each user, and use SVM classification algorithm for friendship prediction. Also, we acquired corresponding classification results for two datasets. (See Table 7) We can notice that the proposed method can predict friendship effectively. Additionally, the

Table 7. Classification Results

	Gowalla	Brightkite
precision	0.923	0.902
recall	0.821	0.753
$F1$	0.869	0.821
AUC	0.879	0.847

ROC curves are illustrated in Fig.5 and Fig.6. And in Table 8 and Table 9, we display the area under the ROC curve. From it, we can find the area of three features fusion reaches 0.879 for Gowalla and 0.847 for Brightkite. And the social relationship makes a greater impact on the friendship prediction.

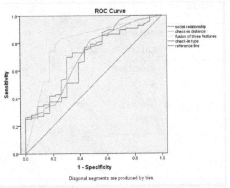
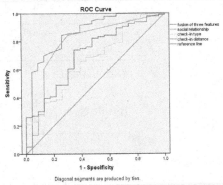

Fig. 5. ROC Curve, Gowalla   **Fig. 6.** ROC Curve, Brightkite

Table 8. Area Under the Curve(Gowalla)

Test Result Variables	Area	Std. Error[a]	Asymptotic Sig.[b]	Asymptotic 95% Confidence Interval	
				Lower Bound	Upper Bound
fusion of three features	.879	.047	.000	.766	.952
social relationship	.787	.060	.000	.669	.905
check-in type	.668	.062	.012	.548	.789
check-in distance	.712	.058	.002	.598	.825

Table 9. Area Under the Curve(Brightkite)

Test Result Variables	Area	Std. Error[a]	Asymptotic Sig.[b]	Asymptotic 95% Confidence Interval	
				Lower Bound	Upper Bound
fusion of three features	.847	.055	.000	.694	.908
social relationship	.731	.059	.001	.615	.847
check-in type	.671	.062	.011	.549	.793
check-in distance	.703	.061	.003	.584	.823

4.5   Comparison with Other Models

Compared with the model for friendship prediction proposed by *Salvatore* , our precision reaches 0.923 in Gowalla dataset and 0.902 for Brightkite dataset, AUC values are 0.879 and 0.847 respectively. Whereas 0.92, 0.98, 0.90 by Model trees, Random forests and J48 algorithm using *Salvatore's* model. Moreover, our calculation process is relatively simple, implicating the effectiveness of our presented approach.

5   Conclusion

This paper presents a SVM based approach for friendship prediction on LBSN. We extract the user social relations, check-in distance and check-in type and establish prediction model. Some experiments show our algorithm valid. But, there still exists some problems, such as whether just the three attributes are enough? We will do more research on mobile datasets to find more latent information to improve the quality of friendship prediction. And travel route recommendation depending on friendship is also the research focus later.

Acknowledgements. Partially supported by the National high Technology Research and Development Program of China (863 Program,2014AA015204), Natural Science Foundation of Shanxi Province of China (Grant No. 2014011022-1), the National Natural Science Foundation of China (Grant No. 61472272), the Open Project Funding of CAS Key Lab of Network Data Science and Technology, Institute of Computing Technology, Chinese Academy of Sciences.

References

1. GigaOM. Foursquare Hits 4 Million Users. http://gigaom.com/2010/10/21/foursquare-hits-4-million-users/
2. Cho, E., Myers, S.A., Leskovec, J.: Friendship and mobility: user movement in location-based social networks. In: Proceedings of the 17th ACM SIGKDD International Conference on Knowledge Discovery and Data Mining, pp. 1082–1090. ACM (2011)
3. Yang, D.N., Shen, C.Y., Lee, W.C., et al.: On socio-spatial group query for location-based social networks. In: Proceedings of the 18th ACM SIGKDD International Conference on Knowledge Discovery and Data Mining, pp. 949–957. ACM (2012)
4. Wakita, K., Tsurumi, T.: Finding community structure in mega-scale social networks:[extended abstract]. In: Proceedings of the 16th International Conference on World Wide Web, pp. 1275–1276. ACM (2007)
5. Huo, Zheng, Meng, Xiaofeng, Hu, Haibo, Huang, Yi: You can walk alone: trajectory privacy-preserving through significant stays protection. In: Lee, Sang-goo, Peng, Zhiyong, Zhou, Xiaofang, Moon, Yang-Sae, Unland, Rainer, Yoo, Jaesoo (eds.) DASFAA 2012, Part I. LNCS, vol. 7238, pp. 351–366. Springer, Heidelberg (2012)
6. Aiello, L.M., Barrat, A., Schifanella, R., et al.: Friendship prediction and homophily in social media. ACM Transactions on the Web (TWEB) 6(2), 9 (2012)
7. Fusco, S.J., Michael, K., Michael, M.G., et al.: Exploring the social implications of location based social networking: an inquiry into the perceived positive and negative impacts of using LBSN between friends. In: Proceedings of the 2010 Ninth International Conference on Mobile Business / 2010 Ninth Global Mobility Roundtable, pp. 230–237. IEEE Computer Society (2010)
8. Li, N., Chen, G.: Multi-layered friendship modeling for location-based mobile social networks. In: Mobile & Ubiquitous Systems Networking & Services Mobiquitous.mobiquitous, pp. 1–10 (2009)
9. Ma, J., Xu, H., Chen, H.: Friendship prediction in recommender system. Journal of National University of Defense Technology 35(1), 163–168 (2013)

10. Quercia, D., Askham, H., Crowcroft, J.: TweetLDA: supervised topic classification and link prediction in Twitter. In: Proceedings of the 4th Annual ACM Web Science Conference, pp. 247–250. ACM (2012)
11. Wang, H., Li, Z., Lee, W.C.: PGT: Measuring Mobility Relationship using Personal, Global and Temporal Factors
12. Crandall, D.J., Backstrom, L., Cosley, D., et al.: Inferring social ties from geographic coincidences. Proceedings of the National Academy of Sciences **107**(52), 22436–22441 (2010)
13. Liben-Nowell, D., Novak, J., Kumar, R., et al.: Geographic routing in social networks. Proceedings of the National Academy of Sciences of the United States of America **102**(33), 11623–11628 (2005)
14. Scellato, S., Noulas, A., Mascolo, C.: Exploiting place features in link prediction on location-based social networks. In: Proceedings of the 17th ACM SIGKDD International Conference on Knowledge Discovery and Data Mining, pp. 1046–1054. ACM (2011)
15. Christopher, M.B.: Pattern recognition and machine learning. Company New York Ny **16**(4), 049901 (2006)
16. Breese, J.S., Heckerman, D., Kadie, C.: Empirical analysis of predictive algorithms for collaborative filtering. In: Proceedings of the Conference on Uncertainty in Artificial Intelligence, pp. 43–52 (1998)
17. Sarwar, B., Karypis, G., Konstan, J., et al.: Item-based collaborative filtering recommendation algorithms. In: Proc International Conference on the World Wide Web, pp. 285–295. ACM (2001)
18. Sutton, R.S., Barto, A.G.: Reinforcement learning: an introduction. IEEE Transactions on Neural Networks **9**(5), 1054 (1998)
19. Cranshaw, J., Toch, E., Hong, J., et al.: Bridging the Gap between Physical Location and Online Social Networks, pp. 119–128. Social Science Electronic Publishing (2010)
20. Zhang, H., Berg, A.C., Maire, M., et al.: SVM-KNN: discriminative nearest neighbor classification for visual category recognition. In: 2006 IEEE Computer Society Conference on Computer Vision and Pattern Recognition, vol. 2, pp. 2126–2136. IEEE (2006)
21. Bergstra, J., Bengio, Y., Bottou, L.: Random search for hyper-parameter optimization. Journal of Machine Learning Research **13**(1), 281–305 (2012)
22. Bergstra, J., Bengio, Y.: Algorithms for hyper-parameter optimization. Nips **24**, 2546–2554 (2011)
23. Fawcett, T.: An introduction to ROC analysis. Pattern Recognition Letters **27**(8), 861–C874 (2006)
24. Hanley, J.A., McNeil, B.J.: The meaning and use of the area under a receiver operating characteristic (ROC) curve. Radiology **143**(1), 29–36 (1982)

Named Entity Recognition
in Vietnamese Tweets

Vu H. Nguyen[1], Hien T. Nguyen[1(✉)], and Vaclav Snasel[2]

[1] Faculty of Information Technology, Ton Duc Thang University,
Ho Chi Minh City, Vietnam
`{nguyenhongvu,hien}@tdt.edu.vn`
[2] Faculty of Electrical Engineering and Computer Science,
VSB-Technical University of Ostrava, Ostrava, Czech Republic
`vaclav.snasel@vsb.cz`

Abstract. Named entity recognition (NER) is a task of detecting named entities in documents and categorizing them to predefined classes such as Person (PER), Location (LOC), Organization (ORG) and so on. There have been many approaches proposed to tackle this problem in both formal texts such as news or authorized web content and short texts such as contents in online social network. However, those texts were written in languages other than Vietnamese. In this paper, we propose a method for NER in Vietnamese tweets. Since tweets on Twitter are noisy, irregular, short and consist of acronyms, spelling errors, NER in those tweets is a challenging task. Our method firstly normalizes tweets and then applies a learning model to recognize named entities using six different types of features. We built a training set of more than 40,000 named entities, and a testing set of 2,446 named entities to evaluate our system. The experiment results show that our system achieves encouraging performance with 82.3% F1 score.

Keywords: Text normalization · Named entity recognition · Spelling error detection and correction

1 Introduction

In recent years, social networks are becoming more and more popular. It is easy for user to share their data using online social networks. Nowaday, one of the most popular social networks is Twitter. According to a statistic in 2011, the number of tweets was up to 140 million per day[1]. With the huge posting everyday, effective extraction and processing of those data will bring great benefit, in particular, to information extraction applications.

Twitter provides an interactive environment allowing users to create their own content through tweets. Since each tweet consists of only 140 characters, a user tends to use acronyms, non-standard words, and social tokens. Therefore,

---

[1] https://blog.twitter.com/2011/numbers

© Springer International Publishing Switzerland 2015
M.T. Thai et al. (Eds.): CSoNet 2015, LNCS 9197, pp. 205–215, 2015.
DOI: 10.1007/978-3-319-21786-4_18

it contains many spelling errors and raises a challenge for NER. For English and others languages, there have been several named entities recognition methods proposed for tweets [1,7,12,13,21]. For Vietnamese language, although there have been many approaches proposed for NER in formal texts, but there is no one for Vietnamese tweets. In this paper, we propose a method for NER in Vietnamese tweets to fill the gap. The system consists of three steps: the first step is to normalize tweets by detecting and correcting spelling errors; the second is capitalization classifier and the last is to recognize named entities.

For example, for the tweet: *xe đón hồ ngọc hà gây tai nạn* **kinhh** *hoàng: sẽ khởi tố tài xế http://fb.me/2MwvznBbj* - (the car picked up ho ngoc ha caused a terrible accident: the driver will be prosecuted); after the first step, the tweet will become: *xe đón hồ ngọc hà gây tai nạn kinh hoàng: sẽ khởi tố tài xế*, where the spelling error word **kinhh** is corrected to **kinh**; And after the second and the third steps, the tweet will become *Xe đón Hồ Ngọc Hà gây tai nạn kinh hoàng: sẽ khởi tố tài xế* and *Xe đón <PER>Hồ Ngọc Hà</PER> gây tai nạn kinh hoàng: sẽ khởi tố tài xế* respectively.

This paper presents the first attempt to NER in Vietnamese tweets and the contribution is three-fold: (1) a method for normalization of Vietnamese tweets based on dictionaries and Vietnamese vocabulary structures in combination with a language model, (2) a learning model for NER in Vietnamese tweets with six different types of features, and (3) a training set of more than 40,000 named entities and a testing set of 2,446 named entities to evaluate NER system of Vietnamese tweets.

The rest of this paper is organized as follows: Section 2 presents related work. Section 3 presents our proposed method. Experiments and results are shown in Section 4. Finally we draw conclusion in Section 5.

2   Related Work

NER has been extensively studied on formal texts, such as news, authorized web content. Several approaches have been proposed using different learning models such as Condition Random Fields (CRF), Maximum Entropy Model (MEM), Hidden Markov Model (HMM), Support Vector Machines (SVM). In particular, [14] used SVM to estimate lattice transition probabilities for NER. [15] applied a feature induction method for CRF to recognize named entities. A combination between a CRF model with latent semantics to recognize named entities was proposed in [8]. A method using soft-constrained inference for NER was proposed in [5]. In [3] and [26], the authors proposed a maximum entropy tagger and a HMM-based chunk tagger to recognize named entities respectively. Unfortunately, those methods gave poor performance on tweets as pointed out in [13].

Regarding to microblog texts written in English, there have been several approaches proposed for NER. Among them, [21] proposed a NER system for tweets, called T-NER, which employed a CRF model for training and

Labled-LDA [20] with the external knowledge base, in particular, Freebase[2] for NER. A hybrid approach to NER on tweets was presented in [13] where a KNN-based classifier was employed together with a CRF model. A combination between heuristics and MEM was proposed in [7]. Since Vietnamese has some specific features presented in [23], it is not able to apply those methods directly to Vietnamese tweets.

For Vietnamese texts, various approaches have been proposed using some learning models such as SVM [23], classifier voting [22], CRF [9,25]. Some other works proposed a rule-based method [16], employed bootstrapping algorithm and a rule-based model [24], combined linguistically motivated and ontological features [17] for NER.

However, until now, there has not been any work focusing on NER in Vietnamese tweets or (short) informal Vietnamese texts. In this paper, we propose a method for NER in Vietnamese tweets to fill the gap. Our method includes three main tasks as follows: (1) a method for normalization of Vietnamese tweets based on dictionaries and Vietnamese vocabulary structures in combination with a language model, (2) a method for detecting and correcting the suitable capital letter, and (3) a model for training and recognizing named entity in Vietnamese tweets. We also conduct experiments to evaluate our NER method focused on three entity types: PERSON, LOCATION and LOCATION.

3   Proposed Method

In this section, we presents our method for NER in Vietnamese tweets. The method is described in Figure 1. We will describe proposed method in details in following subsections.

3.1   The Theoretical Background

Currently, there are several view-points on what is a Vietnamese word. However, in order to meet the goals of automatic error detection, normalization and classification, we follow the view-point in [22]: "A Vietnamese word is composed of special linguistic units called Vietnamese *morphosyllable*". A morphosyllable may be a morpheme, a word, or neither of them [23]. And according to the syllable dictionary of Hoang Phe [19], we split a morphosyllable into two basic parts as follows:

- **Consonant and vowel:**
  - Consonant: Vietnamese language has 27 consonants: "b", "ch", "c", "d", "đ", "gi", "gh", "g", "h", "kh", "k", "l", "m", "ngh", "ng", "nh", "n", "ph", "q", "r", "s", "th", "tr", "t", "v", "x", "p". And in those, there are 8 tail consonants: "c", "ch", "n", "nh", "ng", "m", "p", "t".
  - Vowel: Vietnamese language has 12 single vowels including: "a", "ă","â", "e","ê", "i","o", "ô","ơ", "u","ư", "y".

---

[2] http://www.freebase.com

– **Syllable**: A syllable may be a vowel, or combination of vowels, or combination of vowels and tail consonants. According to the syllable dictionary of Hoang Phe, Vietnamese language has 158 syllables and the vowels in these syllables do not occur consecutively more than once except "ooc" and "oong" syllables.

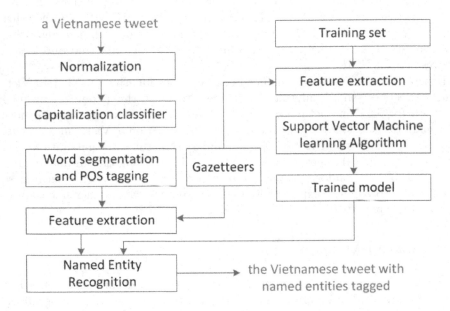

Fig. 1. NER model in Vietnamese tweets

3.2 Normalization

Because Vietnamese tweets contain a lot of spelling errors, we propose a method for normalization them before performing NER. Our normalization method has two steps, the first step is error detection and the second step is error correction.

3.2.1 Error Detection

Before performing this step, tweets must be removed noisy contents such as emotion symbols (e.g: ❤❤,..), hashtag symbols, link url @username, etc. In order to detect errors, we synthesize and build a dictionary for all Vietnamese morphosyllables. This dictionary includes more than 7,300 morphosyllables. A morphosyllable in a text will be identified as an error if it does not appear in the morphosyllable dictionary.

3.2.2 Error Correction

Normally, there are two kinds of errors existing in Vietnamese texts. They are typing errors and spelling errors.

To compose Vietnamese texts, there are two popular typings: Telex typing and VNI typing. Each input typing has a combination of letters and tone-marks to form Vietnamese morphosyllables. Therefore, in order to fix typing errors, we build a set of syllables with their tone-marks and a set of rules to map these syllables to their errors as example follows:

- "án": "asn", "ans", "a1n", "an1"
- "àn": "afn", "anf", "a2n", "an2"
- "ản": "arn", "anr", "a3n", "an3"
- "ãn": "axn", "anx", "a4n", "an4"
- "ạn": "ajn", "anj", "a5n", "an5"

In order to fix spelling errors, we employ the tri-gram model proposed in [18]. Table 1 shows normalization results of Vietnamese tweets with spelling errors and their normalization.

Table 1. Tweets with spelling errors and their normalization.

Spelling error tweets	Normalized tweets
xe dón hồ ngọc hà gây tai nạn **kinhh** *hoàng: sẽ khởi tố tài xế* http://fb.me/2MwvznBbj	xe đón hồ ngọc hà gây tai nạn **kinh** hoàng: sẽ khởi tố tài xế (the car picked up ho ngoc ha caused a terrible accident: the driver will be prosecuted)
hôm nay, **siinh** *viên* **ddaijj** *học tôn* **dduwcss** *thắng được nghỉ học*	hôm nay, **sinh** viên **đại** học tôn **đức** thắng được nghỉ học (today, students of ton duc thang university was allowed to absent)

3.3 Capitalization Classifier

Capitalization is a key orthographic feature for recognizing named entities ([6], [4]). Unfortunately, in tweets, capitalization is much less reliable than in edited texts. Users usually compose and reply message quickly, they do not care much about capitalization. According to [2], a letter is capitalized in the following cases:

1. Capitalize first letter of the first syllable of a complete sentence, after punctuation (.), question mark (?), exclamation point (!), ellipsis (...) and newline.
2. Capitalize name of persons, locations and organizations.
3. Capitalize for other cases: medal name, position name, days of a week, months of a year, holidays, name of books, magazines, etc.

Because our method focuses on three types of entities: person, organization and location; therefore, in capitalization classifier, we take the first and the second cases into account. For the first case, we detect the structure of sentence and correct wrong capitalization letters. In the second case, we use gazetteers of persons, locations and organizations. Table 2 shows the results of capitalization classifier of Vietnamese tweets.

Table 2. Some results of capitalization classifier of Vietnamese tweets.

Tweets before capitalization	Tweets after capitalization classifier
xe dón hồ ngọc hà gây tai nạn kinh hoàng: sẽ khởi tố tài xế	xe dón **Hồ Ngọc Hà** gây tai nạn kinh hoàng: sẽ khởi tố tài xế (the car picked up Ho Ngoc Ha caused a terrible accident: the driver will be prosecuted)
hôm nay, sinh viên đại học tôn đức thắng được nghỉ học	hôm nay, sinh viên **Đại học Tôn Đức Thắng** được nghỉ học (today, students of Ton Duc Thang university was allowed to absent)

3.4 Word Segmentation and POS Tagging

In order to perform word segmentation and POS tagging for normalized tweets, we employ vnTokenizer[3] of [10] for word segmentation and VnTagger[4] of [11] for POS tagging.

3.5 Feature Extraction

This phase aims to convert each word to a vector of feature values. Our system uses IOB model to annotate data in training and classification phases. IOB is expressed as follows:

- B: current morphosyllable is the beginning of a named entity (NE)
- I: current morphosyllable is inside of a NE
- O: current morphosyllable is outside of a NE

Table 3 shows the characteristic value of labels according to IOB model with four classes (PER, LOC, ORG, O).

Selection of specific attributes from the training set plays a key role in identifying the entity type. Since the nature of the Vietnamese language is different from English, we exploit the most appropriate and reasonable features in order to achieve optimum accuracy for the system. Our system uses following features:

- **Word position**: the position of words in a sentence.
- **POS**: POS tag of the current word.
- **Orthographic**: capitalization of first character, capitalization of all letters, lowercase, punctuation, numbers.
- **Gazetteer**: We build several gazetteer lists such as: person, location, organization and prefixes. Those gazetteer lists consist of more than 50,000 person names, nearly 12,000 location names and 7,000 organization names.
- **Prefix, Suffix**: the first and the second character; the last and the next to the last character of the current word.

---

[3] http://mim.hus.vnu.edu.vn/phuonglh/softwares/vnTokenizer
[4] http://mim.hus.vnu.edu.vn/phuonglh/softwares/vnTagger

– **POS Prefix, POS Suffix**: POS tags of two previous words and POS tags of two following words of the current word.

Table 3. The characteristic value of labels according to IOB model

Label	Value	Meaning
O	[1]	Outside a named entity
B-PER	[2]	Beginning morphosyllable of a NE belongs to a Person class
I-PER	[3]	Inside morphosyllable of a NE belongs to Person class
B-LOC	[4]	Beginning morphosyllable of a NE belongs to Location class
I-LOC	[5]	Inside morphosyllable of a NE belongs to Location class
B-ORG	[6]	Beginning morphosyllable of a NE belongs to Organization class
I-ORG	[7]	Inside morphosyllable of a NE belongs to Organization class

4   Evaluation

4.1   Training Set

In Figure 1, before performing feature extraction, we perform word segmentation, POS tagging and assigning labels in Table 3 for each word in the training set. Then the system extracts features of the words and represents each of those word as a feature vector. Finally, a support vector machine learning algorithm is employed to train a model using the training set.

In particular, we assign labels for words in training set by semi-automatic program. It means that we assign labels to those words with self-written program and check in hand. In our self-written program, we consider the noun phrase obtained after tagging step combining with a list of dictionary of text files to label for those words. Text files of dictionary contains:

– The noun prefix for people such as you, sister, uncle, president, etc.
– The noun prefix for organizations such as company, firm, corporation, etc.
– The noun prefix for locations such as province, city, district, etc.
– List of dictionary for states, provinces of Vietnam, etc.

Table 4 shows results of assigning labels to words in Vietnamese tweets. Total number of entities which we assigned label in this phase was presented in Table 5.

Table 4. The results of assigning labels for words in Vietnamese tweets.

Tweets	Tweets after assigning labels
xe đón Hồ Ngọc Hà gây tai nạn kinh hoàng: sẽ khởi tố tài xế	xe đón <**PER**> Hồ Ngọc Hà </**PER**> gây tai nạn kinh hoàng: sẽ khởi tố tài xế (the car picked up Ho Ngoc Ha caused a terrible accident: the driver will be prosecuted)
hôm nay, sinh viên Đại học Tôn Đức Thắng được nghỉ học	hôm nay, sinh viên <**ORG**> Đại học Tôn Đức Thắng </**ORG**> được nghỉ học (today, students of Ton Duc Thang university was allowed to absent)

After assigning labels for words in Vietnamese tweets, we analyze these tweets to build feature vectors for those words. Structure of a feature vector includes: <label> <index1>:<value1> <index2>:<value2> <index3>:<value3>... where:

- <label>: value from 1 to 7 according to 7 labels (O, B-PER, I-PER, B-LOC, I-LOC, B-ORG, IORG).
- <index>:<value>: order of feature and value corresponding to feature of word respectively.

After representing words in training set as feature vectors, we employ lib-SVM[6] to train the model.

Table 5. Total number of named entities in the training set

Entity type	Number of named entities
PER	10,842
LOC	19,037
ORG	12,311

4.2   Experiments

We conduct experiments to evaluate our method. Table 6, Table 7 show our experimental results in the terms of Precision, Recall and F-Measure.

- Precision (P): the number of correctly recognized named entities divided by total number of named entities recognized by the NER system.
- Recall (R): the number of correctly recognized named entities divided by total number of named entities in the testing set.

---

[6] http://www.csie.ntu.edu.tw/cjlin/libsvm/#download

– F-Measure (F1): $F1 = \frac{2*P*R}{p+R}$

We build a testing set including 1,668 Vietnamese tweets and conduct two experiments, with (Case 1) and without (Case 2) normalization and capitalization classifier of tweets.

Table 6. Experiment results

Case	# NEs in testing set	# recognized NEs	# correctly recognized NEs	# wrong recognized NEs	P	R	F1
1	2,446	1,915	1,601	314	**83.6%**	**65.45%**	**73.42%**
2	2,446	2,266	1,939	327	**85.57%**	**79.27%**	**82.3%**

We re-implement the state-of-the-art method proposed in [23] and compare its performance with that of our method.

Table 7. Comparison performance of our method with that of [23]

System	Precision	Recall	F1
Our system	85.57%	79.27%	82.3%
System of [23]	83.20%	76.20	79.55

5   Conclusion

In this paper, we present the first attempt to NER in Vietnamese tweets on Twitter. We propose a method for normalization of Vietnamese tweets based on dictionaries and Vietnamese vocabulary structures in combination with a language model. We also propose a learning model to recognize named entities using six different types of features. To evaluate our system, we build a training set of more than 40,000 named entities and a testing set of 2,446 named entities. The experiment results show that our system achieves encouraging performance with 82.3% F1 score.

References

1. Bandyopadhyay, A., Roy, D., Mitra, M., Saha, S.: Named entity recognition from tweets. In: Proceedings of the 16th LWA Workshops: KDML, IR and FGWM, Aachen, Germany, September 8–10, 2014, pp. 218–225 (2014)
2. Chu, M.N., Nghieu, V.D., Phien, H.T.: Basis of linguistics and Vietnamese. Vietnam educational publisher (2010)

3. Curran, J.R., Clark, S.: Language independent NER using a maximum entropy tagger. In: Proceedings of the Seventh Conference on Natural Language Learning, CoNLL 2003, Held in cooperation with HLT-NAACL 2003, Edmonton, Canada, May 31 - June 1, 2003, pp. 164–167 (2003)

4. Downey, D., Broadhead, M., Etzioni, O.: Locating complex named entities in web text. In: Proceedings of the 20th International Joint Conference on Artificial Intelligence, IJCAI 2007, Hyderabad, India, January 6–12, 2007, pp. 2733–2739 (2007)

5. Fersini, E., Messina, E., Felici, G., Roth, D.: Soft-constrained inference for named entity recognition. Inf. Process. Manage. **50**(5), 807–819 (2014)

6. Florian, R.: Named entity recognition as a house of cards: classifier stacking. In: Proceedings of the 6th Conference on Natural Language Learning, CoNLL 2002, Held in cooperation with COLING 2002, Taipei, Taiwan, 2002 (2002)

7. Jung, J.J.: Online named entity recognition method for microtexts in social networking services: A case study of twitter. Expert Syst. Appl. **39**(9), 8066–8070 (2012)

8. Konkol, M., Brychcin, T., Konopík, M.: Latent semantics in named entity recognition. Expert Syst. Appl. **42**(7), 3470–3479 (2015)

9. Le, H., Tran, M., Bui, N., Phan, N., Ha, Q.: An integrated approach using conditional random fields for named entity recognition and person property extraction in vietnamese text. In: International Conference on Asian Language Processing, IALP 2011, Penang, Malaysia, 15–17 November, 2011, pp. 115–118 (2011)

10. Hông Phuong, L., Thi Minh Huyên, N., Roussanaly, A., Vinh, H.T.: A hybrid approach to word segmentation of vietnamese texts. In: Martín-Vide, C., Otto, F., Fernau, H. (eds.) LATA 2008. LNCS, vol. 5196, pp. 240–249. Springer, Heidelberg (2008)

11. Le-Hong, P., Roussanaly, A., et al.: An empirical study of maximum entropy approach for part-of-speech tagging of vietnamese texts. In: Traitement Automatique des Langues Naturelles-TALN 2010 (2010)

12. Li, C., Sun, A., Weng, J., He, Q.: Tweet segmentation and its application to named entity recognition. IEEE Trans. Knowl. Data Eng. **27**(2), 558–570 (2015)

13. Liu, X., Zhang, S., Wei, F., Zhou, M.: Recognizing named entities in tweets. In: The 49th Annual Meeting of the Association for Computational Linguistics: Human Language Technologies, Proceedings of the Conference, 19–24 June, 2011, Portland, Oregon, USA, pp. 359–367 (2011)

14. Mayfield, J., McNamee, P., Piatko, C.D.: Named entity recognition using hundreds of thousands of features. In: Proceedings of the Seventh Conference on Natural Language Learning, CoNLL 2003, Held in cooperation with HLT-NAACL 2003, Edmonton, Canada, May 31 – June 1, 2003, pp. 184–187 (2003)

15. McCallum, A., Li, W.: Early results for named entity recognition with conditional random fields, feature induction and web-enhanced lexicons. In: Proceedings of the Seventh Conference on Natural Language Learning, CoNLL 2003, Held in cooperation with HLT-NAACL 2003, Edmonton, Canada, May 31 - June 1, 2003, pp. 188–191 (2003)

16. Nguyen, D.B., Hoang, S.H., Pham, S.B., Nguyen, T.P.: Named entity recognition for vietnamese. In: Nguyen, N.T., Le, M.T., Światek, J. (eds.) Intelligent Information and Database Systems. LNCS, vol. 5991, pp. 205–214. Springer, Heidelberg (2010)

17. Nguyen, T.T., Cao, T.H.: Linguistically motivated and ontological features for vietnamese named entity recognition. In: 2012 IEEE RIVF International Conference on Computing & Communication Technologies, Research, Innovation, and Vision for the Future (RIVF), Ho Chi Minh City, Vietnam, February 27 – March 1, 2012, pp. 1–6 (2012)
18. Nguyen, V.H., Nguyen, H.T., Snasel, V.: Normalization of vietnamese tweets on twitter. In: Proceedings of the Second Euro-China Conference on Intelligent Data Analysis and Applications (2015)
19. Phe, H.: syllable Dictionary. Dictionary center, Hanoi encyclopedia Publishers (2011)
20. Ramage, D., Hall, D.L.W., Nallapati, R., Manning, C.D.: Labeled LDA: a supervised topic model for credit attribution in multi-labeled corpora. In: Proceedings of the 2009 Conference on Empirical Methods in Natural Language Processing, pp. 248–256 (2009)
21. Ritter, A., Clark, S., Mausam, Etzioni, O.: Named entity recognition in tweets: an experimental study. In: Proceedings of the 2011 Conference on Empirical Methods in Natural Language Processing, EMNLP 2011, 27–31 July 2011, John McIntyre Conference Centre, Edinburgh, UK, A meeting of SIGDAT, a Special Interest Group of the ACL, pp. 1524–1534 (2011)
22. Thao, P.T.X., Tri, T.Q., Dien, D., Collier, N.: Named entity recognition in vietnamese using classifier voting. ACM Trans. Asian Lang. Inf. Process. 6(4) (2007)
23. Tran, T.Q., et al.: Named entity recognition in vietnamese documents. Progress in Informatics 5 (2007)
24. Le Trung, H., Le Anh, V., Le Trung, K.: Bootstrapping and rule-based model for recognizing vietnamese named entity. In: Nguyen, N.T., Attachoo, B., Trawiński, B., Somboonviwat, K. (eds.) ACIIDS 2014, Part II. LNCS, vol. 8398, pp. 167–176. Springer, Heidelberg (2014)
25. Tu, N.C., et al.: Named entity recognition in vietnamese free-text and web documents using conditional random fields. In: The 8th Conference on Some Selection Problems of Information Technology and Telecommunication (2005)
26. Zhou, G., Su, J.: Named entity recognition using an hmm-based chunk tagger. In: Proceedings of the 40th Annual Meeting of the Association for Computational Linguistics, July 6–12, 2002, Philadelphia, PA, USA, pp. 473–480 (2002)

Beta Current Flow Centrality
for Weighted Networks

Konstantin E. Avrachenkov[1]($\boxtimes$), Vladimir V. Mazalov[2],
and Bulat T. Tsynguev[3]

[1] INRIA, 2004 Route des Lucioles, Sophia-Antipolis, France
k.avrachenkov@sophia.inria.fr
[2] Institute of Applied Mathematical Research, Karelian Research Center,
Russian Academy of Sciences, 11, Pushkinskaya St., Petrozavodsk 185910, Russia
vmazalov@krc.karelia.ru
[3] Transbaikal State University, 30, Aleksandro-Zavodskaya St., Chita 672039, Russia
btsynguev@gmail.com

Abstract. Betweenness centrality is one of the basic concepts in the analysis of social networks. Initial definition for the betweenness of a node in a graph is based on the fraction of the number of geodesics (shortest paths) between any two nodes that given node lies on, to the total number of the shortest paths connecting these nodes. This method has quadratic complexity and does not take into account indirect paths. We propose a new concept of betweenness centrality for weighted network, beta current flow centrality, based on Kirchhoff's law for electric circuits. In comparison with the original current flow centrality and alpha current flow centrality, this new measure can be computed for larger networks. The results of numerical experiments for some examples of networks, in particular, for the popular social network VKontakte as well as the comparison with PageRank method are presented.

Keywords: Beta current flow centrality · Betweenness centrality ·
Pagerank · Weighted graph · Social networks

1 Introduction

The online social networks gave impulse to the development of new graph-theoretical methods for network analysis. Furthermore, social network analysis methods are applied in many other fields such as: economics, physics, biology and information technologies.

One of the basic concepts in the analysis of social networks is betweenness centrality, a measure of centrality that is based on how well a node i is situated in terms of the paths that it lies on [11]:

$$c_B(i) = \frac{1}{n_B} \sum_{s,t \in V} \frac{\sigma_{s,t}(i)}{\sigma_{s,t}}, \qquad (1)$$

© Springer International Publishing Switzerland 2015
M.T. Thai et al. (Eds.): CSoNet 2015, LNCS 9197, pp. 216–227, 2015.
DOI: 10.1007/978-3-319-21786-4_19

where $\sigma_{s,t}$ is the total number of geodesics (shortest paths) between nodes s and t, $\sigma_{s,t}(i)$ is the number of geodesics between s and t that i lies on. The denominator n_B captures that the node i could lie on paths between as many as $n_B = (n-1)(n-2)/2$ pairs of other nodes. The complexity of the fastest algorithm to find $c_B(i)$ is $O(mn)$ where m is the number of edges and presented in [8].

One of shortcomings of the betweenness centrality is that it takes into accounts only the shortest paths, ignoring the paths that might be one or two steps longer, while the edges on such paths can be important for communication processes in the network. In order to take such paths into account, Brandes and Fleischer [9] and Newman [19] introduced the current flow betweenness centrality (CF-centrality, for short). In [9,19] the graph is regarded as an electrical network with edges being unit resistances. The CF-centrality of an edge is the amount of current that flows through it, averaged over all source-destination pairs, when one unit of current is induced at the source, and the destination (sink) is connected to the ground.

However, the modification proposed in [9,19] comes with a cost. In comparison with the original betweenness centrality, the bottleneck in the computation of CF-centrality is the matrix inversion with complexity $O(n^3)$. To mitigate this high complexity, in [2] the authors suggested a modification of CF-centrality, where in addition to the grounded sink, every node is attached to the ground with some small conductance proportional to the node degree.

The proposal in [2] makes the underlying linear system strongly diagonally dominant and reduces the computational cost of CF-centrality significantly but still needs to apply averaging over all source-destination pairs. In the current work, we go further and suggest to ground all nodes equally, which leads to averaging only over source nodes and reduces further computational cost. We refer to our new method as beta current flow centrality (βCF-centrality, for short).

Additionally, in contrast to the works [2,9,19], we consider weighted networks. Of course, the original betweenness centrality can easily be extended to weighted networks with integer weights. Namely, transform each link of the weight k into k parallel links of weight 1. We obtain a multigraph. The shortest path between two nodes is determined the same way as in unweighted graph. But the number of geodesics becomes larger because of the multi-links. For instance, if the nodes i_1 and i_2 are connected by k links and the nodes i_2 and i_3 are connected by l links, then the nodes i_1 and i_3 are connected by $k \cdot l$ paths. Applying the formula (1) to the nodes of multigraph we derive the centrality value for weighted graph, but with a very significant increase in computation cost. In the worst case scenario of k links between any two nodes the complexity of the algorithm to find $c_B(i)$ is $O(mn^k)$. In contrast, we note that our proposed method has the same computational complexity for weighted and non-weighted graphs.

Finally, we would like to note that, due to its relatively small computational cost, the proposed βCF-centrality is very well suited to serve as a characteristic function in the Myerson vector [1,18]. The concept of betweenness centrality via the Myerson vector was proposed in [12–14]. Considering the nodes in the network as players and the links as connections between players they formulate a communication game. The imputation of the general payoff in this cooperative game can be used for ranking of players and, respectively, for nodes of the graph. In [17] for the communication game with special characteristic function it was proposed a fairly simple imputation procedure based on the generating function and was shown that the resulting imputation agrees with the Myerson value. The advantage of the Myerson value is in taking into account the impact of all coalitions. Using the current flow betweenness centrality as a weight of any subset of the network it is possible to determine a new characteristic function and then rank the nodes as the Myerson value. This approach extends the game-theoretic approach from non-weighted to weighted graphs.

2   Beta Current Flow Centrality Based on Kirchhoff's Law

Consider a weighted graph $G = (V, E, W)$, where V is the set of nodes, E is the set of edges, and W is the matrix of weights, i.e.,

$$W(G) = \begin{pmatrix} 0 & w_{1,2} & \dots & w_{1,n} \\ w_{2,1} & 0 & \dots & w_{2,n} \\ \vdots & \vdots & \ddots & \vdots \\ w_{n,1} & w_{n,2} & \dots & 0 \end{pmatrix},$$

where $w_{i,j} \geqslant 0$ is weight of the edge connecting the nodes i and j, $n = |V|$ is the number of nodes. Note that $w_{i,j} = 0$ if nodes i and j are not adjacent. Here we assume that G is undirected graph, i.e. $w_{i,j} = w_{j,i}$. By random walk interpretation, the method can in fact be extended to directed networks.

Next we introduce the diagonal degree matrix:

$$D(G) = \begin{pmatrix} d_1 & 0 & \dots & 0 \\ 0 & d_2 & \dots & 0 \\ \vdots & \vdots & \ddots & \vdots \\ 0 & 0 & \dots & d_n \end{pmatrix},$$

where $d_i = \sum_{j=1}^{n} w_{i,j}$ is the sum of weights of the edges which are adjacent to node i in graph G. The Laplacian matrix $L(G)$ for weighted graph G is defined as follows:

$$L(G) = D(G) - W(G) = \begin{pmatrix} d_1 & -w_{1,2} & \dots & -w_{1,n} \\ -w_{2,1} & d_2 & \dots & -w_{2,n} \\ \vdots & \vdots & \ddots & \vdots \\ -w_{n,1} & -w_{n,2} & \dots & d_n \end{pmatrix}. \tag{2}$$

Let the graph G' be converted from the graph G by extension with an additional node $n + 1$ connected with all nodes of the graph G with the links of constant conductance β. Thus, we obtain the Laplacian matrix for the modified graph G' as:

$$L(G') = D(G') - W(G') = \begin{pmatrix} d_1 + \beta & -w_{1,2} & \cdots & -w_{1,n} & -\beta \\ -w_{2,1} & d_2 + \beta & \cdots & -w_{2,n} & -\beta \\ \vdots & \vdots & \ddots & \vdots & \vdots \\ -w_{n,1} & -w_{n,2} & \cdots & d_n + \beta & -\beta \\ -\beta & -\beta & \cdots & -\beta & \beta n \end{pmatrix}. \tag{3}$$

Suppose that a unit of current enters into the node $s \in V$ and the node $n+1$ is grounded. Let φ_i^s be the electric potential at node i when an electric charge is located at node s. The vector of all potentials $\varphi^s(G') = [\varphi_1^s, \ldots, \varphi_n^s, \varphi_{n+1}^s]^T$ for the nodes of graph G' is determined by the following system of equations (Kirchhoff's current law):

$$L(G')\varphi^s(G') = b_s', \tag{4}$$

where b_s' is the vector of $n+1$ components with the values:

$$b_s'(i) = \begin{cases} 1 & i = s, \\ 0 & \text{otherwise.} \end{cases} \tag{5}$$

The Laplacian matrix (2) is singular. The potential values can be determined up to a constant. Hence, without loss of generality, we can assume that the potential in node $n+1$ is equal to 0 (grounded node). Then, from (3) it follows that

$$\tilde{\varphi}^s(G') = \tilde{L}(G')^{-1} b_s, \tag{6}$$

where $\tilde{\varphi}^s(G')$, $\tilde{L}(G')$ and b_s are obtained from (3) by deleting the last row and column corresponding to node $n+1$. Notice that in $\varphi^s(G')$ and b_s' zero elements are deleted. This yields

$$\tilde{\varphi}^s(G') = [D(G) - W(G) + \beta I]^{-1} b_s, \tag{7}$$

where I is a unity matrix of size n.

Thus we can consider the vector $\tilde{\varphi}^s(G')$ as the vector of potential values for the nodes of graph G, that is,

$$\tilde{\varphi}^s(G) = [L(G) + \beta I]^{-1} b_s.$$

Rewrite (7) in the following form:

$$\tilde{\varphi}^s(G) = [(D(G) + \beta I) - W(G)]^{-1} b_s =$$

$$= [I - (D(G) + \beta I)^{-1} D(G) D^{-1}(G) W(G)]^{-1} (D(G) + \beta I)^{-1} b_s.$$

The matrices $(D(G) + \beta I)^{-1}$ and $(D(G) + \beta I)^{-1} D(G)$ are diagonal with the elements $\frac{1}{d_i + \beta}$ and $\frac{d_i}{d_i + \beta}$, $i = 1, \ldots, n$, denote these matrices as D_1 and D_2, respectively. The matrix $D^{-1}(G) W(G)$ is stochastic. Denote it as P. Consequently, we have

$$\tilde{\varphi}^s(G) = [I - D_2 P]^{-1} D_1 b_s = \sum_{k=0}^{\infty} (D_2 P)^k D_1 b_s. \tag{8}$$

From (8) it follows that the potential vector can be calculated by the recursion:

$$\tilde{\varphi}^s_{k+1}(G) = D_2 P \tilde{\varphi}^s_k(G) + D_1 b_s, \quad \tilde{\varphi}^s_0(G) = 0.$$

Note that the convergence is guaranteed since the matrix $D_2 P$ is substochastic. The rate of convergence can be easily regulated by the value of β.

The current let-through the link $e = (i,j)$ according to Ohm's law is $x^s_e = |\varphi^s_i - \varphi^s_j| \cdot w_{i,j}$. Consequently, given that the electric charge is in node s, the mean value of the current passing through node i is:

$$x^s(i) = \frac{1}{2}(b_s(i) + \sum_{e: i \in e} x^s_e), \qquad (9)$$

where

$$b_s(i) = \begin{cases} 1 & i = s, \\ 0 & \text{otherwise.} \end{cases}$$

Finally, we define beta current flow centrality (βCF-centrality) of node i as follows:

$$CF_\beta(i) = \frac{1}{n} \sum_{s \in V} x^s(i). \qquad (10)$$

We note that the above equation and the law of large numbers can be used to make a simple, light complexity, Monte Carlo type method for quick estimation of βCF-centrality. Specifically, we can take a small subset of nodes, $V_1 \subset V$, chosen independently and uniformly as source nodes in order to approximate βCF-centrality:

$$CF_\beta(i) \approx \frac{1}{|V_1|} \sum_{s \in V_1} x^s(i). \qquad (11)$$

Let us now investigate the limiting cases of large and small values of β. First, assume that β is large. Then, we can derive the following asymptotics for the potential vector.

$$\tilde{\varphi}^s = [L + \beta I]^{-1} b_s = \frac{1}{\beta}[I + \frac{1}{\beta}L]^{-1} b_s = \frac{1}{\beta} b_s - \frac{1}{\beta^2} L b_s + o\left(\frac{1}{\beta^2}\right)$$

From the above asymptotics, we can conclude that $x^s(s) = 1/2(1 + d_s/\beta) + o(1/\beta)$ and $x^s(i) = o(1)$, for $i \neq s$, and consequently,

$$CF_\beta(i) = \frac{1}{2n} + o(1), \quad \text{as} \quad \beta \to \infty,$$

which does not give informative ranking. Now for the other case $\beta \to 0$, we can derive the following asymptotics

$$\tilde{\varphi}^s = [L + \beta I]^{-1} b_s = \left[\frac{1}{\beta}\frac{1}{n}\mathbf{1}\mathbf{1}^T + L^\sharp + O(\beta)\right] b_s = \frac{1}{\beta}\frac{1}{n}\mathbf{1} + L^\sharp_{*,s} + O(\beta),$$

where we have used the Laurent series expansion for inversion of singularly perturbed matrices (see e.g., [4, Chapter 2]) with $\underline{1}$ denoting vector of ones of appropriate dimension, and $L^\sharp = [L - 1/n\underline{1}\underline{1}^T]^{-1} - 1/n\underline{1}\underline{1}^T$ denoting the group inverse of the Laplacian. Thus, we have

$$x_e^s = |L_{i,s}^\sharp - L_{j,s}^\sharp|w_{i,j} + o(1),$$

and hence a well-defined and non-trivial limit for βCF-centrality exists when $\beta \to 0$.

3   Illustrative Examples

3.1   Weighted Network of Six Nodes

Let us start with a simple six nodes network example which nicely explains the properties of the beta current flow centrality (see Fig.1). We compute all main measures of centrality for that weighted graph with six nodes. The results of computation are presented in Table 1. We see that classical betweenness centrality evaluates only the nodes A and D and gives 0 to other four nodes, even though they are obviously also important. The PageRank method ranks all nodes with equal values and thus it is indiscriminatory in this particular case. The current flow betweenness centrality and the βCF-centrality evaluate all nodes in quite similar manner. In particular, they both give rather high values to nodes A and D. As we mentioned in the introduction, the comparative advantage of the βCF-centrality in its small computational costs.

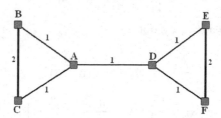

Fig. 1. Weighted network of six nodes

Table 1. Measures of centrality for weighted graph with six nodes

Nodes	A	B	C	D	E	F
Original betweenness centrality	6	0	0	6	0	0
PageRank centrality $\alpha = 0.85$	1/6	1/6	1/6	1/6	1/6	1/6
Current flow betweenness centrality	1.12	0.66	0.66	1.12	0.66	0.66
βCF-centrality $\beta = 1$	0.27	0.19	0.19	0.27	0.19	0.19

3.2   Star Graph

Consider a graph G of n nodes in the form of a star. Let node 1 be the center of the star. The modified Laplacian matrix in this case is given by

$$\tilde{L} = L + \beta I = D(G) - W(G) + \beta I$$

$$= \begin{pmatrix} n-1+\beta & -1 & \cdots & -1 \\ -1 & 1+\beta & \cdots & 0 \\ \vdots & \vdots & \ddots & \vdots \\ -1 & 0 & \cdots & 1+\beta \end{pmatrix}.$$

Its inverse matrix is

$$\tilde{L}^{-1} = (L + \beta I)^{-1}$$

$$= \frac{1}{\beta(1+\beta)(n+\beta)} \begin{pmatrix} (1+\beta)^2 & 1+\beta & 1+\beta & \cdots & 1+\beta \\ 1+\beta & 1+\beta(n+\beta) & 1+\beta & \cdots & 1 \\ 1+\beta & 1 & 1+\beta(n+\beta) & \cdots & 1 \\ \vdots & \vdots & \vdots & \ddots & \vdots \\ 1+\beta & 1 & 1 & \cdots & 1+\beta(n+\beta) \end{pmatrix}.$$

If we take as a source node $s = 1$, we find from (9) that

$$x^s(1) = \frac{1}{2}\left(1 + \frac{n-1}{n+\beta}\right),$$

and

$$x^s(i) = \frac{1}{2(n+\beta)}, \quad i = 2, ..., n.$$

And for a source node $s \neq 1$, we obtain

$$x^s(1) = \frac{2n-3+\beta}{2(1+\beta)(n+\beta)},$$

$$x^s(s) = \frac{1}{2}\left(1 + \frac{n-1+\beta}{(1+\beta)(n+\beta)}\right),$$

$$x^s(i) = \frac{1}{2(1+\beta)(n+\beta)}, \quad i \neq 1, s.$$

The latter yields that the βCF-centrality for the star graph is

$$CF_\beta(1) = \frac{1}{2n}\left(1 + \frac{n-1}{n+\beta} + (n-1)\frac{2n-3+\beta}{(1+\beta)(n+\beta)}\right) = \frac{1}{2n} + \frac{(n-1)(n-1+\beta)}{n(1+\beta)(n+\beta)},$$

$$CF_\beta(i) = \frac{1}{2n}\left(\frac{1}{n+\beta} + 1 + \frac{n-1+\beta}{(1+\beta)(n+\beta)} + (n-2)\frac{1}{(1+\beta)(n+\beta)}\right)$$

$$= \frac{1}{2n} + \frac{n-1+\beta}{n(1+\beta)(n+\beta)}, \quad i = 2, ..., n.$$

In particular, we can conclude from the above expressions that if $\beta \to \infty$ all nodes obtain the same value $1/(2n)$. And if $\beta \to 0$ and n is large, the central node obtains a value very close to one and the other nodes have nearly zero value. This is in agreement with the general asymptotics derived in the previous section.

This example also shows that the βCF-centrality can be viewed as a flexible characteristic function and thus efficiently used in the calculation of the Myerson vector.

3.3   The Results of Computer Experiments with Online Social Network VKontakte

In this subsection we consider the weighted graph extracted from the popular Russian social network VKontakte. The graph corresponds to the online community devoted to game theory. This community consists of 483 participants. As a weight of a link we take the number of common friends between the participants. In fact, the probability that two participants are familiar depends on the number of common friends [14]. This approach is often used in online social networks for link recommendation.

In Fig. 2 we show the principal component of the community Game Theory, which consists of 275 nodes. It is difficult to see from Fig. 2 which nodes are more important with respect to the community connection structure. Then, we have converted this graph to another modified graph by deleting the links whose weights are less than three. This new weighted graph is presented in Fig. 3. The thickness of a link depends on the link weight, i.e. on the number of common friends.

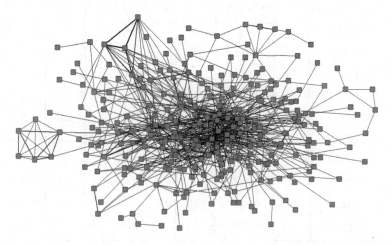

Fig. 2. Principal component of the community Game Theory in the social network VKontakte (number of nodes: 275, number of edges: 805 and mean path's length: 3.36)

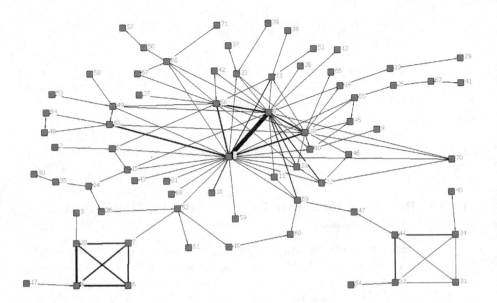

Fig. 3. Principal component of the community Game Theory in the social network VKontakte after deleting the links weighted less than 3 (number of nodes: 71, number of edges: 116 and mean path's length: 3.75)

Table 2. Measures of centrality for top nodes of social network VKontakte

Nodes	βCF-centrality ($\beta = 0.3$)	Nodes	PageRank centrality ($\alpha = 0.85$)	Nodes	Weighted betweenness centrality "tnet" ($\alpha = 1.5$)	Nodes	CF-centrality
1	0.4168	1	0.1359	1	1846	1	0.6406
8	0.3143	8	0.1189	8	1398	8	0.4919
52	0.1463	56	0.0432	52	500	69	0.2946
69	0.1454	28	0.0366	69	494	52	0.2748
28	0.1299	44	0.0277	47	384	28	0.2095
56	0.1273	4	0.0267	44	331	56	0.1942
7	0.1002	32	0.0252	63	331	47	0.1880
15	0.0931	20	0.0244	7	325	44	0.1649
66	0.0922	63	0.0228	55	265	15	0.1645
63	0.0896	6	0.0212	15	228	7	0.1642

Table 3. The results of the Monte Carlo approach with sampling only 10% of the nodes as sources

Nodes	βCF-centrality ($\beta = 0.3$)	Nodes	Monte Carlo approach 10% of the nodes
1	0.4168	1	0.5043
8	0.3143	8	0.4134
52	0.1463	52	0.2468
69	0.1454	23	0.2307
28	0.1299	28	0.2255
56	0.1273	20	0.2003
7	0.1002	7	0.1982
15	0.0931	24	0.1871
66	0.0922	63	0.1789
63	0.0896	10	0.1786
47	0.0889	15	0.1763
24	0.0880	55	0.1756
44	0.0842	36	0.1613
55	0.0801	69	0.1565
49	0.0725	12	0.1457
23	0.0702	39	0.1438
13	0.0699	45	0.1403
10	0.0610	56	0.1397
14	0.0598	3	0.1360
25	0.0564	4	0.1234

The results of computing the βCF-centrality for the social network VKontakte are given in Table 2. Here we take $\beta = 0.3$. It is useful to compare these values of βCF-centrality with the results corresponding to the PageRank and classical notation of centrality using the shortest paths [20] for the parameter $\alpha = 1.5$. We present in the table only the lists of top-10 nodes for each centrality measure.

From Table 2 we find that all four methods ranked two main nodes 1 and 8 in the same order. We can already see that, as in the six node network example,

βCF-centrality is more similar to CF-centrality and betweenness centrality than to PageRank.

On Figure 3 we can see that node 52 connects the subgraph $\{3, 4, 6, 7, 17, 20\}$ with the rest of the graph. Thus, we can expect that node 52 deserves high centrality rank. Similarly, we also expect that node 7 should have high centrality rank. The ranking according to βCF-centrality confirms this intuitive expectation, as they take positions 3 and 7, respectively (See Table 2). We also note that nodes 4, 20, 6, 17 and 3 took positions 22, 24, 36, 68 and 69, respectively. However, PageRank gives to nodes 52 and 7 only positions 22 and 12, respectively. Furthermore, under PageRank ranking nodes 4, 20 and 6 took positions 6, 8 and 10, respectively. Namely, PageRank ranks nodes 4, 20 and 6 higher than node 52. This does not correspond at all to our intuition.

Finally, in Table 3 we present the results of the Monte Carlo approach (see equation (11)) with sampling only 10% of the nodes as sources. Nodes 1 and 8 as before lead the ranking and there are 6 correct elements in the top-10 basket of nodes [3]. Monte Carlo approach also determines correctly the ranks of the key nodes 52 and 7.

4   Conclusion

Betweenness centrality measure is an important tool in the analysis of social networks. The structure of a network is represented by a graph. The original betweenness centrality measure is based on the assumption that the information is transmitted along geodesics (shortest paths) between any two nodes. There is a criticism of this approach that it does not take into account information spread along non-shortest paths. The current flow betweenness centrality based on electric circuit interpretation tries to mitigate this shortcoming. However, this comes with the increase of computational cost. We introduce here the βCF-centrality method which depends on the parameter β. This method is versatile, has lower computational complexity and can be easily used as characteristic function in the Myerson vector.

Acknowledgements. This research is supported by Russian Humanitarian Science Foundation (project 15-02-00352), the Division of Mathematical Sciences of Russian Academy of Sciences, EU Project Congas FP7-ICT-2011-8-317672 and Campus France.

References

1. Aumann, R., Myerson, R.: Endogenous formation of links between players and coalitions: an application of the Shapley value. In: The Shapley value, pp. 175–191. Cambridge University Press (1988)
2. Avrachenkov, K., Litvak, N., Medyanikov, V., Sokol, M.: Alpha current flow betweenness centrality. In: Bonato, A., Mitzenmacher, M., Prałat, P. (eds.) WAW 2013. LNCS, vol. 8305, pp. 106–117. Springer, Heidelberg (2013)

3. Avrachenkov, K., Litvak, N., Nemirovsky, D., Smirnova, E., Sokol, M.: Quick detection of top-k personalized pagerank lists. In: Frieze, A., Horn, P., Prałat, P. (eds.) WAW 2011. LNCS, vol. 6732, pp. 50–61. Springer, Heidelberg (2011)
4. Avrachenkov, K.E., Filar J.A., Howlett, P.G.: Analytic Perturbation Theory and its Applications. SIAM (2013)
5. Borgatti, S.P., Everett, M.G., Freeman, L.C.: Ucinet for Windows: Software for Social Network Analysis. Harvard (2002)
6. Borm, P., Owen, G., Tijs, S.: On the position value for communication situations. SIAM J. on Disc. Math. **5**(3), 305–320 (1992)
7. Borm, P., van den Nouweland, A., Tijs, S.: Cooperation and communication restrictions: a survey. In: Imperfections and Behavior in Economic Organizations. Kluwer (1994)
8. Brandes, U.: A faster algorithm for betweenness centrality. Journal of Mathematical Sociology **25**, 163–177 (2001)
9. Brandes, U., Fleischer, D.: Centrality measures based on current flow. In: Diekert, V., Durand, B. (eds.) STACS 2005. LNCS, vol. 3404, pp. 533–544. Springer, Heidelberg (2005)
10. Calvo, E., Lasaga, J., van den Nouweland, A.: Values of games with probabilistic graphs. Math. Social Sci. **37**, 79–95 (1999)
11. Freeman, L.C.: A set of measures of centrality based on betweenness. Sociometry **40**, 35–41 (1977)
12. Jackson, M.O.: Allocation rules for network games. Games and Econ. Behav. **51**(1), 128–154 (2005)
13. Jackson, M.O., Wolinsky, J.: A strategic model of social and economic networks. J. Econ. Theory **71**(1), 44–74 (1996)
14. Jackson, M.O.: Social and economic networks. Princeton University Press (2008)
15. Jamison, R.E.: Alternating Whitney sums and matchings in trees. Part 1. Disc. Math. **67**, 177–189 (1987)
16. Mazalov, V.: Mathematical Game Theory and Applications. Wiley (2014)
17. Mazalov, V.V., Trukhina, L.I.: Generating functions and the Myerson vector in communication networks. Disc. Math. and Appl. **24**(5), 295–303 (2014)
18. Myerson, R.B.: Graphs and cooperation in games. Math. Oper. Res. **2**, 225–229 (1977)
19. Newman, M.E.J.: A measure of betweenness centrality based on random walks. Social networks **27**, 39–54 (2005)
20. Opsahl, T., Agneessens, F., Skvoretz, J.: Node centrality in weighted networks: generalizing degree and shortest paths. Social Networks **32**, 245–251 (2010)
21. Slikker, M.: Link monotonic allocation schemes. Int. Game Theory Review **7**(4), 473–489 (2005)
22. Slikker, M., Gilles, R.P., Norde, H., Tijs, S.: Directed networks, allocation properties and hierarchy formation. Math. Social Sci. **49**(1), 55–80 (2005)
23. Talman, D., Yamamoto, Y.: Average tree solutions and subcore for acyclic graph games. J. Oper. Res. Soc. Japan **51**(3), 187–201 (2008)

A Method of Link Prediction Based on Betweenness

Pengyuan Zhang[(✉)], Jianping Li, Enming Dong, and Qi Liu

College of Science, National University of Defense Technology,
Changsha 410073, China
pyl1007@163.com

Abstract. Link prediction in complex networks has attracted increasing attention of researchers in many domains. The prediction methods are usually used to find missing information, identify spurious interactions, and reconstruct networks. Inspired by the rich-get-richer mechanism, we propose a novel index on the basis of betweenness. Extensive experiments show that the proposed method performs well on some networks. Especially, on the Adjnoun network and Florida network, it outperforms some mainstream link prediction baselines, such as *CN Index*, *AA Index* and *RA Index*.

Keywords: Link prediction · Complex networks · Betweenness

1 Introduction

Many complex systems can be well described as networks, ranging from biochemical networks, through the Internet, to various social networks. Nodes represent individuals or organizations and edges denote the relations between them [1]. Link prediction as one of the most fundamental problems plays a significant role in understanding the intrinsic evolutionary mechanisms of networks. On protein-protein interaction network, whether a link between two nodes exists must be demonstrated by laboratorial experiments, which are usually very costly. Instead of checking all possible interactions, focusing on those links most likely to exist can sharply reduce the experimental cost if the predictions are accurate enough. On online social networks, the accurate predictions can help users to find new friends and further enhance their loyalties to the web sites [2]. On online shopping networks, with the help of link prediction methods, recommending customers the most likely right goods is welcome to not only the buyers but also the sellers.

Link prediction is a long-standing challenge in complex networks, and a lot of methods have been proposed based on topological features and/or the structural characteristics of networks, like the node attribute, the hierarchical organization [3] and community structure [4]. Mainstream link prediction methods can be classified into two major classes. The first class is the similarity-based algorithms that take into account the topological similarity based on network structures only, such as *CN Index* counting the number of common neighbours [5], *AA*

© Springer International Publishing Switzerland 2015
M.T. Thai et al. (Eds.): CSoNet 2015, LNCS 9197, pp. 228–235, 2015.
DOI: 10.1007/978-3-319-21786-4_20

Index [6] and *RA Index* [1] penalizing the large-degree common neighbours, and so on. This kind of methods usually has a low computational expense but performs not very well on some networks to some extent, like Adjnoun network [15] and Florida network [16]. The second class is proposed based on maximum likelihood estimation, like Stochastic Block Model(BM) [7] and Hierarchical Random Graph(HRG) [8] predicting more accurate with higher computational complexity. Recently, a new measure based on neighbor communities is proposed with good performance [9], meanwhile, a proposed method via convex nonnegative matrix factorization on multiscale blocks gives better results than common neighbours method when the networks have a large number of missing links [10].

However, the network evolutionary mechanisms coincide with our capacity to predict missing links. For example, rich-get-richer mechanism is very popular. When people search papers to read, the higher cited ones are always favoured and then have more popularity to be cited again. When someone has extra money to deposit, larger bank is usually the favourite choice. Does the pair of unlinked nodes with large betweenness also have more probability to be connected? Therefore, we propose an index called $B-Index$ based on betweenness to predict missing links. Empirical results verify that the proposed index improves prediction accuracy, compared with the three mainstream baselines, especially on the Adjnoun network [15], Karate network [14] and Florida network [16].

2   Method

Suppose we have an undirected simple network $G(V, E)$, where V is the set of nodes and E is the set of links.

2.1   Common Neighbours (CN)

Let $\Gamma(i)$ represents the set of neighbours of node v_i. By common sense, two nodes are more possible to have a link if they have many common neighbours. Thus, CN *Index* [5] measures the probability between two nodes with the number of their common neighbours:

$$s_{ij}^{CN} = |\Gamma(v_i) \cap \Gamma(v_j)| \tag{1}$$

where $\Gamma(i)$ denotes the set of neighbours of node v_i, $\Gamma(i) \cap \Gamma(j)$ indicates the set of common neighbours of node v_i and v_j.

2.2   Adamic-Adar Index (AA)

AA Index [6] refines the counting of common neighbours by emphasizing less-connected common neighbours, as:

$$s_{ij}^{RA} = \sum_{z \in \Gamma(i) \cap \Gamma(j)} \frac{1}{\log k_z} \tag{2}$$

in which k_z is the degree of the node v_z.

2.3   Resource-Allocation Index (RA)

RA Index [1] simulates resource transmissions between two nodes, and penalizes common neighbours with large degrees, it is defined as:

$$s_{ij}^{RA} = \sum_{z \in \Gamma(i) \cap \Gamma(j)} \frac{1}{k_z} \tag{3}$$

2.4   B-Index

Different node has different influence, the influence of a node denotes the significance. Obviously, the significance of a node is not only an individual property but also related to other nodes. It determines the ability to capture resource, information, and is always related with network structure. Popularly, the significance of the node is measured by the Betweenness. Betweenness regards a node as being in a favoured position to the extent that the node falls on the shorted paths between other pairs of nodes in the network [11]. Moreover, Betweenness is in proportion to the number of shortest paths from all nodes to the others that pass through that node. If item transfers through the network follows the shortest paths, a node with high Betweenness has a large influence on the transfer behaviour. Motivated by the rich-get-richer phenomenon, we hypothesize that the probability of two nodes to make a link is related to the sum of their betweenness. Therefore, the sum of two nodes' betweenness is bigger, the two nodes are more possible to be connected.

Definition 1. The betweenness [12] of the node v_k is given by the expression:

$$g(v_k) = \sum_{v_i \neq v_k \neq v_j} \frac{\sigma_{ij}(v_k)}{\sigma_{ij}} \tag{4}$$

where σ_{ij} is the total number of shortest paths from node v_i to node v_j, and $\sigma_{ij}(v_k)$ is the number of those paths that pass through node v_k.

We claim that in the network construction, if we make a link between two unconnected nodes v_i and v_j, then the betweenness of v_i and v_j will get bigger, respectively.

The reason is as follows. v_x and v_y are two random nodes but different with v_i and v_j. Take the node v_i for example. Firstly, the distance from node v_x to node v_y will be shorter than or equal to that of node v_i and v_j unconnected. So, it may contribute more to the betweenness of node v_i. Secondly, if a shortest path (between node v_x and v_j) passes through the node v_i, the contribution to the betweenness of node v_i will be 1 if node v_i and node v_j are connected.

The Betweenness is essential in the analysis of complex networks, but costly to compute. Currently, the known algorithms require $O(n^3)$ time, where n is the number of nodes in the network [13].

Thus, we get the $B - Index$ as follows:

$$s_{ij}^{B} = g(v_i) + g(v_j) \tag{5}$$

Table 1. Algorithm of the $B - Index$

Algorithm of $B - Index$
input: $A = (a_{ij})_{n \times n}$ //observed network
output: S //prediction network
1. Compute the betweenness of the node v_k by the equation (4).
2. Calculate s_{ij}^{B} for each v_i, v_j by the equation (5).
3. $S = (s_{ij}^{B})$.

3   Experiments

In general, links between different nodes may have different weights to measure their relative importance in networks. In our experiments, we set all weights to be one. Multiple links and loops are not allowed, and we convert arcs into undirected links.

3.1   Data

In this paper, we consider six representative networks from disparate fields: 1) Karate [14]: the network of friendships between the 34 members of a karate club. 2) Adjnoun [15]: the network of common adjective and noun adjacencies for the novel "David Copperfield". 3) Florida [16]: the food web of Florida ecosystem, the relations of carbon exchanges in the cypress wetlands of South Florida during the wet season. 4) USAir97 [17]: the network of the USA air transportation system, which contains 332 airports and 2126 airlines. 5) ERDOS971 [18]: Erdos collaboration network containing 472 researchers and 1314 papers. 6) Email [19]: the e-mail communication network of a university in Spain.

    Table 2 gives the basic topological features of these networks. Brief definitions of the monitored topological measures can be found in the table caption.

Table 2. The basic topological features of six example networks, $|V|$ and $|E|$ are the number of nodes and edges, respectively. C indicates the clustering coefficient [20], and $< k >$ represents the average degree of the network, ρ denotes the density of the network, defined as $\rho = \frac{2|E|}{|V|(|V|-1)}$, $< d >$ is the average distance. r denotes the assortative coefficient [21], H is the degree heterogeneity, defined as $H = \frac{<k^2>}{<k>^2}$.

Networks	$|V|$	$|E|$	C	$< k >$	ρ	$< d >$	r	H
Karate	34	78	0.588	4.589	0.139	2.408	-0.476	1.693
Adjnoun	112	425	0.190	7.589	0.068	2.536	-0.129	1.815
Florida	128	2,075	0.334	32.422	0.255	1.776	-0.111	1.237
USAir97	332	2,126	0.749	12.807	0.039	2.738	-0.208	3.464
ERDOS971	472	1,314	0.347	5.568	0.012	4.021	0.182	2.442
Email	1,133	5,452	0.254	9.624	0.0086	3.606	0.078	1.941

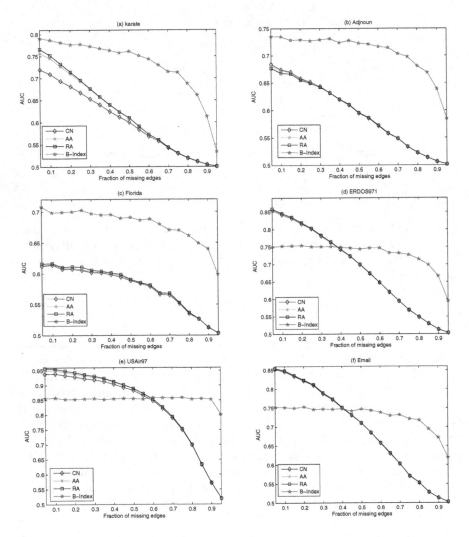

Fig. 1. Comparisons of $B - Index$ with CN, AA and RA on six networks

3.2   Evaluation Method

The set of links E is randomly divided into two parts: the training set E^T and the probe set E^P. The former one is treated as known information, the latter one is used for testing. Clearly, $E = E^T \cup E^P$ and $E^T \cap E^P = \varnothing$.

To quantify the accuracies of link prediction methods, the standard metric we use is AUC [22]. The value of AUC can be interpreted as the probability that a randomly chosen missing link (a link in E^P) is given a higher score than a randomly chosen nonexistent link (a link in $U \backslash E$, where U represents the universal set). In practice, we do n independent comparisons, if there are n^0

times the missing link having a higher score than the nonexistent link and n^* times they are equal, the AUC value can be computed by the formula below:

$$AUC = \frac{n^0 + 0.5 * n^*}{n}. \tag{6}$$

3.3   Results

Firstly, we compare $B-Index$ with the similarity-based algorithms: CN, AA and RA, measured by AUC. As shown in Fig 1, our method outperforms all the time in the networks: Karate, Adjnoun and Florida. On the networks: ERDOS971, USAir97 and Email, although $B-Index$ doesn't do very well at the begining, when the fraction of missing edges is large enough, our method outperforms the other three algorithms. What's more, the AUC of all methods decrease as the fraction of missing edges grows, but our method decreases very slowly.

Note that the AUC of CN, AA and RA are more or less similar in the performance on the six networks, while that of $B-Index$ is not. The reason is twofold. Firstly, the first three methods are all based on common neighbours, yet $B-Index$ is proposed based on betweenness. Secondly, betweenness is a global network attribute, but that three measures are local similarity indices. In addition, as we can see from Fig 1, on the network of Florida of which the average degree is only 1.237 and the degree heterogeneity is only 1.776, the AUC raises almost 0.2 all the time. However when the proposed method is applied on the ERDOS971 network whose average degree is 4.021 and degree heterogeneity is 2.442, it does better only when the fraction of missing edges is bigger than 0.4. Obviously, $B-Index$ does better on the networks that simultaneously have small degree heterogeneity and small average distance. The reason is that in this kind of networks, betweenness determines the ability of capturing the flow of information more perfectly without the turbulence from node's degree.

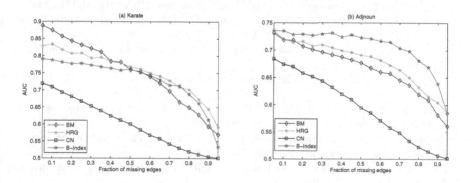

Fig. 2. Comparisons of $B-Index$ with BM, HRG and CN on networks: Karate and Adjnoun

What's more, we compare $B-Index$ with three representative methods: Stochastic Block Model(BM), Hierarchical Random Graph(HRG) and Common neighbors(CN), the AUC results are shown in Fig 2. BM and HRG show better performance in prediction in some networks but suffer from high computational complexity. However, $B - Index$ gets the best performance on the Adjnoun network. Moreover, the AUC of $B - Index$ gets down quickly after a point. This phenomenon is the result that the network is too sparse. Before the point, $B - Index$ shows much more reliability and stability.

4   Conclusion

In this paper, we have proposed a new measure, $B - Index$, motivated by the rich-get-richer mechanism, to predict missing links. The proposed index is calculated by summing up the two ends' betweenness. The experiments on six networks reveal that two nodes are more likely to be connected if the sum of their betweenness is bigger. Besides, the proposed method does well in predicting missing links on monitored network which simultaneously has small average distance and small degree heterogeneity, like Karate, Adjnoun and Florida.

Acknowledgments. The authors would like to thank Zheng Xie for helpful discussions and good ideas, and Xuan Zhao for proofreading this paper.

References

1. Zhou, T., Lü, L., Zhang, Y.-C.: Predicting missing links via local information. Eur. Phy. B **71**, 623–630 (2009)
2. Lü, L., Zhou, T.: Link prediction in complex networks: a survey. Physica A **390**, 1150–1170 (2011)
3. Sales-Pardo, M., Guimer, R., Amaral, L.A.N.: Extracting the hierarchical organization of complex systems. Proc. Natl. Acad. Sci. USA **104**, 15224 (2007)
4. Girvan, M., Newman, M.E.J.: Community structure in social and biological networks. Proc. Natl. Acad. Sci. USA **99**, 7821 (2002)
5. Newman, M.E.J.: Clustering and preferential attachment in growing networks. Phys. Rev. E **64**, 025102 (2001)
6. Adamic, L.A., Adar, E.: Friends and neighbors on the web. Soc. Networks **25**, 211 (2003)
7. Airoldi, E.M., Blei, D.M., Fienberg, S.E., Xing, X.P.: Mixed-membership stochastic blockmodels. J. Mach. Learn. Res. **9**, 1981 (2008)
8. Clauset, A., Moore, C., Newman, M.E.J.: Hierarchical structure and the prediction of missing links in networks. Nature **453**, 98 (2008)
9. Xie, Z., Dong, E., Li, J., Kong, D., Wu, N.: Potential links by neighbor communities. Physica A **406**(C), 244–252 (2014)
10. Dong, E., Li, J., Xie, Z.: Link Prediction via Convex Nonnegative Matrix Factorization on Multiscale Blocks. Journal of Applied Mathematics **2014** (2014)
11. Hanneman, R.A., Riddle, M.: Introduction to Social Network Methods. University of California (2005)
12. Freeman, L.C.: Centrality in networks: I. Conceptual clarification. Social Networks **1**(C), 215–239 (1979)

13. Brandes, U.: A faster algorithm for betweenness centrality. Journal of Mathematical Sociology **25**(2), 163–177 (2001)
14. Zachary, W.W.: An information flow model for conflict and fission in small groups. Journal of Anthropological Research **33**, 452–473 (1977)
15. Newman, M.E.J.: Finding community structure in networks using the eigenvectors of matrices. Preprint physics, 0605087 (2006)
16. Melián, C.J., Bascompte, J.: Food web cohesion. Ecology **85**, 352–358 (2004)
17. Batageli, V., Mrvar, A.: Pajek Datasets. http://vlado.fmf.unilj.si/pub/networks/data/default.htm
18. Davis, T., Hu, Y.: The University of Florida Sparse Matrix Collection. http://www.cise.ufl.edu/research/sparse/matrices/
19. Guimera, R., Danon, L., Diaz-Guilera, A., Giralt, F., Arenas, A.: Self-similar community structure in a network of human interactions. Phys. Rev. E **68**, 065103 (2003)
20. Watts, D.J., Strogatz, S.H.: Collective dynamics of small-worldnetworks. Nature **393**, 440–442 (1998)
21. Newman, M.E.J.: Assortative mixing in networks. Phys. Rev. Lett. **89**, 208701 (2002)
22. Hand, D.J., Till, R.J.: A simple generalisation of the area under the ROC curve for multiple class classification problems. Mach. Learn. **45**, 171–186 (2001)

A Trust Measurement in Social Networks Based on Game Theory

Yingjie Wang[1], Zhipeng Cai[2](✉), Guisheng Yin[3], Yang Gao[1],
and Qingxian Pan[1]

[1] School of Computer and Control Engineering, Yantai University,
Yantai 264005, Shandong, China
[2] Department of Computer Science, Georgia State University,
Atlanta, GA 30303, USA
zcai@gsu.edu
[3] College of Computer Science and Technology,
Harbin Engineering University, Harbin 150001, China

Abstract. In social networks, trust is a complex social relationship between entities. How to calculate the trust degree more accurately is an important research issue. This paper proposes a trust measurement model in social networks based on game theory. The trust degree is calculated from three aspects, service reliability, feedback effectiveness, recommendation credibility, to get more accurate result. In addition, in order to alleviate the free-riding problem, based on game theory, this paper proposes a punishment mechanism according to the specific trust degree and the global trust degree. The simulation results indicate the effectiveness of the proposed trust measurement model. And it can effectively solve the free-riding problem in social networks through the proposed punishment mechanism.

Keywords: Service reliability · Feedback effectiveness · Recommendation credibility · Game theory · Punishment mechanism

1 Introduction

With the current popularity of online social networks, more and more information is distributed through social network services. Participants in online social networks want to share information and experiences with as many reliable users as possible [1,2]. Trust between nodes is the basis of social network services. However, the modeling of trust is complicated and application-dependent [3,4]. Modeling trust needs to consider interaction history, recommendation, user behaviors and so on. Therefore, modeling trust is an important focus for online social networks [5–7].

In social networks, the existing trust models are mainly constructed on the basis of the nodes' global trust. However, these models fail to filter the false feedback and distrust recommendation, which leads to the inaccuracy of

© Springer International Publishing Switzerland 2015
M.T. Thai et al. (Eds.): CSoNet 2015, LNCS 9197, pp. 236–247, 2015.
DOI: 10.1007/978-3-319-21786-4_21

the measurements. Because it is common that nodes intend to be selfish, the free-riding phenomenon often occurs in complex networks, resulting in the decrease of network performance. According to the free-riding, the so-called free-riders attempt to benefit from network resources of others without offering their own resources in exchange [8]. The goal of our work is to build an effective trust measurement model that can benefit social network services, such as controlling feedback, recommendation, and strategy selection, etc. In an effort to solve the above problems, the main contributions of our work are summarized as follows.

1. In order to more accurately measure trust degree of a node, we introduce three novel evaluation factors which are service reliability, feedback effectiveness and recommendation credibility.
2. Another practical problem considered in this paper is the free-riding problem. We propose a punishment mechanism based on the proposed trust model, which is different from the existing works where statistic methods are commonly used. In our punishment mechanism, we employ the evolutionary game theory which is more flexible and effective.

The rest of the paper is organized as follows: Section 2 reviews the related works and presents the motivation for our work. Section 3 introduces the proposed trust measurement model for social networks. Section 4 illustrates our simulation results and analysis of the results. Conclusions and future work are shown in Section 5.

2   Related Work

Much effort has been spent on trust measurement models to depict trust behaviors in complex networks. Trust measurement methods under open network environment and trust measurement methods based on Agent synergy are the most important trust measurement methods.

i. Trust Measurement Methods Under Open Network Environment

Beth et al. [9] first proposed a trust measurement method under open network environment. In their work, trust is regarded as direct trust and recommendation trust, and a probabilistic method is adopted to represent trust. The PeerTrust model [10] uses the transaction and the community background as the source of reputation feedback. It can act as a defense against some of the subtle malicious attacks, e.g., a seller develops a good reputation by being honest for small transactions and tries to make a big profit by being dishonest for large transactions. The EigenRep model [11] assumes that if the direct trust between a node and the destination node is higher, the recommendation trust is more reliable. The model uses direct trust to calculate the global trust. This model can effectively solve the bad effect caused by the malicious recommendation. Wang et al. [12] proposed a trust model based on Bayesian network. This model investigates how to describe different aspects of trust to obtain various properties of entities according to different scenes. Wang et al. [13] solved the problem of recommendation trust based on the Bayesian method. This method calculates recommendation trust based on experts' experience. Lu *et al.* [14] proposed an evaluation method of software reliability. It is a bottom-up calculation process of trust level that can decompose and synthetically derive a parallel structure, so that the trust value of a system can be calculated accurately. However, there are still some shortages about this kind of models. They only adopt probabilistic model to establish subjective trust model. In other words, subjectivity and uncertainty of trust are equivalent to randomness. They also adopt the averaging method to calculate recommendation trust, which cannot reflect the real situations of a trust relationship.

ii. Trust Measurement Methods Based on Agent Synergy

In Agent synergy, trust means that a collaborative Agent can properly and non-destructively predict subjective possibility of a collaborative activity. The source of prediction is the goal service behavior that previous Agent observes. Prediction results are affected by evaluation of important degree from the Agent, such as key collaborative activities, secondary collaborative activities, *etc* [15]. The eBay trust model [16] is one of the most successful cases. In this model, the entities evaluate each other after each transaction. The structure of this system is straightforward, and the computation cost is small. Because trust between Agents is associated with other entities' subjective understanding and fuzziness, it cannot be described and managed by conventional and accurate logic. Subjective trust as a cognitive phenomenon, whose subjectivity and uncertainty present fuzziness, is often managed by Fuzzy Set based methods. It not only reflects fuzziness of Agent trust, but also describes the trust mechanism between Agents with intuitive and concise semantics. Tang *et al.* [17] first proposed the definition and evaluation of trust based on the fuzzy set theory. They gave formalization, and deducted rules of trust to construct a complete subjective trust management model. However, this kind of models fail to consider the cooperative cheating behaviors, which cannot detect the community of cooperative cheating.

In addition, some recent works are also remarkable. Shi *et al.* [18] proposed a dynamic P2P trust model based on the time-window feedback mechanism. The model considers the inherent connection among trust, reputation and incentive and the effect of time factor on the trust computation. Gan *et al.* [19] proposed a reputation-based multi-dimensional trust (RMDT) algorithm which makes use of a self-confident coefficient to synthesize the direct and recommendation trust to evaluate the nodes in a network. A multi-dimensional trust mechanism is also introduced to improve sensitivity of RMDT on a single attribute. Meng *et al.* [20] proposed the @Trust model. Bedi *et al.* [21] proposed a trust based recommender system using ant colony for trust computation. Zhang *et al.* [22] proposed a trust evaluation method based on the cloud model.

These models have promoted the development of trust measurement. However, most of the existing models fail to filter the false feedback and distrust recommendation, which leads to the inaccuracy of the measurements. In addition, the free-riding problem is not comprehensively considered in one trust measurement model. Considering these problems, this paper proposes a new trust measurement model for social networks based on Game Theory. The proposed model introduces three novel evaluation factors which are service reliability, feedback effectiveness and recommendation credibility to more accurately measure the trust degree of a node.

3   The Proposed Trust Measurement Model

In order to describe the trust degree more accurately, this paper divides nodes into four categories, which are service nodes, feedback nodes, recommendation nodes and managed nodes. In particular, each node can become the four different roles in different transactions, *i.e.*, service node, feedback node, recommendation node or managed node. In social networks, *trust* represents the level of confidence about the reliability and correctness of entity's behaviors. *Service reliability* indicates the trustworthiness of service that service nodes provide; *feedback effectiveness* represents the trustworthiness of feedback that feedback nodes return; *recommendation credibility* expresses the trustworthiness of recommendation that recommendation nodes give. In this paper, the global trust of the node i, denoted as T_i, is the probability of i being correct. The service reliability is denoted as ST_i; the feedback effectiveness is denoted as FT_i; and the recommendation credibility is denoted as CT_i.

In this paper, let i be a service node, j be a feedback node and k be a recommendation node; and M_i, M_j, M_k are the managed nodes of i, j, k, respectively. In each transaction, the managed nodes are different, because every node has the probability to become managed node.

When feedback node j requests a specific service s, the managed node M_j searches for the trust node which can provide service s. If there exists such a node i, then node j requests the service from node i. If not, M_j searches for the recommendation node k. Then node k recommends a service node i with the maximum trust degree that can provide service s to node j. If there does not exist a recommendation node k, the transaction fails.

3.1   The Trust Measurement Process

In this model, the specific feedback value $f_{v_{j,i}}$, given by the feedback node j, is known by the system. Therefore, we obtain the calculation method of service reliability based on the specific feedback value $f_{v_{j,i}}$, which is shown by Eq.(1).

$$ST_i = \frac{\sum_{j \in set(i)} f_{v_{j,i}} \cdot \lambda(j,i)}{\sum_{j \in set(i)} \lambda(j,i)}, FT_j \geq \theta \tag{1}$$

In Eq. (1), $set(i)$ is the set of feedback nodes that communicated with service node i, and θ is the threshold of feedback effectiveness. $\lambda(j,i)$ presents the influence effect of node j on node i. In addition, FT_j represents the feedback effectiveness of node j.

In social networks, some feedback nodes may evaluate some trust nodes maliciously, and praise some distrustful nodes. Therefore, we should also evaluate the trust degree of $f_{v_{j,i}}$. In this paper, we calculate the feedback effectiveness based on similarity of specific feedback values. The feedback effectiveness of node j can be derived through a similarity formula as shown by Eq.(2).

$$FT_j = \frac{\sum_{i \in set(j,r)} f_{v_{j,i}} \cdot f_{v_{r,i}}}{\sqrt{\sum_{i \in set(j,r)} f_{v_{j,i}}^2} \cdot \sqrt{\sum_{i \in set(j,r)} f_{v_{r,i}}^2}} \tag{2}$$

In Eq.(2), $set(j,r)$ presents the node-pair set that both nodes communicated with node i. Similar with the calculation method of service reliability, the recommendation credibility of node k is computed by Eq.(3).

$$CT_k = \frac{\sum_{i \in Rset(k)} ST_i \cdot \lambda(k,i)}{\sum_{i \in Rset(k)} \lambda(k,i)} \tag{3}$$

In Eq.(3), $Rset(k)$ is the node set recommended by recommendation node k before. $\lambda(k,i)$ presents the influence effect of node k on node i. There are two factors affecting the value of $\lambda(k,i)$. One is the time interval $T = t_n - t_p$, t_n presents the current time, and t_p presents the time that node k recommend node i. Another is the connection degree $\omega_{k,i}$ of the relationship between node i and node k. Thus, $\lambda(k,i)$ is shown as Eq.(4).

$$\lambda(k,i) = \frac{1}{t_n - t_p} \cdot \omega_{k,i} \tag{4}$$

In this paper, how to determine the connection degree $\omega_{k,i}$ is considered. According to the successful transaction Tr_{suc} and the number of total transactions $|Tr|$ between node k and node i, we determine the connection degree $\omega_{k,i}$, which is shown by Eq.(5). In Eq.(5), successful transaction Tr_{suc} is an indicative function, if $CT > Threshold$, $Tr_{suc} = 1$, otherwise, $Tr_{suc} = 0$.

$$\omega_{k,i} = \frac{\sum_{m=1}^{|Tr|} Tr_{suc}}{|Tr|} \tag{5}$$

According to the above analysis, the calculation method of global trust degree is shown by Eq.(6). In Eq.(6), α, β and γ are weights for service reliability, feedback effectiveness and recommendation credibility, and $\alpha + \beta + \gamma = 1$.

$$T_i = \alpha \cdot ST_i + \beta \cdot FT_i + \gamma \cdot CT_i \tag{6}$$

If a service node provides distrust service, *i.e.*, the service reliability is less than the service threshold ρ, the node will enter the service punishment cycle. In the service punishment cycle, a node should not provide any service. If a feedback node provides distrust feedback, *i.e.*, the feedback effectiveness is less than the feedback threshold θ, the node will enter the feedback punishment cycle. In the feedback punishment cycle, a node should not request any service. If a recommendation node provides distrust recommendation, *i.e.*, the recommendation credibility is less than the recommendation threshold δ, the node will enter the recommendation punishment cycle. In the recommendation punishment cycle, a node should not provide any recommendation for other nodes.

In direct interaction algorithm, the service reliability ST_i, and the feedback effectiveness FT_j will be output.

If there is not a trustful service node i that has interacted with the feedback node j directly, it needs a recommendation node k to recommend a trustful service node i for node j.

3.2   The Punishment Mechanism

Based on the proposed trust model, two punishment mechanisms are proposed according to the specific trust degree and the global trust degree respectively. According to the specific trust degree (service reliability, feedback effectiveness and recommendation credibility), this paper designs three punishment cycles according to different specific trust degree, so that restrain the specific trust behavior of nodes. According to the global trust degree, this paper gives a punishment mechanism based on the game theory [23,24] in order to solve the free-riding problem in social networks.

According to the specific trust degree, we design the specific punishment mechanism, and divide punishment cycles into service punishment cycle, feedback punishment cycle and recommendation punishment cycle. Once a node has selfish behavior, the node will enter the punishment cycle. In the period of punishment cycle, the node must be cooperative and honest in order to restore its reputation. In addition, other nodes reject to provide services for this node. After the punishment cycle, the node can replay transactions. According to the different selfish behavior, this paper gives different punishment strategies.

1. Service punishment cycle. If the service reliability $ST_i < \rho$, node i will enter service punishment cycle. In the service punishment cycle, a node cannot provide service for other nodes, and cannot request any service.
2. Feedback punishment cycle. If the feedback effectiveness $FT_i < \theta$, node i will enter feedback punishment cycle. In the feedback punishment cycle, a node cannot request any service. However, it can provide service for other nodes.
3. Recommendation punishment cycle. If the recommendation credibility $CT_i < \delta$, node i will enter recommendation punishment cycle. In the recommendation punishment cycle, a node cannot recommend any node. However, it can request and provide service for other nodes.

If the global trust degree of a node that stays in punishment cycle, greater a threshold, this node will exit the punishment cycle. And this node can make transactions with other nodes normally. However, if the node behaves distrustfully again, this node will enter the corresponding punishment cycle. In addition, the exiting threshold will be increased, so that strengthen the punishment force.

According to the global trust degree, this paper proposes punishment mechanism based on multi-strategy game to inspire nodes to select the strategies with high trust degree. T_i indicates the whole trust degree of node i. We divide trust degrees into five levels shown as: $[0.8, 1.0] \rightarrow Trust_1$, $[0.6, 0.8) \rightarrow Trust_2$, $[0.4, 0.6) \rightarrow Trust_3$, $[0.2, 0.4) \rightarrow Trust_4$ and $[0.0, 0.2) \rightarrow Trust_5$.

The five-strategy matrix is shown in Table 1. In Table 1, pr_A^{ij} is the profit value that the entity A obtains, if entity A game with entity B that A adopt the strategy i, and entity B adopt the strategy j. And pr_B^{ij} is the profit value that the entity B obtains, if entity B game with entity A that B adopt the strategy i, and entity A adopt the strategy j. Through the game analysis for the entities' behaviors in social networks, we can know that the multi-strategy game matrix is a symmetric matrix. In the analysis of the dynamics model, this game is performed repeatedly. At the end of each stage of the multi-strategy game, any participant's strategy as a historical information can be known by other participants. In addition, all the participants establish the strategy for the next stage of game based the historical information.

Table 1. The five-strategy game matrix with incentive and punishment mechanism

Trust level	Trust 1	Trust 2	Trust 3	Trust 4	Trust 5
Trust 1	pr_A^{11}, pr_B^{11}	pr_A^{12}, pr_B^{12}	pr_A^{13}, pr_B^{13}	pr_A^{14}, pr_B^{14}	pr_A^{15}, pr_B^{15}
Trust 2	pr_A^{21}, pr_B^{21}	pr_A^{22}, pr_B^{22}	pr_A^{23}, pr_B^{23}	pr_A^{24}, pr_B^{24}	pr_A^{25}, pr_B^{25}
Trust 3	pr_A^{31}, pr_B^{31}	pr_A^{32}, pr_B^{32}	pr_A^{33}, pr_B^{33}	pr_A^{34}, pr_B^{34}	pr_A^{35}, pr_B^{35}
Trust 4	pr_A^{41}, pr_B^{41}	pr_A^{42}, pr_B^{42}	pr_A^{43}, pr_B^{43}	pr_A^{44}, pr_B^{44}	pr_A^{45}, pr_B^{45}
Trust 5	pr_A^{51}, pr_B^{51}	pr_A^{52}, pr_B^{52}	pr_A^{53}, pr_B^{53}	pr_A^{54}, pr_B^{54}	pr_A^{55}, pr_B^{55}

In order to prevent the selfish nodes from selecting the strategy with low trust degree to be their preferred strategy for getting more benefits, *i.e.*, to restrain the free-riding phenomenon, a punishment mechanism is established to inspire the nodes to select the strategies with high trust degree based on the multi-strategy game. In the case of $i < j$, the calculation method of the benefits after adding punishment mechanism is shown by Eq.(7). When $i = j$, the calculation method is shown by Eq.(8).

$$PR_A^{ij} = pr_A^{ij} + \mu \cdot (pr_A^{ij} + pr_B^{ij}), \ i < j$$
$$PR_B^{ij} = pr_B^{ij} - \mu \cdot (pr_A^{ij} + pr_B^{ij})$$

(7)

$$PR_A^{ij} = pr_A^{ij}, \ i = j$$

(8)

where PR_A^{ij} represents the benefit of entity A, if entity A selects i strategy and entity B selects j strategy after adding punishment mechanism.

4   Simulations and Performance Analysis

In this section, we present the simulation results to verify the effectiveness of the proposed model. The hardware simulation environment is: Intel Core (TM)

Duo 2.66GHz CPU, 2GB Memory, Windows XP operating system, and Matlab 7.0 simulation platform. There are two kinds of nodes, the normal nodes and the malicious nodes. There are two types of normal nodes: (1) *completely trustful nodes* that can provide trustable service, feedback and recommendation; (2) *mix-type trustful nodes* that provide trustful feedback and recommendation, but random service quality, 20% of files have low quality, *i.e.*, malicious files. The malicious nodes include three types: (1) completely malicious nodes that provide questionable service, feedback and recommendation; (2) random malicious nodes that provide questionable service, feedback and recommendation with a certain probability (in the simulations, the probability is 50%); (3) disguised malicious nodes that provide trustful service and recommendation but questionable feedback. In the simulations, there are 1000 nodes, including 30% completely trustful nodes, 30% mix-type trustful nodes, 10% completely malicious nodes, 20% random malicious nodes, and 10% disguised malicious nodes. The simulation setting is shown in Table 2 where 1 represents completely trustful strategy, 0 represents completely questionable strategy, and ε represents randomly trustful strategy.

Table 2. The simulation setting

The style of nodes/trust	Service	Feedback	Recommendation
Ct	1	1	1
Tt	ε	1	1
Cm	0	0	0
Rm	ε	ε	ε
Dm	1	0	1

We measure the evolution of trust degree according to the service reliability, feedback effectiveness and the recommendation credibility respectively. Fig.1 presents the initial evolution trend of the service reliability without any punishment mechanism. From Fig.1, it can be seen that there exists the free-riding problem. The proportions of completely malicious nodes (Cm) and the random malicious nodes (Rm) increase steadily in the first 50 generations. After that, the system tends to be stable. Therefore, if there is not any punishment mechanism, the malicious nodes will dominate the evolutionary direction of the whole system. Fig.2 shows the ideal condition by adopting the punishment mechanism. From Fig.2, it can be seen that the completely trustful nodes (Ct) will dominant the evolutionary direction of the whole system with the punishment mechanism. The proportions of the completely trustful nodes (Ct) and the disguised malicious nodes (Dm) increase steadily, and tend to be stable at the last. It is because that the disguised malicious nodes (Dm) can provide trustful service so that they can survive in the system.

Fig.3 and Fig.4 show the evolution trend of feedback effectiveness. Fig.3 presents the initial evolution trend of the feedback effectiveness without any

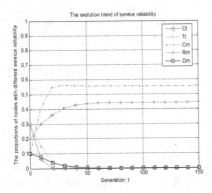

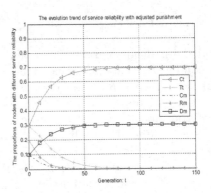

Fig. 1. The initial evolution trend of service reliability without punishment

Fig. 2. The evolution trend of service reliability with punishment

punishment mechanism. From Fig.3, it can be seen that the free-riding problem occurs. The proportions of the completely malicious nodes (Cm), the random malicious nodes (Rm) and the disguised malicious nodes (Dm) increase steadily in the first 50 generations. After that, the system tends to be stable. Since the disguised malicious nodes (Dm) provide distrustful feedback, they will obtain more benefits than the nodes that provide trustful feedback without any punishment mechanism. Fig.4 shows the ideal case by adopting the punishment mechanism. From Fig.4, it can be seen that the proportion of the completely trustful nodes (Ct) and the mix-type trustful nodes (Tt) increase steadily, and tend to be stable at the last. Since the mix-type trustful nodes (Tt) can provide trustful feedback, they will survive in the system.

Fig.5 and Fig.6 show the evolution trend of recommendation credibility. Fig.5 presents the initial evolution trend of the recommendation credibility without

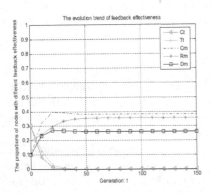

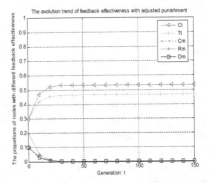

Fig. 3. The initial evolution trend of feedback effectiveness without punishment

Fig. 4. The evolution trend of feedback effectiveness with punishment

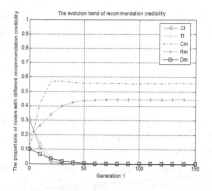
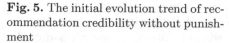

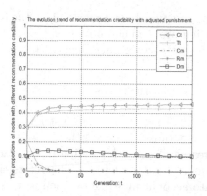

Fig. 5. The initial evolution trend of recommendation credibility without punishment

Fig. 6. The evolution trend of recommendation credibility with punishment

any punishment mechanism. From Fig.5, it can be seen that the free-riding problem generates in social networks. The proportions of the completely malicious nodes (Cm) and the random malicious nodes (Rm) increase steadily in the first 50 generations. After that, the system tends to be stable. Fig.6 shows the ideal case by adopting the punishment mechanism. From Fig.6, it can be seen that the proportions of the completely trustful nodes (Ct) and the mix-type trustful nodes (Tt) increase steadily, and tend to be stable at the last. Because the mix-type trustful nodes (Tt) can provide trustful recommendation, they will survive in the system.

We also verify the effectiveness of the whole punishment mechanism. Through combining the measurement results of service reliability, feedback effectiveness and recommendation credibility, we measure the trust evolution of the whole system. Fig.7 shows the initial evolution results. It can be seen that if there is

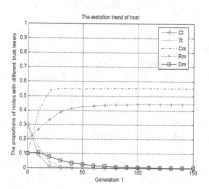

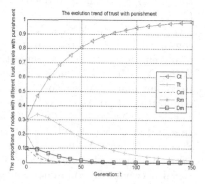

Fig. 7. The initial evolution trend of trust without punishment

Fig. 8. The evolution trend of trust with punishment

not any punishment mechanism, the free-riding phenomenon will occur. The free-riding problem can be solved by employing the punishment mechanism based on multi-strategy game which is shown by Fig.8.

5 Conclusions

In social networks, the trust relationship between nodes is the basis of service transactions. However, the establishment of trust relationship is a complex progressive process depending on interaction history, trust recommendation, trust management and so on. Therefore, modeling trust relationship needs to take into account multiple decision factors. Considering the existing problems of the trust models, this paper proposes a trust measurement model in social networks based on game theory where the trust degree is determined by three aspects, which are service reliability, feedback effectiveness, and recommendation credibility. In addition, based on game theory, we propose punishment mechanisms according to a specific trust degree and the global trust degree respectively in order to solve the free-riding problem. The simulation results show the effectiveness of the proposed trust measurement model. It also shows that the proposed punishment mechanisms can prevent the free-riding phenomenon effectively.

As a future work, we will further investigate more specific trust relationships between entities, e.g., family, best friends, and classmates. We plan to study how to find ordered trust-entity set in social networks.

Acknowledgments. This work is supported by the National Natural Science Foundation of China under Grants No.61170224, No.61272186 and No.61472095.

References

1. Al-Oufi, S., Kim, H.-N., El Saddik, A.: A group trust metric for identifying people of trust in online social networks. Expert Systems with Applications **39**, 13173–13181 (2012)
2. He, Z., Cai, Z., Wang, X.: Modeling propagation dynamics and optimal countermeasures of the social network rumors. In: The 35th IEEE International Conference on Distributed Computing Systems, Columbus, Ohio, USA (2015)
3. Wang, Y., Yin, G., Cai, Z., Dong, Y., Dong, H.: A Trust-based Probabilistic Recommendation Model for Social Networks. Journal of Network and Computer Applications (2015)
4. Zheng, X., Cai, Z., Li, J., Gao, H.: An application-aware scheduling policy for real-time traffic. In: The 35th IEEE International Conference on Distributed Computing Systems, Columbus, Ohio, USA (2015)
5. Wang, Q., Wang, J., Jian, Y., Mei, Y., Zhang, Y.: Trust-aware query routing in P2P social networks. International Journal of Communication Systems **25**, 1260–1280 (2012)
6. Zhang, L., Wang, X., Lu, J., Ren, M., Duan, Z., Cai, Z.: A Novel Contact Prediction Based Routing Scheme for DTNs. Transactions on Emerging Telecommunications Technologies (2014)

7. Zhang, L., Cai, Z., Lu, J., Wang, X.: Spacial Mobility Prediction Based Routing Scheme in Delay/Disruption-Tolerant Networks. Personal and Ubiquitous Computing (2015)
8. Feldman, M., Chuang, J.: Overcoming free-riding behavior in peer-to-peer systems. Newsletter ACM SIGecom Exchanges 5, 41–50 (2005)
9. Beth, T., Borcherding, M., Klein, B.: Valuation of trust in open networks. In: Gollmann, D. (ed.) ESORICS 1994. LNCS, vol. 875, pp. 1–18. Springer, Heidelberg (1994)
10. Xiong, L., Liu, L.: PeerTrust: supporting reputation-base trust in peer-to-peer communities. IEEE Transaction on Data and Knowledge Engineering 16, 843–857 (2004)
11. Kamvar, S., Schlosser, M.: EigenRep: reputation management in P2P networks. In: Proceedings of the 12th International World Wide Web Conference, New York, pp. 123–134 (2003)
12. Wang, Y., Lv, J.: FengXu, Trust Measurement and Evolution Model for Internetware. Journal of Software 17, 1–2 (2006)
13. Wang, Y., Vassileva, J.: Bayesian network-based trust model. In: Proc of the IEEE Computer Society WIC International Conference on Web Intelligence. IEEE Computer Society, Washington DC, pp. 372–378 (2003)
14. Wen, L., Feng, X., Lv, J.: An Approach of Software Reliability Evaluation in the Open Environment. Chinese Journal of Computers 33, 452–462 (2010)
15. Zhu, M., Jin, Z.: Approach for Evaluating the Trustworthiness of Service Agent. Journal of Software 22, 2593–2609 (2011)
16. Yu, Y.: Trust models for open grid network, Masters Thesis of SJTU (2009)
17. Tang, W., Chen, Z.: Research of Subjective Trust Management Model Based on the Fuzzy Set Theory. Journal of Software 14, 1401–1408 (2003)
18. Shi, Z., Liu, J., Wang, Z.: Dynamic P2P trust model based on time-window feedback mechanism. Journal on Communications 31, 120–129 (2010)
19. Gan, Z., Ding, Q., Li, K., Xiao, G., Algorithm, R.-B.M.-D.T.: Reputation-Based Multi-Dimensional Trust Algorithm. Journal of Software 22, 2401–2411 (2011)
20. Meng, X., Ding, Y., Gong, Y.: @Trust: A trust model based on feedback-arbitration in structured P2P network. Computer Communications 35, 2044–2053 (2012)
21. Bedi, P., Sharma, R.: Trust based recommender system using ant colony for trust computation. Expert Systems with Applications 39, 1183–1190 (2012)
22. Zhang, S., Chunxiang, X.: Study on the Trust Evaluation Approach Based on Cloud Model. Chinese Journal of Computers 36, 422–431 (2013)
23. Chen, J., Kiremire, A.R., Brust, M.R., Phoha, V.V.: Modeling online social network users' profile attribute disclosure behavior from a game thoretic perspective. Computer Communications 49, 18–32 (2014)
24. Yin, G., Wang, Y., Dong, Y., Dong, H.: WrightCFisher multi-strategy trust evolution model with white noise for Internetware. Expert Systems with Applications 40, 7367–C7380 (2013)

Whom You Know Matters: Relook Vehicle-to-Vehicle Communications from a Topological Perspective

Syed Fakhar Abbas[1], William Liu[1(✉)], Quan Bai[1], and Adnan Al-Anbuky[2]

[1] School of Computer and Mathematical Sciences, Auckland, New Zealand
{sabbas,william.liu,quan.bai}@aut.ac.nz
[2] Sensor Network and Smart Environment (SeNSe) Research Laboratory,
Auckland University of Technology, Auckland, New Zealand
aalanbuk@aut.ac.nz

Abstract. Vehicular communication networks such as vehicle-to-vehicle (V2V) or vehicle-to-infrastructure (V2I) are a type of network in which vehicles and roadside units are the communicating nodes, providing each other with information such as safety warnings and traffic information. As a cooperative approach, vehicular communications can be more effective in avoiding accidents and traffic congestion than if each vehicle tries to solve these problems individually. Vehicular communications has some distinct characters such as fast moving, short-lived and opportunistic connections. Recently literature in this area is growing rapidly. The main focus in on studying how to advance and evaluate the traditional communication protocols and algorithms so as to be more effective in communicating information among those fast moving vehicles. Unfortunately there is far less work to reveal how the underlie connectivity of wireless network topological can affect the overlay communications behaviors. The vehicles' communications behavior is not merely a function of message transmission or the wireless communications technologies, but also it is a network-wide role and organization. The wireless ties that link a vehicle to other vehicles are also a critical factor. In this paper, we are paving a new line of research on revealing the roles of network topological characters in vehicular communications. This novel research dimension is thought-provoking and opening a new conversation for researchers working in this area to rethink and redesign the communications protocols by also considering the topological connectivity related parameters.

1 Introduction

Vehicular communication networks are a type of network in which vehicles and roadside units are the communicating nodes, providing each other with information such as safety warnings and traffic information. As a cooperative approach, vehicular communications can be more effective in avoiding accidents and traffic congestion than if each vehicle tries to solve these problems individually. Therefore the Intelligent Transportation Systems (ITS) have been developed to address the challenges of safety, security and efficacy of the current transportation systems. The field of Inter Vehicular Communications (IVC), including both Vehicle-to-Vehicle communication (V2V) and

© Springer International Publishing Switzerland 2015
M.T. Thai et al. (Eds.): CSoNet 2015, LNCS 9197, pp. 248–261, 2015.
DOI: 10.1007/978-3-319-21786-4_22

Vehicle-to-Infrastructures (V2I), also known as Vehicular Ad-Hoc Network (VANET). The VANET is recognized as an important component of ITS in various national plans [1]. The VANET communication has become a progressively important research topic in the area of wireless networking as well as the automotive industries. The goal of VANET research is to develop a vehicular communication system to enable cost-effective and fastest communication of data for the benefit of passengers' safety and comfort [2]. VANETs are a class of ephemeral networks [3], in which the nodes have short-lived and opportunistic connections with each other. The density of the network changes continuously as nodes move in and out of the range of each other.

Recently there are rapidly growing literatures in this area but mainly focus on studying how to advance and evaluate the traditional communications protocols and algorithms so as to more effectively transmit information among those fast moving vehicles. Unfortunately there is far less work to reveal how underlie wireless network topological connectivity can affect the overlay communications behaviors. There is a saying that, it is not what you know but who you know in daily communications cases, which argues people get ahead in life based on their connections, not on their skills or knowledge, and every day offers evidence of this 'hiding' law. Actually it is applied to the V2V or V2I communications behavior too. For example, noticing other vehicles for finding alternative paths as earlier as possible by unfolding congestion information due to accident or abnormal traffic down the track. The vehicles' communications behavior is not merely a function of message transmission or the wireless communications technologies, but also it is a network-wide activity. The wireless ties that linking a vehicle to other vehicles are also a critical factor.

The rest of the paper is organized as follows. Section 2 brief some backgrounds of vehicular communications. The novel paradigm of the adaptive network (i.e., the interplay between trust state and topology) is introduced in Section 3. An opening discussion to raise a new research direction, by using the network transitivity as a topological metric example associated with two case studies to validate the transitivity effects on VANET communication performances. The section 5 concludes the findings and also layout future works.

2   Background on VANET

The VANET is one of the emerging wireless communications network areas. It is the upcoming area of mobile ad hoc network (MANET) where vehicles work as portable nodes that are within the network. Comforting passengers and increase in the road users are some of the basic targets of VANET. Its communications takes place through wireless links that are mounted on each and every node i.e. the vehicle [4]. Within VANET, each node always act as the participant and network router as the node communication is through the nodes that are intermediate and also lies in a range within their own transmission. VANET is a network that organizes communications on its own and its reliability is not pegged on any network infrastructure that is fixed although some of the nodes that are fixed work as roadside units. This is to ensure facilitation of the vehicular networking to serve geographical data or to allow access to the internet etc.

The characteristics of VANET include the rapid pattern movement, speed and higher mobility of the node that causes faster changes in topology's network and also opportunist connection [5]. In VANET, vehicles have to follow traffic signals and traffic signs, vehicles move on road that are predefined and the vehicle's velocity always depend on the speed signs unlike in MANET [6]. The stable and reliable communications e.g., end-to-end routing in VANET and some of the challenges to be solved to provide guaranteed services. Therefore, to make VANET more applicable and put it into implementation, some challenging research problems should be addressed. For example, routing is one of most challenging issues because vehicles are dynamic in action, in speed and also high mobility. In addition, there are also much work done in social network aspect of VANET such as trust and reputation management in VANET [7]. Moreover, transport emissions are among the highest environmental pollution sources and the vehicles emissions contain a number of harmful substances which include fine particulate matter, carbon dioxide (CO_2), carbon monoxide (CO), hydrocarbons (HC), nitrogen dioxide and nitrogen monoxide [8]. Refining mobility of transportation and improving the living environment are two challenging issues that require attention for urban traffic.

While many literatures are mainly focusing on studying how to advance and evaluate the traditional communications protocols and algorithms so as to more effectively communicate information among those fast moving vehicles and reduce carbon emissions. Unfortunately there is far less work to reveal how underlie wireless network topological connectivity can affect the overlay network behaviors and associated performance. The wireless ties that linking a vehicle to other vehicles are also a critical factor. A new paradigm named as adaptive networks has been emerged in recently to reveal the co-evolution behavior of dynamic 'on' and 'of' the networks [10, 11]. In other words, how the network entities' state (i.e., behaviors) change affects the underlie network topologies evolves. On the other hand how the change in network topologies can affect the change of over entities' state which we will elaborated in more details in the next section.

3   Adaptive Networks and Topological Robustness Measurement

Adaptive network is a combination of two concepts which are dynamics on networks and dynamics of networks in [12 - 14]. The dynamics on networks is defined as the status or state change on the network entities, and the dynamics of networks is defined as the change of underlie network topology. These two factors are affecting each other which are called as co-evolution. In study [14] of the Internet web, the state (i.e., behaviors) affects how topologies (i.e., structure) change, and on the other hand, the topologies affect how state change as well. For example, the users change their behaviors to do more online shopping which could cause more online shops (i.e., computer servers and their related connectivity to the Internet) are established. On the other hand, the search results associated with positioning could affect user's preference to access various contents and this underlie topology affect overlay state.

3.1   The Interplay Between Network State and Topology

The behavior on a particular node in the network can be considered as the state of that node, once this node's behavior changes which also mean the node's state changes. A typical example is that in the vehicular routing algorithm, if a node is moving out of the communication range of others, the connection state is changed, and the algorithm will detect this change then disconnect this node but look for an alternative node being routed to the destination. This is how the state's change of a node to affect the change of its underlying network topology. On the other hand, the network topology change should affect the state of a node as well. For example, there are two routes between node A and B, so A can select either route to connect with B. However, if one of the routes is disconnected, node A will have no choice but to choose the remaining route to stay connected with B. This can make the remaining route overload as all traffic now come through this route. Some of the nodes might become selfish to preserve energy so as to stay alive longer in the network. This example shows that a reliable node can become unreliable under pressure due to the change of its underlie network topology. Moreover, the change of network topology can affect the efficiency of information dissemination algorithm. The study in [9] found out that the information dissemination speed is much faster that in small world network than in Lattice network. The information dissemination algorithm can be very inefficient in some network topologies, such as scale-free network and star network. In such case, the research question will be stated as "can network topological characteristic become one of the metrics for information dissemination evaluation?" To do that, "how to characterize network topology" is next research question.

3.2   Network Topology Modeling

There are lots of researches on complex network. They are trying to model the network evolution of human social network such as telephone communication, co-authors network , epidemiological network the Susceptible–Infected–Susceptible (SIS) model [12] etc. The extraction is the process deciding which part of the network is going to change, production is decide how this part is going to change, and finally the embedding is embedding the new part into the network. There are two well-known existing models such as Small World and Scale Free networks. Small World effect was introduced by Watts and Strogatz [9]. It is defined as any node in the network can reach any other nodes within k hops. In the real world, a person is very far away from you, but sometimes you will be surprised that he could be your friend's friend. In such case, he is actually very close to you, this is small world effect. Scale-free network was introduced by Barabási and Albert [11] which features as power-law degree distribution. Many studies describe this scale-free network as "robust, yet fragile" [12-14]. Free scale networks are robust against random malicious attacks, but fragile while malicious parties attack its central hops. For example, random attacks need to disable 10 nodes to achieve disconnect the network, but also by disabling one central node can achieve the same damage. The work in [14] found out that the information dissemination speed is faster in Small World network than in Lattice network. This is because the longest distance between two nodes in Small World network is no bigger than k hops. In such case, comparing to lattice network, it obviously has a much longer distance that makes the dissemination speed decrease significantly to Small World network. In addition, the routing algorithm is inefficient in scale-free and star network because such network normally does not have an alternative route between any two nodes in the network. In such case, even though routing algorithm detects the unavailable nodes or links, it cannot avoid them to create a trustable path.

Moreover, a directed or undirected and weighted or unweighted graph G (V, E) is usually used to represent a complex network in [14, 16]. V is node or entity in the network and E is edge or link which is connecting the Vs. A directed graph G is meant the edge has direction like the link is connected from A to B but not necessary vice versa. An undirected graph means the link in the network is reciprocity. The same idea, a weight graph means the links have weight in the network, such as the connection between A and B is motorway, but between B and C is only an alley [18]. In such case, the link weight between A and B is larger. Node strength is the sum of link weights to all its neighbors and a high strength node is normally attracting more nodes to connect, which is "rich get richer".

3.3   Network Robustness Topological Measurement

Networks exist everywhere in our world, like social network, internet network, traffic network etc. Sometimes the network failure can cause significant damage to the companies, society etc. Like a large-scale power outage can cost huge financial loss. Network robustness becomes crucial to prevent this situation happening. What is network robustness? The study in [18] defines the term as "the ability of a network to continue performing well when it is subject to failures or attacks". The study [19] suggests as a network is more robust if the service on the network performs better, where performance of the service is assessed when the network is either (a) in a conventional state or (b) under perturbations, e.g. failures, virus spreading. The research in [17] suggested that robustness has a different definition in different scenarios.

To systematically define network robustness, we need to quantify the robustness. The study [18] has listed four classical graph metrics categories to measure robustness which includes connectivity, distance, betweenness, and clustering. The connectivity is calculated as the percentage of connecting pairs in the network. A fully connected network is having connectivity of 1, and then the completely disconnected network is having the connectivity of 0. There are another two metrics under connectivity, which are vertex connectivity and edge connectivity. The vertex connectivity is the number of nodes needs to remove so as to disconnect the network. Same idea, the edge connectivity is the number of edges needs to be disconnected so as to disconnect the network. Distance has average hop count of all node pair's connection, and the longest hop count is the diameter of the network. Betweenness is calculated as the number of shortest path between any node pair route through a node. Clustering is using the clustering coefficient to measure the percentage of the connected triangle cluster in all connected triples (3 nodes). One of popular metric is algebraic connectivity. It uses the Laplacian matrix to represent the network graph, and the second smallest eigenvalue is the algebraic connectivity. Effective Resistance is another spectral metric to measure the network robustness. It uses the Kirchhoff's circuit laws to calculate the resistance between two vertexes, and the sum of all node pairs' resistance is the effective resistance. The smaller the value is, the network believes more robustness.

The study in [18] has also categorized metrics into Distance class, Connection class, and Spectral class. It is the same as the studies in [19] but more metrics has been listed in each class. It also put the betweenness into the distance class and clustering into the connection class. These additional metrics are similar to that of the given metrics mentioned above. The clustering is first design to describe friends' of my friend is my friends as well in human social network, but here also could measure the backup route as well as the more in clustering, the more in backup routes. Betweenness is the measure of centrality of a particular node in the network. For the algebraic connectivity, studies [18, 19] suggested that if the second smallest eigenvalue are multiple, the algebraic connectivity will not change when additional link added. The effective resistance does not have such problem that it can be more suitable to measure the network robustness by focusing on alternative routes.

4   Directional Discussion and Case Studies

As we can see that, after surveying on the three knowledge domains include vehicular communication networks, social network analysis and adaptive networks associated with robustness measurement, there is a new research gap that can be identified as shown by Fig.1 above. This is to explore the interplay between the underlie network topological

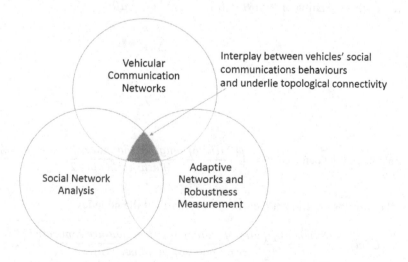

Fig. 1. The research gap identified on the interplay between overlay vehicular social communications behavior and underlie topological connectivity

connectivity and the overlay vehicular communication behaviors. Especially, the concept of using network topological perspective to rethink vehicular communications behavior should become increasingly prominent. This offers a useful approach for examining the connectivity among each network unit and their significant impact on the overall communication behaviors. In this section, we will use a conceptual example of transitivity in a social network to demonstrate the potential of this new direction. In sociology, the transitivity represents a key property of both partial order relations and equivalence relations, in other words, when a friend of my friend become my friend, which means if there is a tie between Vehicle-A and Vehicle-B and one between Vehicle-B and Vehicle-C, then a transitive network Vehicle-A and Vehicle-C will also be connected as shown in Fig.2 below.

Fig. 2. Transitivity in VANET scenario

4.1   Transitivity Model and Its Impact on Communications Behaviors

In reflecting the social network transitivity vehicular network, the tie between Vehicle-A and Vehicle-B and one between Vehicle-B and Vehicle-C will results in Vehicle-A and Vehicle-C also be connected.

For a transitive relation of R: $aRb \land bRc \longrightarrow aRc$

Local Clustering Coefficient $C = \left(\dfrac{|Paths\ of\ length\ 2\ that\ have\ the\ third\ edge|}{|Paths\ of\ Length\ 2|} \right)$

Local clustering coefficient measures transitivity at the node level

$$C\ (vi) = \left(\frac{Number\ of\ pairs\ of\ neighbors\ of\ vi\ that\ are\ connected}{Number\ of\ pairs\ of\ neighbours\ of\ vi} \right)$$

Let G (V, E) be an undirected graph without loops on the vertex set V = {1 ..... n), where n, the order of the graph, is known. A k-subset of V is a subset of V containing k vertices. If we by. $\binom{V}{k}$ denote the class of all k-subsets of V, we have for E, the edge set of G that $E \leq \binom{V}{2}$ The elements of Y, *the corresponding n X n adjacency matrix of G, are given by*

$$y_{ij} = y_{ji} = \begin{cases} 1 \\ 0 \end{cases} if \ \{i,j\} \in E \} . \tag{1}$$

By r we denote the size of the graph, which is the same as the number of edges, thus

$$r = |E| = \sum_{i=1}^{n-1} \sum_{j=i+1}^{n} y_{ij} . \tag{2}$$

Since transitivity and triads are inextricably intertwined, it will be almost inevitable for us to deal with triad counts. Triads can have from 0 to three edges, i.e. be of size 0 to 3. We may use the adjacency matrix Y in order to obtain the count of induced triads of size ℓ as

$$t_{\ell} = \left| \left\{ \{i, j, k\} \in \binom{V}{3} : y_{ij} + y_{ik} + y_{jk} = \ell \right\} \right| . \tag{3}$$

By Δ_{ℓ} we denote the proportion of complete subgraphs of order ℓ out of all possible subsets of order ℓ, thus, for $\ell > 2$,

$$\Delta_{\ell} = \frac{\left| \left\{ C \in \binom{V}{\ell} : \binom{C}{2} \subseteq E \right\} \right|}{\left| \binom{V}{\ell} \right|} = \left(\sum_{C \in \binom{V}{\ell}} \prod_{\{i,j\} \in \binom{C}{2}} y_{ij} \right) \binom{n}{\ell}. \tag{4}$$

If we let $N = \binom{n}{2}$ denote the number of all positions in the graphs where there could be edges, we realize that for $\ell = 2$ we have that

$$\Delta_2 = \left(\sum_{i=1}^{n-1} \sum_{j=i+1}^{n} y_{ij} \right) / \binom{n}{2} = \frac{r}{N} . \tag{5}$$

i.e. Δ_2 is the graph density for $\ell = 3$, we have that

$$\Delta_3 = \left(\sum_{i=1}^{n-2} \sum_{j=i+1}^{n-1} \sum_{k=j+1}^{n} y_{ij} y_{ik} y_{jk} \right) / \binom{n}{3} = t_3 / \binom{n}{3} . \tag{6}$$

In other words, Δ_3 is the proportion of transitive triads out of all triads. This proportion can be used as a transitivity index.

4.2   Case Study 1

In a network where the channel access contention is uniformly distributed (In case of CSMA), or the number of time intermediate node i, and the time that the packet is received at the intermediate node $i + 1$, $where\ i = 1 \dots k$

Table 1. Simulation Parameters

No. of Vehicles	3
Simulation area	0.5 Km
Simulation Time	60 minutes
Vehicle Speed	10 m/sec to 18 m/sec
Channel bandwidth	11Mbps
Transport protocol	UDP
Transmitted power	0.5mW,5mW,15mW,25mW

Fig. 3. Simulation Scenario

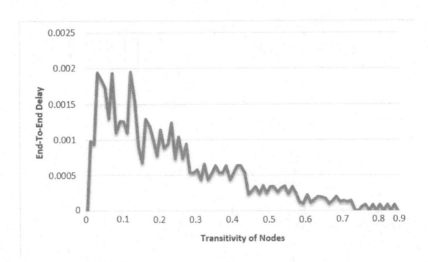

Fig. 4. End-To-End Delay vs. Transitivity of Nodes

From the given graph, it can easily be analyzed that as the transmitted power decreases (i.e., their associated transitivity decreases), the delay increases and vice versa. Moreover, as shown in Fig.5 below, it can be seen that as the transitivity increases (i.e., caused by the transmitted power increases), the overall throughput will increase.

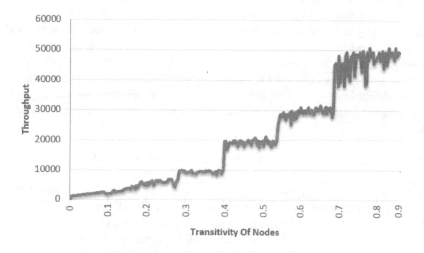

Fig. 5. Throughput vs. Transitivity of Nodes

4.3   Case Study 2

Transport emissions are among the highest environmental pollution sources in cities. The vehicles emissions contain a number of harmful substances. These include fine particulate matter, carbon dioxide (CO_2), carbon monoxide (CO), hydrocarbons (HC) and nitrogen dioxide and nitrogen monoxide. Refining mobility of transportation and bettering the living environment are two essential problems which require to be handled in urban traffic. [22]

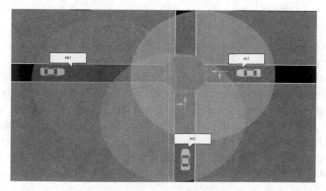

Fig. 6. Simulation Scenario

In general, the factors can be ascribed to a traffic accident, situation of the road (icy, closed), bad driving behavior and traffic signs and lights, as shown in the figure below.

There are two ways to estimate carbon emission. One way is to estimate energy consumption at different vehicular speed in mph. There is one useful model that can be used in our project – Comprehensive Modal Emissions Model. It has 3 main functions: providing detailed fuel consumption, estimating localized emissions and accounting for road grade effects. These 3 functions can act on each individual vehicle. This will be helpful for estimating carbon emission of individual vehicle. [23], [24]. The other way is to evaluate CO_2 emissions based on average speed of trip or trip segment. According to the research, the air pollutant released by burned fuel will increase during acceleration. [26] We can use a formula to calculate the carbon emission.

$$ln(y) = b0 + b1 \cdot x + b2 \cdot x2 + b3 \cdot x3 + b4 \cdot x4 . \tag{7}$$

Where y stands CO_2 emission in g/mi and x stands average travel speed in mph. The concrete number that every letter stands for is in [23], [25].

The total mass of the system, the engine, the speed and acceleration of the calculation, we can produce analog data. This data will be treated as an estimate of the emissions. The emissions include CO_2, CO, hydrocarbon (HC), and nitrous oxide (NOx) and so on. By calculating the two acceleration and deceleration of the vehicle emissions, it can get estimate CO_2 emissions. The tractive power requirement at a vehicle's wheels Ptract is calculated using the following polynomial:

$$P_{tract} = Av + Bv^2 + Cv^3 + Mav + mgv \sin \theta . \tag{8}$$

Based on the tractive power requirement, the gas consumption can be estimated and, consequently, tailpipe emissions of CO2 calculated according to a second polynomial:

$$TP_{CO_2} = \begin{cases} \alpha + \beta v + \delta v^3 + av & if P_{tract} > 0 \\ \alpha' & else \end{cases} . \tag{9}$$

Because road grade is not currently modeled in SUMO, Ptract calculations assumed planar roads and, hence, $\vartheta = 0$. [25]

Also, the numerical value of carbon emission in 0mph to 20mph is highest among 0mph to 90mph. (i.e., relationship between average speed and carbon emission) Meanwhile, the numerical value of carbon emission in 20mph to 0mph will be on the rise. Actually, the vehicle will stay in low speed in the traffic congestion. Which means the vehicle will release more air pollutant including carbon emission in traffic congestion. [26] After traffic congestion ended, the vehicle will accelerate to get up to normal speed. In the acceleration, it will also release more carbon pollutant. In the view of this, the relationship between traffic congestion and carbon emission is that traffic congestion will cause more carbon emission. [27]. Moreover, we found another useful formula to calculate emission of carbon. Like formula below to calculate CO_2 emission.

$$Emissions_{CO_2} = \left(\frac{Liters}{Kilometer}\right)$$
$$\cdot \left(\frac{Mass\ CO_2}{Liter}\right) \cdot (Kilometers\ Traveled)$$

Liter/Kilometer means vehicle-technology-related parameter which is used to describe fuel energy economy of a vehicle. [28]

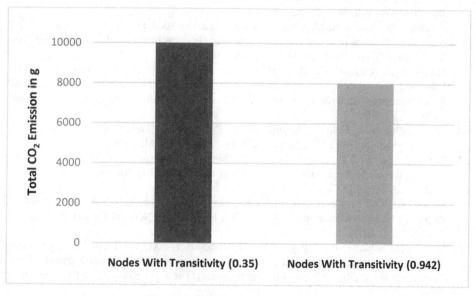

Fig. 7. CO2 Emission vs. Transitivity

5   Conclusion

In this paper, we have surveyed three domains of knowledge include the vehicular communications network, social network analysis and adaptive networks (include robustness topological measurement). It can be seen that, there are rich literature on vehicular communications protocols and algorithms which mainly focus on advancing the traditional communication protocols to handle new challenges in vehicular networks, but unfortunately there is limited work on how underlie network topological connectivity can affect the overlay vehicular social communications behaviors. Like our daily sayings, it is not what you know but who you know. This means that you can only get the right information through the right connectivity. This universal rule embed in the networks can be applied to the vehicular communication networks. The vehicular communications behavior is not merely a function of message transmission, but also network-wide activity. Finally, we need to notice that, in this paper, we are more focusing on the rationales to establish a new developing line of research on exploring the roles of underlie network topological connectivity in vehicular communications. The current state of our research in this direction does not yet lend itself to sweeping prescription. Nevertheless, we believe the topic is thought-provoking and opens a new conversation for researchers to rethink and redesign the vehicular communications protocols from the network topological perspective.

References

1. U.S. Department of Transportation, Intelligent Transportation Systems (ITS) Home, July 2013. http://www.its.dot.gov/index.htm
2. Rylander, D., Axelsson, S.: Using wireless communication to improve road safety and quality of service at road construction work sites (Poster). In: Vehicular Networking Conference (VNC), pp. 14–16. IEEE, November 2012
3. Manvi, S.S., Kakkasageri, M.S., Mahapurush, C.V.: Performance analysis of AODV, DSR, and swarm intelligence routing protocols in vehicular Ad hoc network environment. In: International Conference on Future Computer and Communication, pp. 21–25, April 2009
4. Wex, P. Breuer, J., Held, A., Leinmuller, T., Delgrossi, L.: Trust issues for vehicular Ad Hoc networks. In: IEEE, VTC Spring 2008, pp. 2800–2804, May 2008
5. Taleb, T., Sakhaee, E., Jamalipour, A., Hashimoto, K., Kato, N., Nemoto, Y.: A stable routing protocol to support its services in vanet networks. IEEE Transactions on Vehicular Technology **56**(6), 3337–3347 (2007)
6. Huang, Z., Ruj, S., Cavenaghi, M.A., Stojmenovic, M., Nayak, A.: A Social Network Approach to Trust Management in VANETs. Peer-to-Peer Networking and Applications **7**(3), 229–242 (2012)
7. Lin, S., De Schutter, B., Zegeye, S.K., Hellendoorn, H., Xi, Y.: Integrated urban traffic control for the reduction of travel delays and emissions. In: 2010 13th International IEEE Conference on Intelligent Transportation Systems (ITSC), pp. 677–682. IEEE, September 2010
8. Gross, T., Sayama, H.: Adaptive Networks. In: Adaptive Networks, pp. 1–8. Springer, Heidelberg (2009)
9. Watts, D.J., Strogatz, S.H.: Collective dynamics of 'small-world' networks. Nature **393**(6684), 440–442 (1998)
10. Sayama, H., Pestov, I., Schmidt, J., Bush, B.J., Wong, C., Yamanoi, J.: Modeling complex systems with adaptive networks. Computers & Mathematics with Applications **65**(10), 1645–1664 (2013)
11. Barabási, A.-L., Albert, R.: Emergence of Scaling in Random Networks. Science **286**(5439), 509–512 (1999)
12. Gross, T., Sayama, H.: Adaptive Networks. In: Adaptive Networks, pp. 1–8. Springer, Heidelberg (2009)
13. Sayama, H., Pestov, I., Schmidt, J., Bush, B.J., Wong, C., Yamanoi, J.: Modeling complex systems with adaptive networks. Computers & Mathematics with Applications, 1645–1664, May 2013
14. Martin-Hernandez, J., Van Mieghem, P.: TU Delft: Electrical Engineering, Mathematics and Computer Science: Intelligent Systems (INSY), and TU Delft, Delft University of Technology, Measuring Robustness of Complex Networks, October 28, 2013

15. Sydney, A., Scoglio, C., Gruenbacher, D.: Optimizing algebraic connectivity by edge rewiring. Applied Mathematics and Computation **219**(10), 5465–5479 (2013)
16. Granovetter, M.S.: The Strength of Weak Ties. American Journal of Sociology **78**(6), 1360–1380 (1973)
17. Ellens, W., Kooij, R.E.: Graph measures and network robustness arXiv:1311.5064 (physics), November 2013
18. Wang, H., Van Mieghem, P.: TU Delft: Electrical Engineering, Mathematics and Computer Science: Telecommunications, and TU Delft, Delft University of Technology, Robustness of networks (2009)
19. Varga, A.: OMNeT++ Chapter in the book. In: Klaus, W., Mesut, G., James, G. (Eds.) Modeling and Tools for Network Simulation. Springer (2010). ISBN: 978-3-642-12330-6
20. SUMO Simulation of Urban Mobility. http://sumo.sourceforge.net/ (accessed: May 25, 2015)
21. Veins. http://veins.car2x.org/ (accessed: May 25, 2015)
22. Lin, S., De Schutter, B., Zegeye, S.K., Hellendoorn, H., Xi, Y.: Integrated urban traffic control for the reduction of travel delays and emissions. In: 2010 13th International IEEE Conference on Intelligent Transportation Systems (ITSC), pp. 677–-682. IEEE, September 2010
23. Dissanayake, H., Koggalage, R.: A cost effective intelligent solution to reduce traffic congestion. In: Wireless and Optical Communications Networks (WOCN) (2010)
24. Kraschl-Hirschmann, K., Zallinger, M., Luz, R., Fellendorf, M., Hausberger, S.: A method for emission estimation for microscopic traffic flow simulation. In: Integrated and Sustainable Transportation System (FISTS) (2011)
25. Barth, M., Boriboonsomsin, K.: Real-world carbon dioxide impacts of traffic congestion. Transportation Research Record: Journal of the Transportation Research Board **2058**(1), 163–171 (2008)
26. Li, H.: Calculation of additional pollutant gas emissions and their social cost from transport congestion. In: Mechanic Automation and Control Engineering (MACE) (2011)
27. Li, Q., Li, Q.: Low carbon transportation in Japan and its developmental analysis. In: Transportation of China, 6th Advanced Forum (2010)
28. Chang, X., Chen, B.Y., Li, Q., Cui, X., Tang, L., Liu, C.: Estimating real-time traffic carbon dioxide emissions based on intelligent transportation system technologies. IEEE Transactions onIntelligent Transportation Systems **14**(1), 469–479 (2013)

To Sleep or Not to Sleep: Understanding the Social Behavior of Lifetime-Aware Networks

Luca Chiaraviglio[1]([⊠]), Syed Fakhar Abbas[2], and William Liu[2]

[1] University of Rome Sapienza, Rome, Italy
luca.chiaraviglio@uniroma1.it
[2] Auckland University of Technology, Auckland, New Zeland
{sabbas,william.liu}@aut.ac.nz

Abstract. Network lifetime is one of the key characteristics for evaluating wireless sensor networks (WSNs) in an application-specific way based on the availability of sensor nodes, wireless radio coverage, and wireless connectivity. Basically it shows in a resource constrained environment the consumption of every limited resource must be considered. A large number of energy efficient protocols and algorithms have been proposed in WSNs, mainly by introducing a sleep mode (SM) state to prolong the lifetime of a sensor network. The network nodes or links can be switched between working and sleep modes dynamically according to the real-time traffic situations. While there are far less critical discussions on what can be the negative effects of SMs on network lifetime in terms of hardware reliability such as failure rate. The duration of SMs tends to increase hardware lifetime, while the frequency of power state transitions tends to decrease it. In this paper, we extend the lifetime concepts in WSNs to wired network to reveal the side-effects of SMs on the hardware reliability. We have extensively studied the lifetime behavior of network links in a backbone network scenario as well as identified the sensitive social factors impacting the network lifetime. This novel research dimension is thought-provoking and opening a new conversation for researchers who are working in the areas of sustainable communications and computing to rethink and redesign the energy efficient approaches so as to address their possible side-effects on hardware reliability for the next stage of their implementation.

Keywords: Lifetime-aware networking · Energy efficiency · Reliability · Failure rate and analysis · Social behavior

1 Introduction

Lifetime-aware network is one of most important research topics in sensor networks which are under the resource-constrained environments and the consumption of every limited resource must be considered. It is a key characteristic for

This work is supported by the Sapienza Awards LIFETEL.

© Springer International Publishing Switzerland 2015
M.T. Thai et al. (Eds.): CSoNet 2015, LNCS 9197, pp. 262–272, 2015.
DOI: 10.1007/978-3-319-21786-4_23

evaluating wireless sensor networks (WSNs) in an application-specific way based on the availability of sensor nodes, the wireless radio coverage, and the wireless connectivity. Even the quality of service (QoS) measures can be related to lifetime considerations[1]. The network lifetime can be a measure of energy consumption occupies the exceptional position since it decides the upper time bound for the utility of a sensor network. The network can only fulfill its communications purpose as long as it is considered alive but not after that. It is therefore an indicator for the maximum utility and service of which a sensor network can provide. Lifetime studies first came up because the recharging or replacement of batteries is not feasible in many scenarios e.g., too many nodes, hostile environment etc., and thus the lifetime of network cannot be extended infinitely. If network lifetime metric is used in an analysis on a real life deployment, the estimated network lifetime can also contribute to moderate the cost of the deployment. Lifetime is also considered as a fundamental but very critical parameter in the context of availability and security in networks[2]. Network lifetime strongly depends on the lifetimes of single nodes or links that constitute the network. This fact does not depend on how the network lifetime is defined. Each definition can finally be interpreted to the fundamental question of when the individual node or link fails. Thus, if the lifetime of single node or link are not predicted accurately, it is possible that the derived network lifetime metric will deviate in an uncontrollable manner. It should therefore be clear that accurate and consistent modeling of single node or link is very important. The lifetime of a sensor node basically depends on two factors according to the literature[3]: how much energy it consumes over time, and how much energy is available for its use.

In the literature, we can find a great number of relevant publications that address the problem of sensor network lifetime and how to prolong it by introducing a sleep mode (SM) state and exploiting the possibility to enable the sensor node with SMs as much as possible to save energy consumption. When a SM state is set for a device in a network, the other nodes that remain powered on have to sustain the traffic flows between the source and destination nodes. This promising SMs approach has been applied into the wired networks and different works have investigated the management of backbone networks by adopting sleep modes (SMs)[4–7]. The main outcome of these works is that networks with SM enabled are able to save a significant amount of energy, due to the fact that the traffic experiences high fluctuations between the day and the night, resulting in a large number of resources that can be put in SMs during off peak hours. However, the impact of SMs on the reliability of network devices is an open issue[8,9]. In particular, there are two opposite effects influencing the lifetime of network devices[10]: the duration of SMs tends to increase the lifetime, but the change frequency of the power state (i.e., from SM to full power and vice versa) has the opposite effect, i.e., lifetime decrease. When a device experiences a failure, the device may not be available any more to accommodate traffic forwarding, resulting in a Quality of Service (QoS) degradation for users. Additionally, the reparation costs are incurred, which may involve even the replacement of the whole device. In particular, the reparation costs may even exceed the monetary savings gained from SMs[11]. These all facts suggest that the lifetime of the devices plays a crucial role in determining the effectiveness of SMs, showing that the energy saving may be not the only metric to prove the effectiveness of a SM-based approach.

The remainder of the paper is organized as follows. The mathematical model adopted for evaluating the network lifetime is presented in Sec. 2. Sec. 3 defines and formulates the lifetime aware network problems. A backbone network scenario is introduced for setting up the numerical studies in Sec. 4, and Sec. 5 has conducted the simulation studies and then the results analysis is presented to explain the different network lifetime behavior of each network resource. Discussion of our work is reported in Sec. 6. The final conclusion is drawn and future work is laid out in Sec. 7.

2   The Network Lifetime Model

We first review the model of [9,10] to compute the network lifetime. Here we report the main intuitions, while we refer the reader to [10] for the complete models. In particular, our focus is on links of a backbone network. The generic failure rate for a link at full power is defined as γ^{on}. The failure rate is the inverse of the lifetime. When SM is applied into the link, the new failure rate γ^{tot} is defined as:

$$\gamma^{tot} = \underbrace{\gamma^{on}\left(1 - \frac{\tau^{off}}{T}\right) + \gamma^{off}\frac{\tau^{off}}{T}}_{\text{Failure Rate Decrease}} + \underbrace{\frac{f^{tr}}{N^F}}_{\text{Failure Rate Increase}} \tag{1}$$

where τ^{off} is the total time in SM during time period T, γ^{off} is the failure rate in SM, f^{tr} is the power switching rate between full power and SM, and N^F is a parameter called number of cycles to failures. Thus, the total failure rate is composed by two terms: the first one tends to decrease the failure rate, while the second one has the opposite effect.

In order to evaluate the lifetime increase or decrease w.r.t. the always on solution, we define a metric called Acceleration Factor (AF). The AF is lower than one if the link lifetime is increased compared to the always on solution. On the contrary, a value higher than one means that the lifetime is decreased compared to the always on case. More formally,

$$AF = \frac{\gamma^{tot}}{\gamma^{on}} = 1 - \underbrace{(1 - AF^{off})\frac{T^s}{T}}_{\text{Lifetime Increase}} + \underbrace{\chi f^{tr}}_{\text{Lifetime Decrease}} \tag{2}$$

where AF^{off} is defined as $\frac{\gamma^{off}}{\gamma^{on}}$, which is always lower than one since the failure rate in SM γ^{off} (by neglecting the negative effect due to power state transitions) is always lower than the failure at full power γ^{on}. Moreover, χ is defined as $\frac{1}{\gamma^{on} N^F}$, which acts as a weight for the power frequency rate f^{tr}. The AF is then composed of two terms: the first one which tends to increase the lifetime (longer periods of SMs tends to increase this term which is negative), instead the second one tends to decrease the lifetime (the more often power state transitions occur, the higher this term will be). Moreover, the model is composed by parameters AF^s and χ, which depend solely on the hardware components used to build the device, while parameters τ^s and f^{tr} depend instead on the specific SM strategy. In the following, we detail the optimization model for minimizing the AF of a set of links.

3   Problem Definition

We consider an Internet Service Provider (ISP) network, where nodes are sources and destinations of aggregated traffic requests generated by users. We also assume that the links capacity and the traffic demand by all source/destination node pairs for each time period are given. Our objective is to find the set of links that must be powered on so that the total AF in the network is minimized, subject to flow conservation and maximum link utilization constraints. More formally, we report the formulation of the problem of [12]. More in depth, let $G = (V, E)$ be the graph representing the network infrastructure. Let V be the set of the network nodes, while E the set of the network links. We assume $\mid V \mid = N$ and $\mid E \mid = L$. Let $c_{i,j} > 0$ be the capacity of the link (i, j) and $\alpha \in (0, 1]$ the maximum link utilization that can be tolerated in order to avoid congestion and to guarantee over-provisioning. Let us denote as T the total amount of time under consideration. T is divided in time slots of period δ_t. Finally let $t^{s,d}(k) \geq 0$ be the traffic demand from node s to node d during slot k.

Focusing on the variables, we denote with $f_{i,j}^{s,d}(k) \geq 0$ the amount of flow from s to d that is routed through link (i, j) during slot k. Similarly, we denote as $f_{i,j}(k) \geq 0$ the total amount of flow on link (i, j) during slot k. Moreover, let $\tau_{i,j}^{off} \geq 0$ be the total time in sleep mode of link (i, j). Finally, let us denote with $AF_{i,j} \geq 0$ the AF for link (i, j).

In the following, we consider the integer variables. Let us denote with $x_{i,j}(k)$ a binary variable which takes value one if the link (i, j) is powered on during slot k, zero otherwise. Moreover, let us denote with $\xi_{i,j}(k)$ a binary variable which takes value one if link (i, j) has experienced a power state transitions from slot $k - 1$ to slot k, zero otherwise. Additionally, $C_{i,j} \geq 0$ are integer variables counting the number of power state transitions for link (i, j).

Given the previous notation, the objective is to minimize the total AF in the network:

$$\min \frac{1}{L} \sum_{(i,j) \in E} AF_{(i,j)} \qquad (3)$$

We then consider the constraints. In particular, traffic has to be routed in the network:

$$\sum_{j:(i,j)\in E} f_{i,j}^{s,d}(k) - \sum_{j:(j,i)\in E} f_{j,i}^{s,d}(k) = \begin{cases} t^{sd}(k) \text{ if } & i = s \\ -t^{sd}(k) \text{ if } & i = d \\ 0 \text{ if } & i \neq s,d \end{cases} \quad \forall i \in V \quad \forall k \quad (4)$$

We then compute the total amount of traffic on each link:

$$f_{i,j}(k) = \sum_{s,d} f_{i,j}^{s,d}(k) \quad \forall (i,j) \in E \quad \forall k \tag{5}$$

And we impose the maximum link capacity constraint:

$$f_{i,j}(k) \leq \alpha c_{i,j} x_{i,j}(k) \quad \forall (i,j) \in E \quad \forall k \tag{6}$$

Additionally, links have to assume the same power state in both directions:

$$x_{i,j}(k) = x_{j,i}(k) \quad \forall (i,j) \in E \quad \forall k \tag{7}$$

We then count the number of power state transitions for each link:

$$\begin{cases} x_{i,j}(k) - x_{i,j}(k-1) \leq \xi_{i,j}(k) \\ x_{i,j}(k-1) - x_{i,j}(k) \leq \xi_{i,j}(k) \end{cases} \quad \forall (i,j) \in E \quad \forall k \tag{8}$$

Moreover, we count the total number of transitions for each link:

$$C_{i,j} = \sum_{k} \xi_{i,j}(k) \quad \forall (i,j) \in E \tag{9}$$

And also the total time in SM for each link:

$$\tau_{i,j}^{off} = \sum_{k} (1 - x_{i,j}(k)) * \delta_t \quad \forall (i,j) \in E \tag{10}$$

Finally, we compute the total AF for each link:

$$AF_{i,j} = [1 - (1 - AF_{(i,j)}^s) \frac{\tau_{i,j}^{off}}{T} + \chi_{(i,j)} \frac{C_{i,j}}{2}] \quad \forall (i,j) \in E \tag{11}$$

The total number of transitions is divided by two since we assume that a power cycle is always composed of two transitions.

4 Scenario Description

We adopt the Orange - France Telecom (FT) scenario of [13]. Tab. 1 reports the main network characteristics. We refer the reader to [13] for more details. Here we report a brief summary. In brief, the FT network comprises the core level of the network. The topology, reported in Fig. 1, is composed of 38 nodes and 72 bidirectional links. Additionally, link capacities and routing weights are provided.

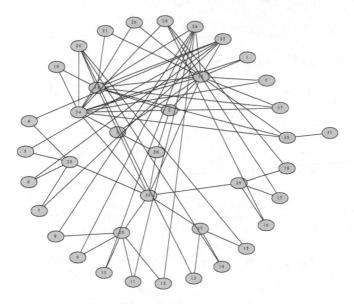

Fig. 1. Orange-FT network topology

More in depth, the FT network provides specific weights used to balance the load in the network. Finally, focusing on power consumption, we have adopted the same power model of [14], in which each link consumes an amount of power corresponding to a pair of transponders and a pair of IP interface ports. Each 10 Gbps transponder consumes 37W and each 1 Gbps port consumes 10W. Finally, we assume that when a link is in sleep mode, the power consumption is negligible.

Together with the topology, we have considered different sets of traffic matrices provided by the operator, together with the source and destination nodes. A total of 289 matrices is provided, which covers a working day. The total traffic (normalized to one) over time is reported in Fig. 2. As expected, traffic exhibits a strong day-night trend, with a peak during the day and an off peak during the night.

Finally, the maximum link utilization α is set to 50%, as recommended by the operator.

5   Case Studies

Since the presented problem is very hard to be solved due to the high number of links in the FT topology, as well as the number of TMs, we have followed a heuristic approach. In particular, we have applied the Most Power (MP) heuristic [6] for each TM, and we have then computed as a post-processing phase the resulting AF. The main idea of the MP heuristic is to put in SM the largest number of links, by ordering them in descending value of power to selectively

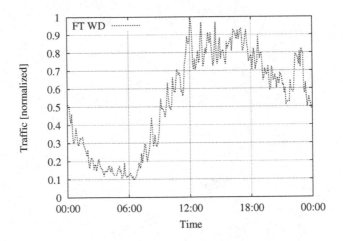

Fig. 2. Traffic profile for the Orange-FT scenario

Table 1. Network Characteristics

Parameter	FT
Type	Core Level
Number of Nodes	38
Number of Links	72
Average Degree	3.78
Routing Weights	Provided by Operator
Routing Algorithm	Min. Cost Path
Traffic Variation	1 working day

put them in SM. For each link put in SM, the connectivity and maximum link utilization constraints are verified. If they are not met, the link is put at full power, otherwise it is left in SM.

In our scenario, we have considered a time period of 15 days, and the repetition of the same traffic profile over the days. Moreover, we have set the hardware parameters as $\chi_{(i,j)} = 0.5$ and $AF^{off} = 0.2$ for all the links. In this way, the gain in terms of lifetime from putting in SM links is high, but we consider also a penalty $\chi_{(i,j)}$ not negligible for the transitions.

The Fig. 3 reports the values of the AF for each link in the network at the end of the 15 days period. Interestingly, we can see that the observed AF is not the same for all the links, with some links having an AF close to one and others instead with larger AF (i.e. more than four). This suggests that the lifetime behavior is not the same for all components in the network, with some links that decrease the lifetime and others instead which tends to keep it similar compared to the full power solution. Thus, we can clearly see that the lifetime depends on the particular device in the topology. In particular, it is a metric that depends on the global conditions of the network (i.e. guaranteeing connectivity

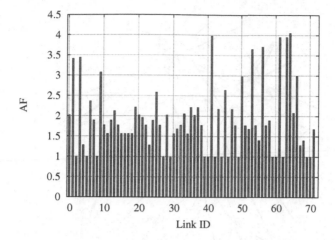

Fig. 3. AF for each link in the topology

and maximum link utilization), but also on the local policy adopted to decide when to put in SM the network device.

To give more insight, the Fig. 4 reports the network topology with the link width proportional to the measured AF. Interestingly, the links with the largest value of AF (i.e., consequently lower lifetime) are the ones that are connected to at least one node with high degree (e.g., 30,32,33,34). In particular, these links are frequently put in SMs to save energy and then powered on to support the traffic flows. Thus, the resulting lifetime is significantly decreased due to the fact that there are a lot of power state transitions occurred. This suggests that also the position of the link in the network topology strongly influences its AF too. In other words, the ways of links being connected play a crucial role for its lifetime.

In the following, we consider the evolution of AF over the time. Fig. 5 reports the AF vs. time for four different links in the topology. Interestingly, also the AF tends to vary notable vs. time. In particular, links 4-23, 19-33 and 20-32 experience an AF less than one in the first two days, meaning that at the beginning their lifetime tends to be increased compared to full power solution. However, in the following days, their AF is higher than one, meaning that the lifetime is reduced. This suggests also that time needs to be considered as one of sensitive factors for the lifetime.

In the last part of our work, we have considered the variation of the hardware parameters χ and AF^{off}. Fig. 6 reports the variation of the average AF in the network vs. the hardware parameters. In particular, when AF^{off} is increased, the AF is increased too: this is due to the fact that the gain for putting in SM devices is smaller compared to low values of AF^{off}. Additionally, the lifetime tends to be decreased when the penalty for power state transitions is increased. This suggests also that the lifetime management should take into account the hardware parameters, which depends on how the single devices are being built.

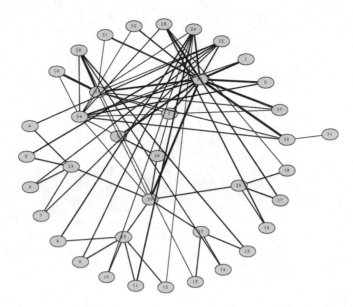

Fig. 4. FT/Orange Topology. The link width is proportional to the AF value.

6   Discussion

The presented results pointed out some interesting insights about the social behavior of lifetime-aware networks. First of all, the lifetime is not the same for all devices in the network, with some devices increasing the lifetime and other keeping it almost constant. This is due to the following factors: i) the specific algorithm used to select the links in sleep mode, ii) the traffic variation over time,

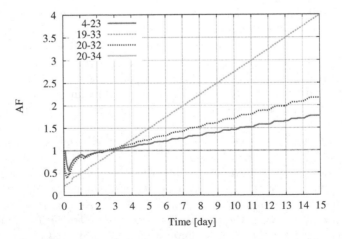

Fig. 5. AF vs. time for a set of links in the topology

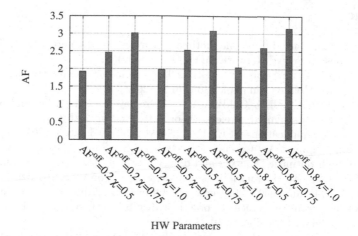

Fig. 6. AF values vs. the variation of HW parameters χ and AF^{off}

iii) the network topology under consideration, and iv) the HW parameters. Thus, we may claim that there exists a social behavior in this scenario: each link has to optimize its own lifetime, but this metric depends on both endogenous (e.g. the HW parameters) and exogenous parameters (e.g. the power state of the other links in the network). Additionally, also the link position in the topology tends to influence the lifetime, i.e., the links connected to the nodes with the highest degree (i.e. the highest number of "connections") tend to vary more notably their lifetime. Moreover, we have seen that the lifetime changes over time, passing from the situation in which it is clearly increased (e.g. during the initial days) to the case in which it is decreased (e.g. during the last days under consideration). Clearly, also this issue should be considered as future direction of investigation for our work.

7   Conclusions and Future Work

We have studied the impact of applying a sleep mode based algorithm on a telecom backbone network, with an emphasis on its social behavior. We have first proposed a simple model to evaluate the lifetime increase or decrease of network links as a consequence of sleep mode. After optimally formulating the problem of maximizing the lifetime of a set of links in a backbone network, we have conducted an extensive case study to validate the lifetime behavior of network links. Our results show that the link lifetime is influenced by the position of the link in the topology, as well as endogenous (e.g. the HW components used to build the link) and exogenous parameters (e.g. the current set of links in SM, and the traffic in the network). Moreover, we have shown that the lifetime varies over time. As next step, we plan to develop an algorithm that is able to consider the aforementioned social effects, as well to study the impact of traffic matrix on the lifetime.

References

1. Dietrich, I., Dressler, F.: On the lifetime of wireless sensor networks. IEEE/ACM Transactions on Sensor Networks **5**(1), 1–39 (2009)
2. Khan, M., Misic, J.: Security in IEEE 802.15.4 cluster based networks. Wireless Networks and Mobile Communications **6**(1) (2008)
3. Madan, R., Cui, S., Lall, S., Goldsmith, A.: Cross-layer design for lifetime maximization in interference-limited wireless sensor networks. In: 24th IEEE Conference on Computer Communications (IEEE INFOCOM 2005), vol. 3, pp. 1964–1975 (2005)
4. Addis, B., Capone, A., Carello, G., Gianoli, L., Sansò, B.: Energy management through optimized routing and device powering for greener communication networks. IEEE/ACM Transactions on Networling (ToN) **22**, 313–325 (2014)
5. Giroire, F., Mazauric, D., Moulierac, J., Onfroy, B.: Minimizing routing energy consumption: from theoretical to practical results. In: Proc. of IEEE GreenCom, Hangzhou, China, December 2010
6. Chiaraviglio, L., Mellia, M., Neri, F.: Minimizing ISP network energy cost: Formulation and solutions. IEEE/ACM Transactions on Networking (TON) **20**, 463–476 (2012)
7. Chabarek, J., Sommers, J., Barford, P., Estan, C., Tsiang, D., Wright, S.: Power awareness in network design and routing. In: Proc. of IEEE INFOCOM, Phoenix, USA, April 2008
8. Wiatr, P., Monti, P., Wosinska, L.: Energy efficiency versus reliability performance in optical backbone networks. Journal of Optical Communications and Networking **7**, 482–A491 (2015)
9. Chiaraviglio, L., Wiatr, P., Monti, P., Chen, J., Lorincz, J., Idzikowski, F., Listanti, M., Wosinska, L.: Is Green Networking Beneficial in Terms of Device Lifetime? IEEE Communications Magazine (2015) (to appear)
10. Chiaraviglio, L., Wiatr, P., Monti, P., Chen, J., Wosinska, L., Lorincz, J., Idzikowski, F., Listanti, M.: Impact of energy-efficient techniques on a device lifetime. In: Online Conference on Green Communications (GreenCom) (2014)
11. Wiatr, P., Chen, J., Monti, P., Wosinska, L.: Energy efficiency and reliability tradeoff in optical core networks. In: Proc. of OSA OFC, San Francisco, USA, March 2014
12. Amorosi, L., Chiaraviglio, L., DellOlmo, P., Listanti, M.: Sleep to stay alive: optimizing reliability in energy-efficient backbone networks. In: 11th ICTON Workshop on Reliability Issues in Next Generation Optical Networks (RONEXT), Budapest, Hungary, July 2015
13. Idzikowski, F.: Trend d3.3 final report for the ira energy-efficient use of network core resources, June 2012. http://www.fp7-trend.eu/system/files/private/71-wp3/d33-final-report-ira.pdf
14. Idzikowski, F.: Power consumption of network elements in ip over wdm networks. TU Berlin, TKN Group, Tech. Rep. TKN-09-006 (2009)

The Structure and Evolution of Large Cascades in Online Social Networks

Jiang Li[1](✉), Jiagui Xiong[2], and Xiaojie Wang[1]

[1] Beijing University of Posts and Telecommunications, Beijing, China
{li_jiang,xjwang}@bupt.edu.cn
[2] Www.kaixin001.com, Beijing, China
jiagui@corp.kaixin001.com

Abstract. The emergence of online social network services allows user to share photo, video or other content with their social friends. The content is transmitted from person to person, and a diffusion cascade form. Many recent works have discovered that the vast majority of cascades are small and only a tiny fraction of content can spread widely. In this paper, we focus on the structure of these rare but large cascades in online social networks. We introduce the concept of combined graph which not only contains the diffusion links but also includes relevant friendship edges. We find that the characteristics of combined graph provide a deep understanding of how information flows and reaches a large population on social network.

We investigate over 45000 large cascades whose sizes range from thousands to hundreds of thousands. We show the temporal dynamics of cascade tree and combined graph, and find that the combined graph is sparse, less clustering and lack of a dense core. In addition, we analyze the phenomenon from a microscopic perspective. Finally, we examine the correlations between structural properties and summarize four structural patterns.

Keywords: Social network · Information diffusion · Large cascades

1 Introduction

The online social networks, such as Facebook, Twitter and KaiXin, provide a mechanism for posting and sharing information. In these sites, photos, videos, or other piece of information are transmitted from person to person, creating long chains or "cascades". On one hand, large cascades capture a lot of attention, ranging from viral marketing [6] to social sites [3,10]. However, on the other hand, many recent works have indicated that large cascades in the real world are rare [5], and quantitative analysis of large cascades is scarce.

In this paper, we focus on the structure of large cascade and its evolution in online social network. We analyze a large dataset from KaiXin, with over 45000

J. Li—Most of research was performed while the author worked at KaiXin.

M.T. Thai et al. (Eds.): CSoNet 2015, LNCS 9197, pp. 273–284, 2015.
DOI: 10.1007/978-3-319-21786-4_24

cascades of which sizes range from the thousands to hundreds of thousands. We do not only explore the dynamics of the tree structure over time, but also concentrate on the corresponding underlying social network.

1.1   Summary of Findings and Contributions

The first contribution of this paper is the introduction of a new concept called combined graph. It gives a more effective way to describe cascades of information diffusion. In addition, we examine several structural properties, such as cascade depth, the average depth of node, the ratio of leaf node, the community structure of combined graph, etc. We show detailed observations of these properties and how they evolve over time. The most striking result is that, for most cascades, the combined graphs are sparse, less clustering and without a dense core.

We attribute this phenomenon to two reasons. One reason is the "low infection rate": each resharer can only affects a small fraction of its neighbours; another reason is the "persistent adoption probability": the adoption probability is relatively stable and does not increase with multiple exposures.

Next, we seek insight into the correlations between structural properties. These correlations indicates a strong relationship between cascade tree structure and the combined graph. In addition, we summarize four structure patterns: the "long chains" cascades spread via the viral mechanism and generate a lot of long chains; the "one-step broadcast" cascades always include a high-degree root node and form a radial shape, whereas the "multi-step broadcast" cascades rely on multiple hub nodes, and both two kinds can reach a large population in short distance; the "combination" cascades benefit from both viral and broadcast mechanisms.

2   Related Work

In recent years, the emergence of blogs and social networks has offered opportunities to collect massive amounts of user interaction data, and a number of studies have used these online data to empirically observe the information diffusion process [1,6,12]. Several recent papers [3,5] have analyzed and cataloged properties of online information cascades, and find that the vast majority of cascades are small and large cascades are rare.

Some recent studies focus on the mutual effects between the diffusion process and underlying network: [2] discusses the causal effect of content exposure on reshare activities and the influences of strong and weak ties; [15] stats that the contagion process is affected by the community structure of underlying network; [10] finds that information diffusion process often result in the bursts of new connections; [12] finds that repeated exposures have a much less marginal effect on the adoption; These works give a rich understanding of the mechanism behind diffusion process, while our study provides a global view of large cascades.

The most relevant work is [3], which use the data of photos in Facebook to study the characteristics of cascades. The study shows that large cascades have remarkable difference in time evolution, reshare depth distribution and the demographics of individuals. However, they study only two typical large cascades, and our dataset contains tens of thousands.

3   Preliminaries

3.1   Concepts and Terminology

Let the undirected graph $G = \{V, E\}$ denote the whole underlying social network. The undirected edge $e = \langle v_i, v_j \rangle \in E$ represents the friendship between node v_i and v_j, as $v_i, v_j \in V$.

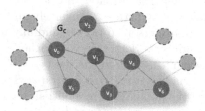

Fig. 1. A cascade starts at node v_0 and spreads through the solid arrow lines. The shadow area illustrates the corresponding combined graph, where the dashed lines represent the friendship edges between cascade nodes.

We define the **cascade** as comprising a "seed" individual, who create or post some content independently of any other individual, followed by other individuals who are influenced either directly or indirectly by the seed and take a reshare action. Under this definition, a cascade can be abstracted as tree structure $T = \{V_T, E_T\}$. V_T denotes the nodes participating in the cascades, $V_T \subset V$, and the directed edge $e_T = \langle v_i, v_j \rangle \in E_T$ represents that the spreading direction of the content is from node v_i to node v_j, as $v_i, v_j \in V_T$. For example, in Figure 1, the node $\{v_0, v_1, \ldots, v_6\}$ with the solid arrow lines form a cascade tree, the v_0 is the root node and the cascade size is 7.

Meanwhile, these nodes in cascade tree have friendships with each other, and form a **subgraph** in the underlying social network. The subgraph provides information about the social connections between cascade nodes. Formally, we denote the *subgraph* as $G_S = \{V_T, E_S\}$, where $E_S = \{e_S = \langle v_i, v_j \rangle \mid v_i, v_j \in V_T \wedge e_S \in E\}$.

It is important to note that, a small fraction of reshares occur between social strangers because of the external influence [11]. Such observation implies that the diffusion links do not necessarily belong to the friendship edges, as $E_T \not\subset E_S$. As a result of this situation, the subgraph of some cascades might not be a connected graph.

Due to the defects of subgraph, we introduce another revised concept: **combined graph**. The combined graph is the combination of both cascade tree and the corresponding subgraph, including all the diffusion links and friendship edges between cascade nodes. Formally, we denote the combined graph as $G_C = \{V_T, E_C\}$, and the edges satisfy $E_C \equiv E_T \cup E_S$. The shadow area in Figure 1 shows an example of combined graph, where the dash lines represent the friendship between cascade nodes. In contrast to the subgraph with uncertainty in structure, the definition above guarantees that the combined graph is a single connected network.

Some studies [3,4,9] have found that, despite achieving comparable diffusion sizes, the large cascades have significant distinctions in structure. In general, cascades form two main shapes, which are always referred as **star cascade** and **chain cascade**. In the star cascades, information are originated from a high-degree node, and many reshares are made directly from the source. It often creates a wide and shallow tree structure. Conversely, in the chain cascades, the content spread naturally from person to person and any node can only directly infects a few others. It produces a deep tree with richer structure.

[4] argues that the mechanisms behind the two types of cascades are different. The chain cascades grow via the **viral mechanism** which is analogous to the virus spreading and reach the population by multi-step interpersonal contagion, whereas the star cascades grow via the **broadcast mechanism** which affects the majority of individuals by one or few super hubs. Our next study will shows that the different mechanisms not only determine the shape of cascades but also influence the properties of underlying combined graph.

3.2 Data Description

KaiXin(www.kaixin001.com) is one of the largest online social network services in China, which was founded in 2008 and has hundreds of millions of registered users. Similar to Facebook, individuals can reshare the content which is posted by others in KaiXin. By rough estimating, the majority of these reshared contents are hot news and funny videos.

Our data consists of 45358 large cascades which posted during from June 1, 2009 to November 30, 2009. The sizes of all cascades are larger than 3000, among them, sizes of 7013 cascades are larger than 30000 and the largest cascade includes 448643 cascade nodes.

An unexpected discovery is, comparing with the photo re-share activities in Facebook[3], the cascades in KaiXin is striking larger. A plausible explanation is that, compare with private photos, people are more willing to share interesting news and funny clips.

4 Structure of Large Cascade and Its Evolution

In this section, we show our observations and analysis on the cascade structure and its evolution in detail. For each cascade, we study the structural properties

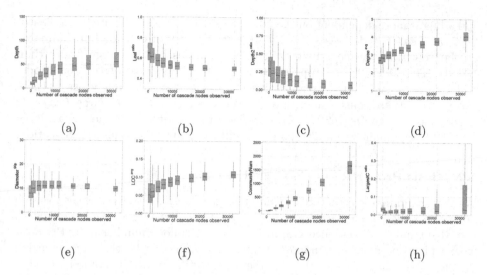

Fig. 2. Structure properties as a function of the number of cascade nodes observed

by observing a certain number (k) of first "infected" nodes. Our plots show how the structural properties change with the increase of k.

4.1   Temporal Dynamics of Tree Structure

Here we study the temporal evolution of several properties in cascade tree. The first natural property we consider is the *Depth* of the cascade. We define the depth of cascade as the largest number of edges from any node to the root node plus one, and the depth of root node is 1. We calculate the depth of all large cascades with different number of observed nodes, and show the statistics in Figure 2a. It is not surprising that, as the cascade grows over time, the depth increases sub-linearly.

One problem of the *Depth* is that this measurement is sometimes not robust and a single long chain can dramatically affect the result [4]. Therefore, we also examine some alternative measures. For example, we use the 90th percentile cascade node depth $Depth^{90p}$ as a metric, which will not affect by the long tail and more effective than *Depth*. Similarly, we also calculate the average node depth $Depth^{avg}$ and the median of node depth $Depth^{50p}$. In addition, prior work [4] suggests that the *WienerIndex*, defined as the average distance between all pairs of nodes in the cascade tree, provides a measure of "structural virality". Thus, a cascade with small Wiener index indicates a pure "broadcast" and the large Wiener index implies a highly "viral". It will be demonstrated in later section that these measures are all highly correlated with each other.

Other features have also been studied, such as the ratio of leaf nodes in the tree ($Leaf^{ratio}$), and how many nodes reshare the content directly from the root node ($Depth2^{ratio}$). A cascade with large fraction of leaves indicates that most of adopters receive the content from a small number of branching nodes and the spreading is more dependent on the "broadcast" mechanism. Figure 2b indicates that the $Leaf^{ratio}$ decreases rapidly with the growth of cascade but maintains a relatively stable value (about 0.5) when cascades become large (more than 10000). The $Depth2^{ratio}$ property describes the infectivity of the root node, and Figure 2c shows it decreases with the cascade size. In many cases, these two properties are consistent with each other, as a cascade with large $Leaf^{ratio}$ always has a large $Depth2^{ratio}$. However, we find a kind of cascades with large $Leaf^{ratio}$ but small $Depth2^{ratio}$, and we will return to this issue in next section.

4.2   Basic Properties of Combined Graph

We now examine the structure of combined graph. First, we look at the average degree of nodes in combined graph ($Degree^{avg}$). The $Degree^{avg}$ is defined as the edge number of combined graph divided by the node number in it. Previous work [7] has observed that, in most of the real social networks, the average degree increases over time. We can also observe that, the $Degree^{avg}$ of combined graph increases with the growth of cascade size, as shown in Figure 2d. The result indicates that the "Densification Laws" do also apply to the combined graph. The phenomenon is not surprising, since larger and deeper the cascade grows, the new infected node is likely to have more friendships with preceding nodes. Thus, the new joined node might not only have larger degree but also augment the degree of old ones.

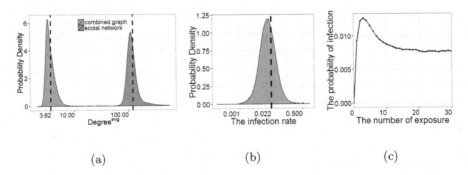

(a)   (b)   (c)

Fig. 3. (a) The significant difference between $Degree^{avg}$ of combined graph and $Degree^{avg}$ in the underlying social network. Note that the x axis is log scale. (b) The probability density of "infection rate". (c) The probability of infection with the repeated number of exposures.

On the other side, however, we can observe that, the value of $Degree^{avg}$ in combined graph is unexpectedly small, which implies that the combined graph is extremely sparse. In order to better illustrate the point, we also calculate the average degree of cascade nodes in the underlying social network. Therefore, each cascade has two related $Degree^{avg}$: one is the average degree in combined graph and the other is the average degree of these cascade nodes in the social network. The comparison of both $Degree^{avg}$ is demonstrate in Figure 3a, and note that the x axis is log scale. We can see that, compared with the $Degree^{avg}$ of combined graph, the $Degree^{avg}$ in social network is one order of magnitude larger and the mean value is more than 100. The remarkable difference between two $Degree^{avg}$ indicates that the density of combined graph is mush smaller than the underlying social network.

Another basic property of combined graph is the diameter. The diameter of a network is the maximum distance between any two nodes. However, just like the depth, the diameter will greatly affected by a occasional long chain. Thus, we use the 90th percentile diameter ($Diameter^{90p}$) as the alternative measure which equals the minimum value that greater than the distance of 90% node pairs. Figure 2e shows that, as the number of observed nodes increases, the $Diameter^{90p}$ increases in the early stage and decreases slowly at the later period. The similar phenomenon of "shrinking diameter" was observed by [7], and our study shows that diameter shrinking happens in combined graph as well.

4.3   The Community Structure

Here we study the community structure of the combined graph in detail. We first compute the average local clustering coefficient (LCC^{avg}) which is often used to measure how close a node's neighbors are to being a clique [14]. Figure 2f shows that the LCC^{avg} increases monotonically with the cascade growth. Moreover, we can see that the LCC^{avg} maintains a stable relatively small value, and the mean is about 0.1. In previous subsection we have indicated that the $Degree^{avg}$ in combined graph is small and the mean value is 3.6. Therefore, on average, the ego network of a node in combined graph only includes 0.6 triangles. It implies that, for most of the cascade nodes, there is zero or only one friendship links between its neighbors.

In order to obtain more detailed community structures, we detect the communities of combined graph and examine two community-related properties: one is the number of communities in combined graph ($CommunityNum$), the other is the size ratio of the largest community ($LargestC^{ratio}$). The $LargestC^{ratio}$ is defined as the size of combined graph divided by the size of largest community. Figure 2g shows that the $CommunityNum$ is on a linear growth as the cascade size. Moreover, most combined graph are composed by a large number of communities. Figure 2h indicates that, despite the significant growth of the cascades, the largest community invariably contains a small faction of cascade nodes. Even if when the cascades grow very large (about 50000) and the $LargestC^{ratio}$ seems to be larger in many cascades, the mean value still smaller than 0.1. Thus, it is difficult to find a large dense core in those cascades.

4.4   Microscopic Perspective

The small $Degree^{avg}$, LCC^{avg} and $LargeC^{ratio}$ in combined graph indicate that, for most cascades, the underlying combined graphs are sparse, less clustering and without a large dense core. Our finding is somewhat surprising because plenty of previous studies [8,13] have examined the structure of social network can roughly described by the "jellyfish" model. That is, the center of network is a tight "core" which contains a large proportion of nodes, and then there are a large number of relatively small "whiskers" connected to the "core". So the question is raised: why the combined graph, which is sparse and less clustering, is generated on top of a dense network substrate?

We argue that this phenomenon can be ascribed to two reasons. The first reason is that, even for the popular event which can obtain a large population, the "infection rate" is still very small. In other words, an infected node can only affect a tiny fraction of its neighbors. To illustrate our point, in each cascade, we computer the infection probability of node which equals the number of node's neighbours in social network divided by the number of followers in the cascade tree. We show the distribution in Figure 3b. The mean value of infection probability is about 0.022, which implies that, for a node with 100 social friends, only 2 of its neighbours will reshare the content on average. The "low infection rate" guarantees the small sparsity of combined graph.

Another reason is that, the likelihood of "adoption" does not increases linearly but be somewhat persistence with the number of repeated exposures. It means that a large proportion of "infected" social neighbours will not motivate the user to "adopt" the content. We demonstrate the adoption probability with the number of repeated exposures in Figure 3c, and the exposure curve is amazingly consistent with the result in [12]. The figure shows that the adoption probability increases and reaches the peak early but declines rapidly and keeps a small value. The adoption probability of multiple exposures (> 10) is even smaller than the probability of first exposure. Therefore, multiple exposures has almost no impact on the adoption, and the "persistent adoption probability" ensures that it is difficult to form a large dense core in combined graph.

5   The Structural Patterns of Cascades

In this section, we first examine the correlations between structural properties. In order to achieve the target, these large cascades are pruned and only retain the first 3000 resharers for each. As a consequence, all pruned cascades have the same size, and the structure properties become comparable. In addition, we summarize four patterns for large cascades. Each pattern represents a typical cascade structure and reflects different diffusion mechanism.

5.1   Correlations Between Structure Properties

First, We study the correlations between the depth-related properties, such as $Depth$, $Depth^{50p}$, $Depth^{90p}$, $Depth^{avg}$ and $WienerIndex$. Table 1 shows the Pearson correlation coefficients between the five properties. We can see that these properties are highly correlated with each other. In addition, we find that the $Depth^{90p}$ has the largest average correlation coefficients with other features, and we use the $Depth^{90p}$ as the representative in our next study.

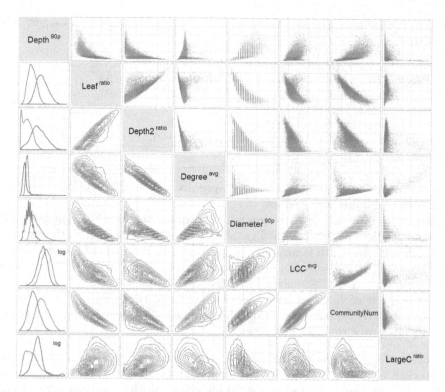

Fig. 4. 1. The upper right plots illustrate the correlations between pairs of structure properties, and the point in plots represents a large cascade; 2. The left lower plots illustrate the two-dimensional probability densities for property pairs; 3. The plots in the first left column show the one-dimensional probability density for the corresponding properties; 4. The cascades are divided into two groups with different colors, red for the deep cascades and blue for the shallow ones; 5. The "log" label in plots mean that the horizontal axis is log scale.

Next, we focus on the correlations between eight structure properties: $Depth^{90p}$, $Leaf^{ratio}$, $Depth2^{ratio}$, $Degree^{avg}$, $Diameter^{90p}$, LCC^{avg}, $CommunityNum$ and $LargeC^{ratio}$. We display a scatter plot for each property pair on the upper right part of Figure 4, and every point in each plot represents a large cascade. The meaning of colors in plots will be explained in next subsection, and here we only place emphasis on correlations between variables:

- $Leaf^{ratio}$ generally has a increasing relationship with $Depth2^{ratio}$, and both two properties trend to be inversely proportional to the $Depth^{90p}$. The phenomenon is easy to understand that, if the cascade is deeper, more nodes will grow on the deep layers and become branching nodes.
- $Leaf^{ratio}$ shows a negative correlation with $Degree^{avg}$ and LCC^{avg}. As we have mentioned before, a large ratio of leaves indicates that the cascade is more attributed to the broadcast mechanism. So the negative correlation

Table 1. Correlation coefficients between $Depth$, $Depth^{50p}$, $Depth^{90p}$, $Depth^{avg}$ and $Wiener Index$

	$Depth$	Dep^{50p}	Dep^{90p}	Dep^{avg}	$WIdx$
$Depth$	1	0.87	0.97	0.92	0.94
$Depth^{50p}$	0.87	1	0.93	0.99	0.90
$Depth^{90p}$	0.97	0.93	1	0.97	0.96
$Depth^{avg}$	0.92	0.99	0.97	1	0.93
$WIndex$	0.94	0.90	0.96	0.93	1

implies that, while most combined graphs are sparse, the combined graph of star cascades are even sparser.

- $Leaf^{ratio}$ also has a negative correlation with $Diameter^{90p}$. The reason is that, the diameter of combined graph is always in proportion to the cascade depth, but the depth is inversely proportional to the ratio of leaves.
- Because of the decreasing relationships with $Degree^{avg}$, LCC^{avg} and $Diameter^{90p}$, $Leaf^{ratio}$ shows a negative correlation with $CommunityNum$.
- While a few of cascades have high $LargeC^{ratio}$, the $LargeC^{ratio}$ for most cascades is very small. In addition, the $LargeC^{ratio}$ does not show any strong relationships with other properties.

These strong correlations indicate that the shape of cascade tree and the structure of combined graph are not independent with each other. In addition, we find that the $Leaf^{ratio}$ is a fairly good indicator, which has strong correlations with most other properties. This property not only describes the shape of cascade tree but also provides insight on the combined graph.

5.2   Structural Patterns

As we mentioned before, the large cascades can be roughly categorized into the star cascades and the chain cascades. However, besides the two categories, is

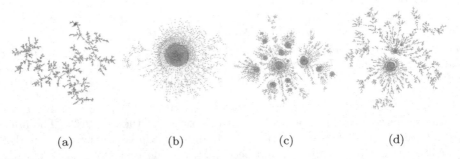

(a)   (b)   (c)   (d)

Fig. 5. Examples of different structure patterns. The large red circle represents the initial node, the small red circles represent the leaves of cascades, and the blue ones are branching nodes. All the four cascades have the same size of 3000.

there other specific shape which can describe a specific collection of cascades? Another question is, does there exist some more structural difference between the cascades in the same category?

In order to examine the structural patterns in more detail, we divide our cascades into two groups according to their $Depth^{90p}$, where the "deep" group is labeled by red color and the "shallow" group is labeled with blue color in Figure 4. In addition, we plot two dimensional probability densities of property pairs for each group in the left lower part, and demonstrate the one dimensional density of corresponding structure properties in the first left column of Figure 4.

We find that each group has specific structural characteristics, and the difference in structure between two groups is significant. The deep group, which represents the chain cascade, statistically contains less leaves and only a few nodes followed directly from the root node. Moreover, the combined graph is denser, has larger diameter and clustering coefficient, and is made up of more tiny clusters. We refer this structural pattern as *"long chains"*.

A typical "long chains" cascade is visualized in Figure 5a. The cascade forms exceedingly long branches, where the $Depth^{90p}$ is 122, and only 0.1% of nodes follow directly from the root. A distinctive characteristic of the cascade is that the root is not in the center of the tree structure. In addition, there are no apparent hub nodes. It implies that, even without huge hubs, information might spread widely and reach a large population.

On the other side, the structures of cascades in shallow group is slightly complicated. A large part of these cascades contain only one super-hub node (is always the root node), and the majority of other nodes receive the information directly from the super-hub. In the cascade tree, the depth is very short, and most of nodes surround the hub. In addition, the combined graph is sparser and covers fewer tiny clusters. We refer this kind of cascades as *"one-step broadcast"*. A typical example is shown in Figure 5b.

However, instead of the only one hub node in cascade tree, there exists a part of cascades in which the spreading of information relies on multiple hub nodes. These cascades have small $Depth2^{ratio}$ but also infect massive individuals in short distance. We show a example in Figure 5c. This kind of cascades are not extreme cases, and we find thousands of similar instances in our dataset. We refer this pattern as *"multi-step broadcast"*.

In reality, a sizable majority of cascades are somewhere in between the pure "long chains" and extreme "broadcast". These cascades form via both broadcast and viral mechanisms. Correspondingly, they consist of not only a high degree root node but also multiple long chains. A typical cascade is visualized in Figure 5d and we refer this pattern as *"combination"*.

Besides the four common structural patterns, there are also some unfrequent shapes. For example, we observe some cascades which form exact star topology with depth of 2. All those cascades are artificial and produced by Sybils. Furthermore, we have mentioned previously that a few cascades contain a large community in their combined graph. However, the tree structure of these cascades shows no significant differences with the ones visualized in Figure 5, so they are not discussed here.

6 Conclusion

In this paper we focus on the structure and its evolution of large cascade. We introduced the concept of combined graph and found that the combined graphs of most large cascades are fairly sparse and less clustering. We analyzed this phenomenon from a microscopic view. In addition, we examined the correlations between structural properties and summarized four structural patterns.

More structural properties should be considered in our future work. For example, the border nodes and edges surrounding cascade may have effects on the cascade structure. The four structural patterns are coarse-grained, and more particular fine-grained patterns need to be discovered. Another direction is to develop a diffusion model which can simulate these structural characteristics.

References

1. Adar, E., Adamic, L.A.: Tracking information epidemics in blogspace. In: Web Intelligence, pp. 207–214 (2005)
2. Bakshy, E., Rosenn, I., Marlow, C., Adamic, L.A.: The role of social networks in information diffusion. In: WWW, pp. 519–528 (2012)
3. Dow, P.A., Adamic, L.A., Friggeri, A.: The anatomy of large facebook cascades. In: ICWSM (2013)
4. Goel, S., Anderson, A., Hofman, J., Watts, D.: The structural virality of online diffusion. Preprint **22**, 26 (2013)
5. Goel, S., Watts, D.J., Goldstein, D.G.: The structure of online diffusion networks. In: EC, pp. 623–638 (2012)
6. Leskovec, J., Adamic, L.A., Huberman, B.A.: The dynamics of viral marketing (2007)
7. Leskovec, J., Kleinberg, J.M., Faloutsos, C.: Graphs over time: densification laws, shrinking diameters and possible explanations. In: KDD, pp. 177–187 (2005)
8. Leskovec, J., Lang, K.J., Dasgupta, A., Mahoney, M.W.: Statistical properties of community structure in large social and information networks. In: WWW, pp. 695–704 (2008)
9. Leskovec, J., McGlohon, M., Faloutsos, C., Glance, N.S., Hurst, M.: Patterns of cascading behavior in large blog graphs. In: SDM, pp. 551–556 (2007)
10. Myers, S.A., Leskovec, J.: The bursty dynamics of the twitter information network. In: WWW, pp. 913–924 (2014)
11. Myers, S.A., Zhu, C., Leskovec, J.: Information diffusion and external influence in networks. In: KDD, pp. 33–41 (2012)
12. Romero, D.M., Meeder, B., Kleinberg, J.M.: Differences in the mechanics of information diffusion across topics: idioms, political hashtags, and complex contagion on twitter. In: WWW, pp. 695–704 (2011)
13. Tauro, S.L., Palmer, C.R., Siganos, G., Faloutsos, M.: A simple conceptual model for the internet topology. In: GLOBECOM, pp. 1667–1671 (2001)
14. Watts, D., Strogatz, S.: Collective dynamics of 'small-world' networks. Nature **393**, 440–442 (1998)
15. Weng, L., Menczer, F., Ahn, Y.Y.: Virality prediction and community structure in social networks (2013)

Breaking Bad: Finding Triangle-Breaking Points in Large Networks

Thang Dinh[1]([envelope]) and Ravi Tiwari[2]

[1] Department of Computer Science, Virginia Commonwealth University,
Richmond, VA 23284, USA
tndinh@vcu.edu
[2] Ebay Inc, San Jose, CA, USA
ravtiwari@ebay.com

Abstract. Many online social networks (OSN), such as Facebook, Twitter can quickly become popular, but many such as Friendster or MySpace can also suffer catastrophic decline in traffics and users. Understanding the capability of OSNs to withstand perturbation and changes, termed social resilience, is a matter of the uttermost importance. In this paper, we investigate the resilience of OSNs under nodes and links removals, where the robustness of the network is measured through *the number of triangles*, a fundamental property in many networks. Specifically, we strive to discover critical nodes and links whose failures will critically break most triangles in the network, changing the network's organization and (possibly) leading to the unpredictable dissolving of the network. We formulate this vulnerability analysis as optimization problems, and provide proofs of their intractability. Given the intractability of the problem, we also investigate approximation algorithms and their efficient implementations.

Keywords: Triangle breaking · Social networks · Approximation algorithm

1 Introduction

With massive amount of users, online social networks (OSNs) have fundamentally changed the way people communicate and interact nowadays. While many OSNs, such as Facebook, Twitter witness rapid expanding in terms of number of users and user-engagement, the other sites such as Friendster and MySpace suffered catastrophic decline in traffics and users. For example, Friendster had over 100 million users at its peak but most of its users have fled to other networks such as Facebook by the end of 2009. Thus, understanding the capability of OSNs to withstands perturbation and changes, termed social resilience, is a matter of the uttermost importance.

Network resilience to perturbation and changes is a growing concern nowadays. Roughly speaking, network resilience evaluates how much the network's normal function and capacity is affected in case of external perturbation [1].

© Springer International Publishing Switzerland 2015
M.T. Thai et al. (Eds.): CSoNet 2015, LNCS 9197, pp. 285–295, 2015.
DOI: 10.1007/978-3-319-21786-4_25

Complex and social systems that can sustain their organizational structure, functionality and responsiveness under such unexpected perturbation are considered more robust than those that fail to do so.

While many measures are proposed for resilience of technological and biological systems, there are little understanding on the resiliency of social networks. Most studies in the literature focus on how the network behaves under perturbation using the measures such as the pair-wise connectivity [2], natural connectivity [3], or using centrality measures, such as degrees, betweeness [4], the geodesic length [1], eigenvector [5], etc. However, these measures are not suitable for assessing social network resilience.

In this paper, we investigate the resilience of OSNs under nodes and links removals, where the robustness of the network is measured through *the number of triangles*, a fundamental properties in many networks including social networks, communication networks, biological networks and more. There are many advantages of number of triangles over other structural measures. First, it is a popular and fundamental metric for evaluating network clustering: a high number of triangles in a network facilitates more cooperative behaviors and interactions among its users. Second, a high number of triangles positively correlate with other structural properties such as modular structure (or community structure), small diameter, and clustering coefficient.

Our goal is to discover critical nodes and links whose failures will critically break most triangles in the network, changing the network's organization and (possibly) leading to the unpredictable dissolving of the network. We formulate this vulnerability analysis as optimization problems, and provide proofs of their intractability. Given the intractability of the problem, we propose efficient and scalable approximation algorithms to identify triangle-breaking points (i.e. nodes or links) in the networks. The proposed algorithms guarantee the solutions to be close to the optimal solutions by a constant factor.

Related Works. Many metrics and approaches have been proposed to account for network robustness and vulnerability [6–10]. While each of these measures has its own emphasis and rationality, they often come with several shortcomings that prevent them from capturing desired characteristics of network connectivity and resilience. For example, measures based on shortest path are rather sensitive to small changes (e.g. removing edges or nodes); algebraic connectivity and diameter are not meaningful for disconnected graphs (all disconnected graphs have the same values); number of connected components and component sizes, arguably, do not fully reflect level of network connectivity.

Network structural vulnerability in social networks, has so far been an untrodden area. In a related work [11], the authors introduced the community structure vulnerability to analyze how the communities are affected when top k vertices are excluded from the underlying graphs. They further provided different heuristic approaches to find out those critical components in modularity-based community structure. [12] suggested a method based on the generating edges of a community to find out critical the components.

Table 1. List of Symbols

Notation	Meaning
n	Number of vertices/nodes ($N = |V|$)
m	Number of edges/links ($M = |E|$)
d_u	The degree of u
$N(u)$	The set of u's neighbors
$Tri(u)$	The set of triangles on a node u
$T(u) = |Tri(u)|$	The number of triangles on u
$Tri(u, v)$	The set of triangles on an edge (u, v)
$Tri(S) = \cup_{u \in S} Tri(u)$	The set of triangles on $S \subseteq V$
$Tri(F) = \cup_{(u,v) \in F} Tri(u, v)$	The set of triangles on a subset of edges $F \subseteq E$

Counting and listing triangles in a graph is an important problem, motivated by applications in variety of areas. The problem of counting triangles on a graph with n vertices and m edges can be performed in a straightforward manner in $O(mn)$. This has been improved to $O(m^{3/2})$ in [13] and $O(m^{\frac{2w}{w+1}})$ where $w < 2.376$ is the exponent of matrix multiplication [14]. Recently, the k-triangle-breaking-node and k-triangle-breaking-edge problems are investigated in [15]. The authors provides NP-completeness proofs and greedy algorithms for the problems. Unfortunately, the NP-completeness proofs contains fundamental flaws that cannot be easily fixed.

2 Model and Problem Definition

2.1 Problem Definition

We abstract a social network using an undirected graph $\mathcal{G} = (V, E)$ with $|V| = n$ nodes and $|E| = m$ edges. Given a graph $G = (V, E)$, we study attack models in which the attackers attempt to minimize the number of triangles in the graph by removing nodes and edges.

We define four different versions of the problems as follows. In the first two, the attacker seeks to remove k nodes/edges from the graph to break as many triangles in G as possible. Here a triangle is broken if one of its edges or nodes is removed. The k-triangle-breaking-node problem is defined as

Definition 1 (k-Triangle-Breaking-Node). *Given an undirected graph $G = (V, E)$ and budget size k, find a subset S of k nodes that removal will break the maximum number of triangles in G*

$$\max \quad |Tri(S)| \tag{1}$$
$$s.t. \quad |S| \leq k, \tag{2}$$
$$S \subseteq V, \tag{3}$$

where $Tri(S)$ is the set of triangles with at least one node in S, i.e.,

$$Tri(S) = \{(u, v, w) \mid (u, v), (v, w), (w, u) \in E \text{ and } \{u, v, w\} \cap S \neq \emptyset\}.$$

Note that we can formulate the above problem as an Integer Linear Programming problem (ILP). Define for each $u \in V$, $x_u \in \{0, 1\}$ that satisfies

$$x_u = \begin{cases} 1 & \text{if node } u \text{ is removed,} \\ 0 & \text{otherwise.} \end{cases}$$

and for each triangle $(u, v, w) \in Tri(V)$ define an integral variable $y_{uvw} \in \{0, 1\}$ that satisfies

$$y_{uvw} = \begin{cases} 1 & \text{if triangle } (u, v, w) \text{ is broken,} \\ 0 & \text{otherwise.} \end{cases}$$

The k-triangle-breaking-node problem is to delete at most k nodes, i.e., $\sum_{u \in V} x_u \leq k$, to break the maximum number of triangles, i.e., to maximize the objective function $\sum_{(u,v,w) \in Tri(V)} y_{uvw}$. Since the triangle (u, v, w) is only broken if at least one node among u, v, w is removed, we have the constraint

$$x_u + x_v + x_w \geq y_{uvw}.$$

In summary, we have the following equivalent ILP formulation.

$$\max \quad \sum_{(u,v,w) \in Tri(V)} y_{uvw} \tag{4}$$

$$\text{s.t.} \quad \sum_{v \in V} x_v \leq k, \tag{5}$$

$$x_u + x_v + x_w \geq y_{uvw}, \quad \forall (u, v, w) \in Tri(V), \tag{6}$$

$$x_u, y_{uvw} \in \{0, 1\}. \tag{7}$$

Observe that the ILP in (13) is a special case of the Max-k-Coverage[16] problem in which the set of universe is $\mathcal{U} = Tri(V)$ (i.e. all the triangles) and the collection of subsets is $\mathcal{S} = \{Tri(v) \mid v \in V\}$. This special case of Max-k-Coverage also satisfies the condition that *all the elements have the same frequency three*, as each triangle involves exactly three nodes.

Similarly, the k-triangle-breaking-edge problem is defined as

Definition 2 (k-Triangle-Breaking-Edge). *Given an undirected graph $G = (V, E)$ and budget size k, find a subset F of k edges that removal will break the maximum number of triangles in G*

$$\min \quad Tri(F) \tag{8}$$

$$\text{s.t.} \quad |F| \leq k,$$

$$F \subseteq E,$$

where $Tri(F)$ is the set of triangles with at least one edge in $F \subset E$. The equivalent ILP of k-triangle-breaking-edge is

$$\max \quad \sum_{(u,v,w) \in Tri(V)} y_{uvw} \tag{9}$$

$$\text{s.t.} \quad \sum_{(u,v) \in E} x_{uv} \leq k, \tag{10}$$

$$x_{uv} + x_{vw} + x_{wv} \geq y_{uvw}, \quad \forall (u,v,w) \in Tri(V), \tag{11}$$

$$x_{uv}, y_{uvw} \in \{0,1\}, \tag{12}$$

where

$$x_{uv} = \begin{cases} 1 & \text{if edge } (u,v) \text{ is removed,} \\ 0 & \text{otherwise.} \end{cases}$$

for all $(u,v) \in E$.

Again, k-triangle-breaking-edge is an another special case of Max-k-Coverage in which the elements to be covered are the triangles in G, and the collection of subsets includes the set of triangles involve each edge $(u,v) \in E$. As each triangle consists of three edges, *the frequency of each element in this instance is also three*. Moreover, any two *subsets have at most one triangle in common*.

Also, we studies the converse version in which we wants to break a certain number of triangles by removing the least number of nodes/edges from the graph.

Definition 3 (min-triangle-breaking-node). *Given an undirected graph $G = (V, E)$ and an integer $|Tri(V)| \geq p > 0$, find a minimum size subset of nodes S that removal will break at least p triangles in G.*

Definition 4 (min-triangle-breaking-edge). *Given an undirected graph $G = (V, E)$ and an integer $|Tri(V)| \geq p > 0$, find minimum size subset of edges F that removal will break at least p triangles in G.*

The ILP for min-triangle-breaking-node

$$\min \quad \sum_{v \in V} x_v \tag{13}$$

$$\text{s.t.} \quad \sum_{(u,v,w) \in Tri(V)} y_{uvw} \geq p,$$

$$x_u + x_v + x_w \geq y_{uvw},$$

$$x_u, y_{uvw} \in \{0,1\}.$$

The ILP for min-triangle-breaking-edge

$$\max \quad \sum_{(u,v) \in E} x_{uv} \tag{14}$$

$$\text{s.t.} \quad \sum_{(u,v,w) \in Tri(V)} y_{uvw} \geq p,$$

$$x_{uv} + x_{vw} + x_{wv} \geq y_{uvw},$$

$$x_{uv}, y_{uvw} \in \{0,1\}.$$

As shown through the ILP formulations, min-triangle-breaking-node and min-triangle-breaking-edge are special cases of the partial set-cover problem [17], in which each element is in exactly three subsets and the intersection of any three subsets contains at most one element.

Table 2. Summary of Complexity and Best Approximation Guarantees

Problem	Complexity	Best approximation ratio
k-triangle-breaking-node	NP-complete	19/27 [18]
min-triangle-breaking-node	NP-complete	3 [17]
k-triangle-breaking-edge	NP-complete	19/27 [18]
min-triangle-breaking-edge	NP-complete	3 [17]

2.2 Hardness and Approximability

We summarize the complexity and approximability results for the studied problems in Table 2. We first discuss the complexity, then present the best approximation guarantees for those problems.

NP-Hardness. Proofs of NP-completeness of k-triangle-breaking-node and k-triangle-breaking-edge are presented in [15]. Unfortunately the proofs have fundamental flaws. Specifically, the proof of Theorem 2.1 [15] relies on a weaker constraint of the set system: "the intersection of any *three* subsets in S has at most one element". Indeed for the k-triangle-breaking-edge, the correct (and stronger) condition should be: the intersection of any *two* subsets in S has at most one element. Moreover, the proof relies on another wrong assumption (2nd paragraph) that if a problem is not NP-hard then there is a polynomial-time algorithm to solve it. Indeed, we do not know if there exist NP-intermediate problems between NP and P. Consequently, the validity of the reduction cannot be confirmed.

We show that all the four problems are indeed NP-complete problems. First, we present simple NP-completeness proofs of min-triangle-breaking-node and k-triangle-breaking-node via reduction from the Vertex-Cover problem [16]. The decision versions of k-triangle-breaking-node and min-triangle-breaking-node can be polynomial-time reducible from the following decision problem, called *Node-Triangle-Free*:

"*Given a undirected graph $G = (V, E)$ and a number k, can we delete k nodes from G so that there is no more triangles in G (aka G is triangle-free)?*".

In turn, we shall show that *Node-Triangle-Free* is polynomial-time reducible from the decision version of Vertex-Cover:

"*Given a graph $G = (V, E)$ and an integer $0 < k < |V|$, is there a vertex-cover of size k?*".

Let $\Phi =< G = (V, E), k >$ is an instance of the vertex-cover problem. For each edge $(u, v) \in E$, we add to G a new node t_{uv} and connect t_{uv} to both u and v. Let G' be the resulted graph. We shall reduce ϕ to an instance $\Lambda =< G', k >$ of *Node-Triangle-Free*. Obviously, if we have a vertex-cover $S \subset V$ of size k in G then we can delete the same set of nodes S in G' to obtain a triangle-free graph. In the reverse direction, we can assume w.l.o.g. that t_{uv} will never be removed. The reason is that we can always remove u or v and break an equal or greater number of triangle(s). Thus a subsets of size k that removal makes G' triangle-free must induce a vertex-cover of size k in G.

Theorem 1. *The problems k-triangle-breaking-node and min-triangle-breaking-node are NP-complete.*

Similarly, both k-triangle-breaking-edge and min-triangle-breaking-edge can be polynomial-time reducible to the following problem:

"Can we delete k edges from a graph $G = (V, E)$ to make the graph triangle-free?".

That problem is known to be NP-complete according to [19]. Hence, we have the following result.

Theorem 2. *The problems k-triangle-breaking-edge and min-triangle-breaking-edge are NP-complete.*

Approximability. Since the min-triangle-breaking-node and min-triangle-breaking-edge problems are special cases of the partial set cover problem with bounded frequencies $f = 3$ [17], the primal-dual algorithm in [17] provides a 3-approximation algorithm for both the problem.

Theorem 3. *There exist 3-approximation algorithms for min-triangle-breaking-node and min-triangle-breaking-edge.*

The k-triangle-breaking-node and k-triangle-breaking-edge problems are special cases of Max-k-Coverage and the pipage-rounding method in [18] results in an approximation algorithm with ratio $1 - (1 - 1/3)^3 = 19/27$. Note that both the primal-dual method in [17] and the pipage-rounding algorithm in [18] have high time-complexity and are not scalable for large networks. Thus we will propose faster algorithm for the studied problems with slightly worse approximation ratios.

Theorem 4. *There exist 19/27-approximation algorithms for k-triangle-breaking-node and k-triangle-breaking-edge.*

3 Finding Triangle-Breaking Nodes

In this section, we present a typical Greedy Algorithm and our Improved Greedy Algorithm (IGAN) to solve the k-triangle-breaking-node and min-triangle-breaking-node problems. The first algorithm (Alg. 1) is a naive and simple greedy algorithm serving as a baseline for comparison purpose. Our main algorithm (Alg. 2) employs more techniques to provide faster running-time

The first algorithm (Alg. 1) select in each step the node that break the most number of triangles. That node $u = \arg\max_{v \in V \setminus S} \Delta_S(v)$ is then added to the solution S. The algorithm continues until *meeting requirement*: either k nodes

---

Algorithm 1. Greedy Algorithm (Simple_Greedy)

1: $S \leftarrow \emptyset$;
2: Repeat
3:    $S \leftarrow S + \arg\max_{v \in V \setminus S} \Delta_S(v)$
4: Until (*meeting requirement*)
5: **return** S

Algorithm 2. Improved Greedy Algorithm (Node) (IGAN)

1: Number nodes from 1 to n such that $u < v$ implies $d(u) \leq d(v)$.
2: $S \leftarrow \emptyset$;
3: **for each** $u \in V$ **do** $T(u) \leftarrow 0$;
4: **for each** $(u,v) \in E$ **do** $tr(u,v) \leftarrow 0$;
5: **for** $u \leftarrow n$ to 1 **do**
6:      **for each** $v \in N(u)$ with $v < u$ **do**
7:          **for each** $w \in A(u) \cap A(v)$ **do**
8:              Increase $T(u), T(v)$ and $T(w)$ by one;
9:              Add u to $A(v)$;
10: **Repeat**
11:      $u_{max} \leftarrow \arg\max_{u \in V \setminus S}\{T(u)\}$;
12:      Remove u_{max} from G and add u_{max} to S;
13:      **for each** $(v,w) \in E$ **do**
14:          **if** $v, w \in N(u_{max}) \setminus S$ **then**
15:              Decrease $T(v)$ and $T(w)$ by one;
16: **Until** (*meeting requirement*)
17: **return** S

---

are selected (for k-triangle-breaking-node) or when the number of triangles broken reach p (for min-triangle-breaking-node).

Since the k-triangle-breaking-node and min-triangle-breaking-node problems are special cases of Max-k-Coverage and partial set cover, respectively, the typical greedy algorithm provide performance guarantees $(1 - 1/e)$ and $\ln p$ for those problems, respectively. The $(1 - 1/e)$ approximation ratio for k-triangle-breaking-node is also shown in [15] by proving that the objective function (the number of broken triangles) is a monotone submodular function.

The complexity of Alg. 1 is $O(kmn)$ assuming k nodes are selected in the solution. Later the time complexity for Alg. 1 is brought down to $O(km^{3/2})$ in [15] using the fast triangle computation method in [13]. For large value of $k = \theta(n)$, the time-complexity of the algorithm in [15] could be as high as $O(nm^{3/2})$. In next part, we will present our improved implementation of greedy with time complexity $O(m^{3/2} + km)$ which is up to $m^{1/2}$ times faster than the algorithm in [15].

3.1 Improved Greedy Algorithm

Our Improved Greedy Algorithm (IGAN - Alg. 2) speeds up significantly the simple greedy algorithm. For small k's, this algorithm requires as much time as counting the number of triangles.

In principle, IGAN employs an adaptive strategy in computing the marginal gains (the number of broken triangles) when nodes are removed one after another. At each round, the node v that breaks the most number of triangles is selected into the solution. Node v is then excluded from the structure and the procedure repeats itself on the remaining nodes and *recomputes* efficiently the new marginal gain for each node u.

We structure IGAN into two phases. The first phases (lines 1–9) extends the algorithm in [13] to compute the number of triangles that are incident with each node in the graph. This algorithm was proved to be time-optimal in $\theta(m^{3/2})$ for triangle-listing, and has been shown to be very efficient in practice. The second phase (lines 10–16) repeats the vertex selection for k rounds. In each round, we select the node u_{max} with the highest value of $\Delta_S(u) = T(u)$ into the solution. The algorithm then removes u_{max} from the graph, and performs the necessary update for $T(u)$ for all $u \in V \setminus S$. The algorithm repeats until meeting requirement: either k nodes are selected (for k-triangle-breaking-node) or when the number of triangles broken reach p (for min-triangle-breaking-node).

The key efficiency of IGAN algorithm is in its update procedure for $\Delta_S(u) = T(u)$. The update for all $O(n)$ values of $\Delta_S(u)$ after removing u_{max} can be done in *linear time*. This is made possible due to the information on the number of triangles involving each node.

Time-complexity: The first phase takes $O(m^{3/2})$ as in [13]. The second phase takes a linear time in each round and has a total time complexity $O(k(m + n))$. Thus, the overall complexity is $O(m^{3/2} + km)$. For $k < m^{1/2}$, the algorithm has an effective time-complexity $O(m^{3/2})$, dominated by the counting triangles procedure.

Obviously, IGAN is an implement of the greedy method and retains the approximation guarantees of the greedy method for Max-k-Coverage and partial set cover.

Theorem 5. *The IGAN algorithm is an $(1 - 1/e)$-approximation algorithm for k-triangle-breaking-node and an $H(p) - 1/2$ approximation algorithm for min-triangle-breaking-node, where $H(p)$ denotes the harmonic function $H(p) = 1 + 1/2 + \ldots + 1/p$.*

4 Finding Triangle-Breaking Edges

We present IGAE, the improved greedy algorithm for finding triangle-breaking edges in Alg. 3. IGAE is faster than its node-version, IGAN, and possesses a time-complexity $O(m^{3/2} + kn)$.

Unlike IGAN, IGAE maintains on each edge the number of triangles incident on that edge and updated the measure efficiently when removing nodes from G.

Algorithm 3. Improved Greedy Algorithm (Edge) (IGAE)

1: Number nodes from 1 to n such that $u < v$ implies $d(u) \leq d(v)$.
2: $F \leftarrow \emptyset$;
3: **for each** $u \in V$ **do** $T(u) \leftarrow 0$;
4: **for each** $(u, v) \in E$ **do** $tr(u, v) \leftarrow 0$;
5: **for** $u \leftarrow n$ to 1 **do**
6:      **for each** $v \in N(u)$ with $v < u$ **do**
7:           **for each** $w \in A(u) \cap A(v)$ **do**
8:                Increase $tr(u, v), tr(v, w)$ and $tr(u, w)$ by one;
9:                Add u to $A(v)$;
10: **Repeat**
11:      $e_{max} \leftarrow \arg\max_{(u,v)\in E \setminus F}\{tr(u, v)\}$;
12:      Remove e_{max} from G and add e_{max} to F;
13:      Let $(u', v') = e_{max}$;
14:      **for each** $w \in N(u') \cap N(v')$ **do**
15:                Decrease $tr(w, u')$ and $tr(w, v')$ by one;
16: **Until** (*meeting requirement*)
17: **return** F

After removing an edge (u', v') we only needs to consider only $|N(u') \cap N(v')|$ updates to discount the triangles incident on (u', v') from the corresponding edges. Thus the overall time-complexity in each iteration is on finding the edge that breaks the maximum number of triangles. Similar to the node version, we also have the same approximation guarantees for the edge-deletion problems.

Theorem 6. *The IGAE is an* $(1 - 1/e)$-*approximation algorithm for* k-*triangle-breaking-edge and an* $H(p) - 1/2$ *approximation algorithm for* min-*triangle-breaking-edge.*

5   Conclusion

In this paper, we study the problems of finding critical nodes and links whose failures will critically break most triangles in the network, changing the network's organization and (possibly) leading to the unpredictable dissolving of the network. We formulate this vulnerability analysis as optimization problems, and provide proofs of their intractability and their best approximation guarantees: 19/27-approximation for k-triangle-breaking-node and k-triangle-breaking-edge as well as 3-approximation for min-triangle-breaking-node and min-triangle-breaking-edge. Since the methods to obtain the best approximation guarantees are not scalable, we provide efficient implementations for the greedy approaches with worsen approximation ratios. In the future, we aim to bridge the gaps between theory and practice to design the scalable approximation with best possible approximation ratios.

References

1. Holme, P., Kim, B.J., Yoon, C.N., Han, S.K.: Attack vulnerability of complex networks. Phys. Rev. E **65**, 056109 (2002)
2. Dinh, T.N., Xuan, Y., Thai, M.T., Pardalos, P.M., Znati, T.: On new approaches of assessing network vulnerability: hardness and approximation. IEEE/ACM Trans. Netw. **20**(2), 609–619 (2012)
3. Chan, H., Tong, H., Akoglu, L.: 37. In: Make It or Break It: Manipulating Robustness in Large Networks. SIAM, pp. 325–333 (2014)
4. Albert, R., Jeong, H., Barabsi, A.L.: Error and attack tolerance of complex networks. Nature **406**, 200 (2000)
5. Allesina, S., Pascual, M.: Googling food webs: Can an eigenvector measure species' importance for coextinctions? PLoS Comput Biol **5**(9), e1000494 (2009)
6. Grubesic, T.H., Matisziw, T.C., Murray, A.T., Snediker, D.: Comparative approaches for assessing network vulnerability. Inter. Regional Sci. Review **31** (2008)
7. Murray, A., Matisziw, T., Grubesic, T.: Multimethodological approaches to network vulnerability analysis. Growth Change (2008)
8. Neumayer, S., Zussman, G., Cohen, R., Modiano, E.: Assessing the vulnerability of the fiber infrastructure to disasters. IEEE/ACM Trans. Netw., 1610–1623 (2011)
9. Dinh, T.N., Thai, M.T.: Precise structural vulnerability assessment via mathematical programming. In: Proc. of IEEE MILCOM (2011)
10. Dinh, T., Thai, M.: Network under joint node and link attacks: Vulnerability assessment methods and analysis. IEEE/ACM Transactions on Networking (2015)
11. Nguyen, N.P., Alim, M.A., Shen, Y., Thai, M.T.: Assessing network vulnerability in a community structure point of view. In: Proceedings of the 2013 IEEE/ACM International Conference on Advances in Social Networks Analysis and Mining. ASONAM 2013, pp. 231–235. ACM, New York (2013)
12. Alim, M.A., Nguyen, N.P., Thang, D.N., Thai, M.T.: Structural vulnerability analysis of overlapping communities in complex networks. In: Proceedings of the 2014 IEEE/WIC/ACM International Conference on Web Intelligence. WI 2014, pp. 231–235. ACM, New York (2014) (to appear)
13. Schank, T., Wagner, D.: Finding, counting and listing all triangles in large graphs, an experimental study. In: Nikoletseas, S.E. (ed.) WEA 2005. LNCS, vol. 3503, pp. 606–609. Springer, Heidelberg (2005)
14. Alon, N., Yuster, R., Zwick, U.: Finding and counting given length cycles. Algorithmica **17**(3), 209–223 (1997)
15. Li, R., Yu, J.: Triangle minimization in large networks. Knowledge and Information Systems, 1–27 (2014)
16. Vazirani, V.: Approximation Algorithms. Springer (2001)
17. Gandhi, R., Khuller, S., Srinivasan, A.: Approximation algorithms for partial covering problems. Journal of Algorithms **53**(1), 55–84 (2004)
18. Ageev, A., Sviridenko, M.: Pipage rounding: A new method of constructing algorithms with proven performance guarantee. Journal of Combinatorial Optimization **8**(3), 307–328 (2004)
19. Yannakakis, M.: Edge-deletion problems. SIAM Journal on Computing **10**(2), 297–309 (1981)

Shifts in Collective Attention and Stock Networks

Evidence from Standard & Poor's 100 Corporations and Firm-Level Google Trends Data

Raphael H. Heiberger[✉]

Institute for Sociology, University of Bremen,
Mary-Somervile-Str. 3, 28205 Bremen, Germany
raphael.heiberger@uni-bremen.de

Abstract. In this paper, we combine network analytical methods to understand the structure of financial markets with recent research about collective attention shifts by utilizing massive social media data. Our main goal, hence, is to investigate whether changes in stock networks are connected with collective attention shifts. To examine the relationship between structural market properties and mass online behavior empirically, we merge company-level Google Trends data with stock network dynamics for all S&P 100 corporations between 2004 and 2014. The interplay of massive online behavior and market activities reveals that collective attention shifts *precede* structural changes in stock market networks and that this connection is mostly carried by companies that already *dominate* the development of the S&P 100.

Keywords: Collective attention · Stock networks · Econophysics · Financial crisis · Information Theory · Computational Science

1   Introduction

Recent economic turmoil has underlined the need to understand modern financial markets as complex systems. Especially in times of crisis, the classical models of rational agents and efficient markets turn out to be insufficient. One interdisciplinary challenger of the dominant paradigm is often dubbed Econophysics [1]. A major branch of this field is the statistical analysis of stock interaction networks and their structural dynamics. This kind of analysis was first conducted by Mantegna [2], using the correlation between price fluctuations of single stocks to construct hierarchical networks and reproduce the topological properties of a market. The main idea is to decrease the immense complexity of financial markets to facilitate investigation, and, at the same time, retain the markets' core information. Several filtering methods have been successfully used to achieve such a reduction, for instance, minimal spanning trees [2–5], dynamic spanning trees and asset graphs [6–8] or a winner-take-all approach [9], which is applied in [10, 11]. Independent of the construction technique, it can be stated that major economic events cause dramatic changes in the network structure of stock markets [12].

© Springer International Publishing Switzerland 2015
M.T. Thai et al. (Eds.): CSoNet 2015, LNCS 9197, pp. 296–306, 2015.
DOI: 10.1007/978-3-319-21786-4_26

Another innovative research line trying to explain complex economic systems focuses on collective attention shifts and their influence on markets by utilizing massive social media data [13–16]. The publicly available service Google Trends seems to be especially fruitful for scientists to comprehend collective financial behavior. Therein, Google provides access to aggregated information on the volume of queries for specific search terms over time. These search query data have delivered useful information to predict trading volumes [17], to diversify portfolio risks [18] and to quantify trading behavior [19].

In this paper, we combine both research approaches. Our main goal is to investigate *whether structural changes in stock networks are connected to collective attention shifts*. For this purpose, we interpret Google Trends as an approximation for collective informational needs and the stock network topologies as representation of the structure of financial markets. The subsequent results indicate a general pattern in modern investment decision making regarding the importance of collective attention shifts for stock formations and, hence, the importance of understanding the link between mass internet behavior and price movements.

2   Data and Methods

2.1   Stock Networks

To examine the relationship between structural market properties and mass online behavior, we merge two large data sets in the period between 2004 and 2014 for all companies listed in the Standard & Poor´s 100 in August, 2014 [20]. The S&P 100 index composition is based on one hundred large and well established "blue chips". Their raw stock data was retrieved from Yahoo [21]. The basic information consists of N assets with price $P_i(\tau)$ for asset i at the first trading day of week τ. The logarithmic return for this specific week is then given as the difference to the first trading day of the previous week, $r_i = \ln P_i^\tau - \ln P_i^{\tau-1}$. In order to investigate the dynamics of the stock market, we divide the individual stock data into M windows, denominated $t = 1, 2, \dots M$ of width T, that is, the number of weekly returns in M. The windows overlap and shift further at length δT, which is also measured in trading weeks to match the second data set on Google Trends queries discussed below. We use one year (i.e. 52 trading weeks) as window width. Following [6–8] we can now quantify the degree of similarity between assets i and j for the given window around t with the correlation coefficient

$$\rho_{ij}^t = \frac{[\langle r_i^t r_j^t \rangle - \langle r_i^t \rangle \langle r_j^t \rangle}{\sqrt{[\langle r_i^{t^2} \rangle - \langle r_i^t \rangle^2][\langle r_j^{t^2} \rangle - \langle r_j^t \rangle^2]}}, \tag{1}$$

where $\langle \dots \rangle$ indicates a time average over the consecutive trading weeks t that are contained in the return vector r_i^t. Finally, we can now the $N \times N$ correlation matrix C^t, which is completely characterized by $N(N-1)/2$ correlation coefficients.

From the moving stock price correlation matrices we construct dynamic networks by using the "winner-take-all"-approach discussed in [9]. According to Tse and colleagues, only those correlations between stocks are used that lie above a certain connection criterion z. Here, the condition is set to 0.6, as a lower bound of considerable correlations. Please note that different levels of correlations have no big impact on the network construction [9, 11]. Thus, to be part of the stock network the correlation (i.e. the weight of the relation) between stock i and j has to satisfy the condition $\rho_{ij}^t > |z|$.

In our view, there exist two major advances of the threshold approach compared with other reduction techniques: (a) The constructed networks loose no essential information. Both minimal spanning trees and planar graphs remove edges with high correlation if the respective nodes fit certain topological conditions and are, on these grounds, already within the reduced graph. (b) There is no fixed upper bond, i.e. the number of nodes included in the network is not mandatory but dependent on the specific period and its topology.

After the construction of the dynamic stock network, we apply two longitudinal measures to quantify its stability and structure on its macro level. To study the systemic stability we use a well-established measure for biological systems. May [22, 23] introduced three parameters for which random networks are almost certainly stable: the size of the network (N), the density of connections (D), and the average interaction strength (α) at a given time t. The stability condition can therefore be formalized as

$$S_t < 1, with \ S_t = \sqrt{ND} \ \alpha . \tag{2}$$

This condition challenged the established relationship between ecological diversity and the stability of ecosystems. In fact, May´s model turned it upside down. According to (2), an increasing network complexity decreases the stability of a system, not the other way round. This relation proofs to be valid in a multitude of ecological networks [24]. Even more important for our undertaking, the stability criterion also displays the state of financial markets rather well [11].

A second measure to comprehend the development of whole networks is their modularity [25, 26], which displays their compartmentalization, i.e. its community structure and community division. For weighted networks it is defined as

$$Q_t = \frac{1}{2x}\Sigma_{i,j}\left[\alpha_{ij} - \frac{k_i k_j}{2x}\right]f(c_i, c_j), \tag{3}$$

with α_{ij} being the interaction strength (weight) of the edge between i and j, k_i being the number of degrees of node i, and c_i being the community of node i. The function is 1 if $c_i = c_j$ and 0 otherwise, and $x = \frac{1}{2}\Sigma_{ij}\alpha_{ij}$. We calculated the modularity for each dynamic network by using the algorithm of Blondel et al. [27].

Regarding the global structure of financial networks, hence, modularity Q_t addresses the partitioning of a stock network and decreases in times of crisis due to common influence factors that concern all stocks. On the other hand, ecological stability S_t follows a reversed numerical logic, since high values indicate an unstable market with large risks and high uncertainties.

To complement these structural indicators with an analysis of individual company positions, we apply the Eigenvector Centrality [28, 29]. This measure does not only account for the number of connections of each node (as done, for instance, by the degree centralization), but considers also the "prestige" of each relationship partner. In other words, nodes are more central if they are connected to neighbors that are themselves more central. This makes it the most useful measure for economic networks according to properties discussed in [30, 31]. Formally, the Eigenvector Centrality E_i of node i is defined as the product of i's relation to j and j's respective Eigenvector. This set of parallel equations is elegantly solved by

$$E_i = \alpha_{i1}E_1 + \alpha_{i2}E_2 \dots + \alpha_{in}E_n = \frac{1}{\mu}\Sigma_j^n \alpha_{ij}E_j , \tag{4}$$

with μ being the largest positive eigenvalue. Thus in the case of stock networks, E_i measures how much the price movements of company i influences other stocks, and how central and influential those other stocks are themselves. Since stock networks are known to be scalefree [9], it is expected that relatively few stocks are exerting much of the influence over the majority of stocks.

2.2 Collective Attention

As a second data set we gathered search volumes provided by Google Trends for each of the S&P 100 companies. Changes in search queries are interpreted as collective attention shifts [17–19]. We retrieved search volume data for all companies in the S&P 100 index from Google Trends website (http://www.google.com/trends) between August 23, 2014 and August 29, 2014. Search volume data are restricted to requests of users localized in the USA, the home location of all companies contained in the S&P 100. We used the full corporate name in combination with "company" as search term to avoid semantic ambiguities. Since only five search terms can be looked up simultaneously, we retrieved the data for each company separately. The scores produced by Google Trends consist of the volume of each search query relative to the total number of searches carried out at each week t. They are normalized by Google with a maximum of 100, serving as a scaling factor for the rest of the series.

Each series is reported weekly on a Sunday to Saturday frequency. In order to investigate the connection between shifts in collective attention and stock networks, the two data sets are matched by following rule: Trends reported on Sunday of week τ correspond to the stock network of *the next week* that is derived from the subsequent correlation matrix $\mathbf{C}^{t+1}$. In doing so, the information about collective attention published on Sunday would be available for potential investors on the subsequent Monday and correspond to the respective stock network of week $\tau + 1$ (and thus time window $t+1$).

To demonstrate the development of attention for the whole index the arithmetic mean for each week over all company search queries is used. Due to the data-inherent normalization of Google Trends scores between 100 and 0, the simple average gives us a broad picture of collective information gathering processes, i.e. if there was a relatively high or low interest in S&P 100 companies in a certain week. To examine the connection between shifts in collective attention and structural dynamics in the stock networks empirically, we utilize time-lagged Pearson cross correlation coefficients [14]. According to the matching rule mentioned above, hence, the correlation is calculated for each week between Google Trends time series and the measures derived for the whole network of the next week, S_{t+1} and Q_{t+1}, respectively.

3   Empirical Results

3.1   Collective Attention and Network Structure

The results are discussed on two levels: the global macro-structure of the stock networks and the individual position of each corporation within this network. Fig. 1 shows the general connection between the development of both global network measures (i.e. May´s ecologic stability criterion and Newman´s modularity) and the relative change of Google search volumes. To facilitate the interpretation, we also picture the course of the S&P 100 index.

If only the course of the lines are considered initially, both network measures express a rather abrupt shift in the market state at the beginning of recent financial turmoil. Systemic stability S_t breaks down, indicated by a steep rising curve in the beginning of 2008. Simultaneously, the previously compartmentalized network becomes cohesive, i.e. less partitioned in subgroups and communities. In other words, the relations get more instable and centralized during the enfolding crisis – a mechanism that is well-known in stock networks [6, 7, 11]. A second eruption in both measures can be observed during 2011, when the "Euro crisis" led to high economic uncertainties that are also reflected in a volatile S&P 100. Here again, the stability criterion is heavily violated and the regained structuration in divisible "modules" is dispersed.

Additionally, each curve in Fig. 1 is dyed by a color code corresponding to the average search volumes of each Standard and Poor´s 100 company. Red sections indicate increasing collective attention in terms of search queries and green parts illustrate declining numbers. If we concentrate on the two points in time when both network measures change drastically, we see that collective attention changes *before* those alterations take place. In both periods, the relatively high search volumes start previous to the breakdown of systemic stability and the de-compartmentalization of relations in terms of separated communities. Moreover, in the beginning crisis in 2007/08 the uprising of search volumes is succeeded by a period of relatively low collective interest in the S&P 100 corporations.

However, Fig. 1 only provides "weak evidence" through descriptive visualization. To test the proposed relationship more thoroughly we calculate regression models. The models allow us to check the robustness of our results on a weekly basis and explore the direction (i.e. the sign) of the relationships between the change in search volume (denominated *GT*) and its influence on the structural network measures in

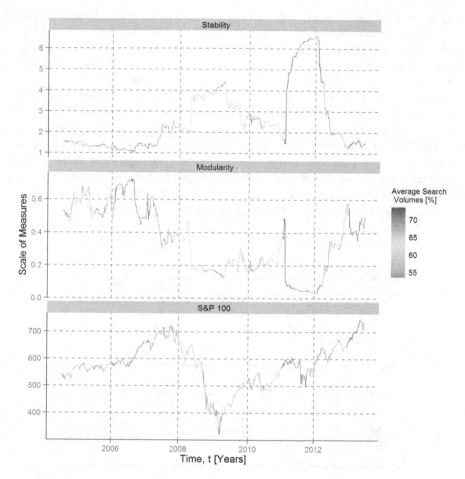

Fig. 1. The S&P 100 network over time and its stability, modularity, and index values matched with average Google search volumes

the following week. This is done (A) for the whole time period as well as (B) for the period from 2008 onward to account for the particular market upheavals in the financial crisis. To control for general market developments we also include the S&P 500 Volatility Index (VIX) and the S&P 100 (SPY) in the regression

$$m_{t+1} | Q_{t+1} = \beta_0 + \beta_1 GT_t + \beta_2 VIX_t + \beta_3 SPY_t + \varepsilon_t, \tag{5}$$

where ε_t is an error term.

The highly significant results in Tab. 1 for Google Trends scores confirm the suggested relationship between stability and collective attention for the whole observation period as well as for the shortened timeframe. The model also approves the evidence provided in Fig. 1 regarding the positive direction of that relationship. In contrast, the influence of collective attention on the modularity score Q_{t+1} is not significant for the whole period of time (cf. Tab. 2). Looking back to Fig. 1, the reason may lie in the relative high volatility of the measure in the early observation period. However, if we only focus on the period after 2008 the effect of search volumes is both highly significant and negative. Thus the suggested preceding influence of collective attention processes can also be confirmed for the modularity scores and, hence, the de-compartmentalization of the network structure in times of crisis.

On the level of the whole network, in summary, the information about Google Trends search volumes provides evidence for the connection of structural changes in stock networks and those in collective attention. Especially in times of great economic uncertainties company-level search queries increase strongly *before* major network alterations happen. This result is supported by time-lagged regression models for each structural measure. As a consequence, the findings provide evidence that collective attention shifts precede the rearrangement of market structures before financial crisis hit the economy (and media coverage, supposedly) with full force.

Table 1. Regression results. Dependent Variable: Stability S_{t+1}

(A) Timeframe: *2004-2014*				
	Estimate	SE	*t* statistic	Pr ($>|t|$
GT	0.0637	0.0185	3.44	0.0006
VIX	0.0781	0.0074	10.59	0.0000
SPY	-0.0004	0.0010	-0.38	0.7012
Adj. R^2	0. 3044			
N	424			

(B) Timeframe: *2008-2014*				
GT	0.1229	0.0228	5.39	0.0000
VIX	0.0300	0.0108	2.78	0.0059
SPY	-0.0036	0.0014	-2.53	0.0121
Adj. R^2	0.2094			
Obs.	260			

Table 2. Regression results. Dependent Variable: Modularity Q_{t+1}

(A) Timeframe: *2004-2014*				
	Estimate	SE	*t* statistic	Pr ($>|t|$
GT	-0.0004	0.0024	-0.18	0.8544
VIX	-0.0106	0.0010	-10.97	0.0000
SPY	-0.0000	0.0001	-0.31	0.7601
Adj. R^2	0.3076			
Obs.	424			

(B) Timeframe: *2008-2014*				
GT	-0.0105	0.0019	-5.67	0.0001
VIX	0.0006	0.0009	0.73	0.4689
SPY	0.0008	0.0001	7.38	0.0001
Adj. R^2	0.3013			
Obs.	260			

3.2   Collective Attention and Network Positions

On the micro level of stock markets we investigate individual company network positions and their correlation with collective attention processes as approximated with Google Trends. After establishing a general relationship between changes in public information gathering behavior and shifts in stock network structures in the previous chapter, the question now is for which corporations these connection is especially strong and in which positions the respective firms are located in the network. Fig. 2 illustrates the time-lagged Pearson cross correlation coefficients between dynamic Eigenvector centralities for each company and their Google search volume. The respective significance level is displayed on the x-axis. In addition, each circle size corresponds to the individual Eigenvector centrality.

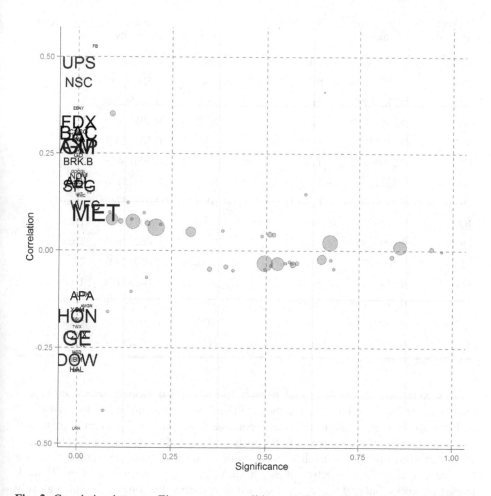

Fig. 2. Correlation between Eigenvector centralities and Google Trends scores of S&P 100 companies, plotted against their significance levels. Circle size is relative to the Eigenvector centrality of each firm.

For all companies with correlations that have significance levels lower than 0.05 the ticker symbols are drawn, whereby the size of the names are also corresponding to the Eigenvector centrality values. The results shown in Fig. 2 provide evidence that the highest correlation scores between attention shifts and network measures emerge for the most influential corporations, i.e. for those with the largest Eigenvector centralities. Unsurprisingly, firms like MetLife, General Motors, FedEx or General Electric are located in the center of the network. For many of those prominent corporations we find that changes in their position is significantly correlated to shifts in "Googling" them, i.e. to their search volumes.

Table 3. Top 30 Eigenvector centrality scores of the S&P 100 and the significance level of their correlation to Google Trends volumes

Ticker Symbol	Eigenvector Centrality	Significance < 5	Ticker Symbol	Eigenvector Centrality	Significance < 5
MET	0.31760607	True	GS	0.15254205	False
DD	0.29576174	False	SPG	0.15070847	True
GM	0.2700878	True	ALL	0.14350536	True
JPM	0.26691977	False	UNP	0.105724	False
MS	0.25837733	False	NSC	0.10512067	True
GE	0.25099612	True	WFC	0.1001642	True
AXP	0.2327536	True	HD	0.09815718	False
DIS	0.22479677	False	APA	0.0887574	True
BAC	0.21162135	True	CVX	0.06912528	True
HON	0.20062998	True	BA	0.06228806	True
FDX	0.18942757	True	BRK.B	0.05955707	True
BK	0.18442774	False	NOV	0.04284174	True
USB	0.18432278	False	MMM	0.03856179	False
DOW	0.16941818	True	TGT	0.03338286	False
UPS	0.16332159	True	HAL	0.03225848	True

This connection between central network positions and Google search queries is underlined in Tab. 2. Therein, the more detailed numbers emphasize the visualized relationship between network influence (operationalized via the Eigenvector centrality) and the strength of correlation. Of the 30 largest centrality scores we find for 19 a significant correlation between changes in position and related search volumes. Thus, we observe not only that collective attention shifts precede structural changes in stock market networks, but that this connection is mostly carried by companies which already dominate the development of the S&P 100.

4   Conclusion

Overall, company-level Google Trends data allow us to combine recent research efforts about massive online behavior with the structural analysis of stock market networks, rather than to analyze them separately. Taking both into account reveals that collective attention shifts *precede* structural changes in stock market networks and that this connection is mostly carried by companies that already *dominate* the development of the S&P 100. As a consequence, both mechanisms point to (further) "Matthew effects" [32] on stock markets. Existing structures and hierarchies are reproduced through collective information gathering behavior, since the impact of collective attention is especially strong on stocks with already influential positions, which strengthens their position further. Practically, however, the significant correlation between collective attention and network structure indicates that shifting – and publicly available – search queries could be a very valuable estimator for changing financial structures and, hence, financial crisis.

References

1. Yakovenko, V.M., Rosser Jr., J.B.: Colloquium: Statistical mechanics of money, wealth, and income. Rev. Mod. Phys. **81**, 1703 (2009)
2. Mantegna, R.N.: Hierarchical structure in financial markets. Eur. Phys. J B-Condens. Matter Complex Syst. **11**, 193–197 (1999)
3. Bonanno, G., Caldarelli, G., Lillo, F., Mantegna, R.N.: Topology of correlation-based minimal spanning trees in real and model markets. Phys. Rev. E. **68**, 046130 (2003)
4. Dias, J.: Sovereign debt crisis in the European Union: A minimum spanning tree approach. Phys. Stat. Mech. Its Appl. **391**, 2046–2055 (2012)
5. Gilmore, C.G., Lucey, B.M., Boscia, M.W.: Comovements in government bond markets: A minimum spanning tree analysis. Phys. Stat. Mech. Its Appl. **389**, 4875–4886 (2010)
6. Onnela, J.-P., Chakraborti, A., Kaski, K., Kertész, J.: Dynamic asset trees and Black Monday. Phys. Stat. Mech. Its Appl. **324**, 247–252 (2003)
7. Onnela, J.-P., Chakraborti, A., Kaski, K., Kertesz, J., Kanto, A.: Dynamics of market correlations: Taxonomy and portfolio analysis. Phys. Rev. E. **68**, 056110 (2003)
8. Onnela, J.-P., Chakraborti, A., Kaski, K., Kertesz, J., Kanto, A.: Asset trees and asset graphs in financial markets. Phys. Scr. **T106**, 48 (2003)
9. Tse, C.K., Liu, J., Lau, F.: A network perspective of the stock market. J. Empir. Finance **17**, 659–667 (2010)
10. Liu, J., Tse, C.K., He, K.: Fierce stock market fluctuation disrupts scalefree distribution. Quant. Finance **11**, 817–823 (2011)
11. Heiberger, R.H.: Stock network stability in times of crisis. Phys. Stat. Mech. Its Appl. **393**, 376–381 (2014)
12. Tumminello, M., Lillo, F., Mantegna, R.N.: Correlation, hierarchies, and networks in financial markets. J. Econ. Behav. Organ. **75**, 40–58 (2010)
13. Mathiesen, J., Angheluta, L., Ahlgren, P.T.H., Jensen, M.H.: Excitable human dynamics driven by extrinsic events in massive communities. Proc. Natl. Acad. Sci. **110**, 17259–17262 (2013)
14. Bordino, I., Battiston, S., Caldarelli, G., Cristelli, M., Ukkonen, A., Weber, I.: Web Search Queries Can Predict Stock Market Volumes. PLoS ONE **7**, e40014 (2012)

15. Moat, H.S., Curme, C., Avakian, A., Kenett, D.Y., Stanley, H.E., Preis, T.: Quantifying Wikipedia Usage Patterns Before Stock Market Moves. Sci. Rep. **3** (2013)
16. Bollen, J., Mao, H., Zeng, X.: Twitter mood predicts the stock market. J. Comput. Sci. **2**, 1–8 (2011)
17. Preis, T., Reith, D., Stanley, H.E.: Complex dynamics of our economic life on different scales: insights from search engine query data. Philos. Trans. R. Soc. Lond. Math. Phys. Eng. Sci. **368**, 5707–5719 (2010)
18. Kristoufek, L.: Can Google Trends search queries contribute to risk diversification? Sci. Rep. **3** (2013)
19. Preis, T., Moat, H.S., Stanley, H.E.: Quantifying Trading Behavior in Financial Markets Using Google Trends. Sci. Rep. **3** (2013)
20. S&P 100 - S&P Dow Jones Indices. http://us.spindices.com/indices/equity/sp-100
21. Yahoo Finance. https://finance.yahoo.com/
22. May, R.M., Levin, S.A., Sugihara, G.: Complex systems: Ecology for bankers. Nature **451**, 893–895 (2008)
23. May, R.M.: Will a Large Complex System be Stable? Nature **238**, 413–414 (1972)
24. Dobson, A., Allesina, S., Lafferty, K., Pascual, M. (eds.): Food-web assembly and collapse: mathematical models and implications for conservation. Philos. Trans. R. Soc. B Biol. Sci. **364**, 1643–1806 (2009)
25. Newman, M.E.: Modularity and Community Structure in Networks. Proc. Natl. Acad. Sci. **103**, 8577–8582 (2006)
26. Girvan, M., Newman, M.E.: Community structure in social and biological networks. Proc. Natl. Acad. Sci. **99**, 7821–7826 (2002)
27. Blondel, V.D., Guillaume, J.-L., Lambiotte, R., Lefebvre, E.: Fast unfolding of communities in large networks. J. Stat. Mech. Theory Exp. **2008**, P10008 (2008)
28. Freeman, L.: Centrality in social networks: Conceptual clarification. Soc. Netw. **1**, 215–239 (1979)
29. Bonacich, P.: Factoring and weighting approaches to status scores and clique identification. J. Math. Sociol. **2**, 113–120 (1972)
30. Ahern, K.R.: Network Centrality and the Cross Section of Stock Returns. Social Science Research Network, Rochester (2013)
31. Borgatti, S.P.: Centrality and network flow. Soc. Netw. **27**, 55–71 (2005)
32. Merton, R.K.: The Matthew Effect in Science The Reward and Communication Systems of Science Are Considered. Science **159**, 56–63 (1968)

Author Index

Printed in the United States
By Bookmasters

Printed in the United States
By Bookmasters